Ergebnisse der Mathematik und ihrer Grenzgebiete 93

A Series of Modern Surveys in Mathematics

Editorial Board: P. R. Halmos P. J. Hilton (Chairman)
R. Remmert B. Szőkefalvi-Nagy

Advisors: L. V. Ahlfors R. Baer F. L. Bauer
A. Dold J. L. Doob S. Eilenberg K. W. Gruenberg
M. Kneser G. H. Müller M. M. Postnikov
B. Segre E. Sperner

Arthur L. Besse

Manifolds all of whose Geodesics are Closed

With 71 Figures

Springer-Verlag
Berlin Heidelberg New York 1978

AMS Subject Classification (1970): Primary: 53-04, 53A05, 57A65, 34B25, 53B20, 53B99, 53C20, 53C25, 53C99, 58G99
Secondary: 57A65, 53B20, 53B99, 53C20, 53C25, 53C35, 53C99, 50D20, 50D25, 53D15, 47G05, 35L05

ISBN 3-540-08158-5 Springer-Verlag Berlin Heidelberg New York
ISBN 0-387-08158-5 Springer-Verlag New York Heidelberg Berlin

Library of Congress Cataloging in Publication Data. Berger, Marcel, 1927.—Manifolds all of whose geodesics are closed. (Ergebnisse der Mathematik und ihrer Grenzgebiete; Bd. 93). Bibliography: p. Includes index. 1. Geometry, Differential. 2. Manifolds (Mathematics, 3. Topological dynamics. I. Title. II. Series. QA649.B459. 516'.362. 77-21336

Printing and binding: Brühlsche Universitätsdruckerei, Lahn-Gießen
2141/3140-543210

Préface

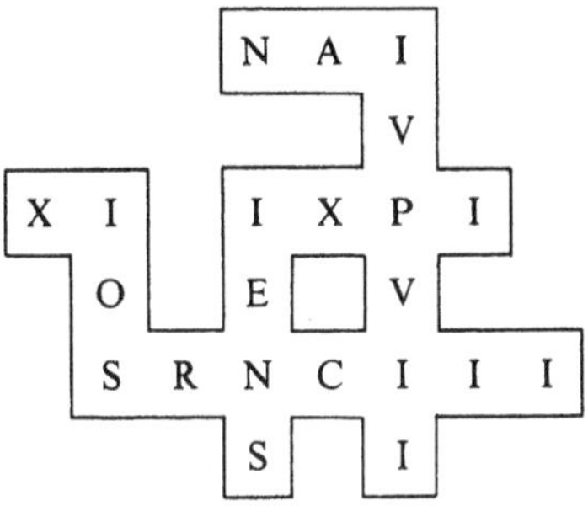

Cher lecteur,

J'entre bien tard dans la sphère étroite des écrivains au double alphabet, moi qui, il y a plus de quarante ans déjà, avais accueilli sur mes terres un général épris de mathématiques. Il m'avait parlé de ses projets grandioses en promettant d'ailleurs de m'envoyer ses ouvrages de géométrie.

Je suis entiché de géométrie et c'est d'elle dont je voudrais vous parler, oh! certes pas de toute la géométrie, mais de celle que fait l'artisan qui taille, burine, amène, gauchit, peaufine les formes. Mon intérêt pour le problème dont je veux vous entretenir ici, je le dois à un ami ébéniste.

En effet comme je rendais un jour visite à cet ami, je le trouvai dans son atelier affairé à un tour. Il se retourna bientôt, puis, rayonnant, me tendit une sorte de toupie et me dit: «Monsieur Besse, vous qui calculez les formes avec vos grimoires, que pensez-vous de ceci?» Je le regardai interloqué. Il poursuivit: «Regardez! Si vous prenez ce collier de laine et si vous le maintenez fermement avec un doigt placé n'importe où sur la toupie, eh bien! la toupie passera toujours juste en son intérieur, sans laisser le moindre espace.» Je rentrai chez moi, fort étonné, car sa toupie était loin d'être une boule. Je me mis alors au travail ...

Après quelques recherches, je compris que, sans être une boule, sa toupie était une surface de révolution dont toutes les géodésiques avaient (sensiblement) la même longueur. En consultant mes grimoires, comme disait mon ami l'ébéniste, je m'aperçus que Darboux avait déjà posé ce problème d'équations différentielles en 1894. En 1903 Zoll avait donné un exemple explicite d'une telle surface et Funk avait trouvé en 1913 une méthode pour en construire une famille continue (mais V. Guillemin n'a rendu cette construction rigoureuse que de nos jours). Gambier, quant à lui, a consacré de multiples études dans les années 20 à l'extension de ces constructions, en particulier au cas des polyèdres. Il prit aussi plaisir à questionner certain de mes amis, mais ceci est une autre histoire.

Tout au long de ces années, les mathématiciens travaillaient aussi à formaliser leur art... Depuis le début du siècle ils réussirent à abstraire les caractéristiques essentielles de la notion de surface plongée dans notre espace de tous les jours, créant les variétés différentielles, riemanniennes et autres choses savantes. Mieux, parmi les espaces riemanniens symétriques d'E. Cartan, une nouvelle classe d'espaces à géodésiques fermées de même longueur apparut: les espaces projectifs sur les nombres réels, sur les nombres complexes et sur les quaternions (ainsi que l'ésotérique plan projectif des octaves de Cayley).

Voilà un sujet comportant bien peu d'exemples, me direz-vous! C'est cela qui incita Blaschke à conjecturer que, sur les espaces projectifs, seule leur structure symétrique pouvait être à géodésiques toutes fermées de même longueur.

A propos de ce problème bien spécifique de géométrie, les techniques les plus variées des mathématiques ont eté mises en œuvre, d'où l'idée directrice de ce livre: rassembler tous les outils permettant une approche de ce problème géométrique. Aussi j'expose les résultats de G. Reeb (1950) repris en 1974 par A. Weinstein se fondant sur la géométrie symplectique; le théorème de R. Bott (1954) utilise la topologie algébrique, celui de L. Green (1963) la géométrie intégrale du fibré unitaire, ceux de R. Michel (1972) outre la géométrie riemannienne l'analyse géométrique, enfin ceux de J. Duistermaat et V. Guillemin (1975) l'analyse des distributions.

Mais, malgré tous ces résultats partiels, on ne peut pas dire que le problème ait beaucoup avancé. J'ai donc été très heureux de pouvoir accueillir en Mai 1975 dans mes terres, au pays que hantent toujours mouflons et cumulo-nimbus, une table ronde consacrée à l'étude des espaces à géodésiques fermées* et autour de laquelle la cuisine locale a été fort appréciée. C'est à cette occasion que mes amis m'ont convaincu qu'il fallait que je fasse un livre**. Mes réceptions du mardi après-midi ont vite pris un caractère animé. Pourtant je crus plusieurs fois que je n'arriverais pas à atteindre le but que je m'étais assigné.

Il m'a été possible d'ordonner ce que je connais de ce problème et de préciser les nombreuses questions non résolues qui s'y rattachent à l'occasion de nombreuses discussions avec entre autres: Pierre Bérard, Lionel Bérard Bergery, Marcel Berger, Jean-Pierre Bourguignon, Yves Colin de Verdière, Annie Deschamps, Jacques Lafontaine, René Michel, Pierrette Sentenac. Le texte de ce livre est trop formalisé à mon goût, mais vous savez comment sont devenus les mathématiciens! De plus, écrire dans la langue du poète W. Hamilton a été pour moi une dure contrainte. Je crains, cher lecteur, que vous ne pâtissiez des conséquences.

En espérant tout de même avoir pu vous faire partager mon goût pour la géométrie, je reste Géodésiquement

Vôtre,

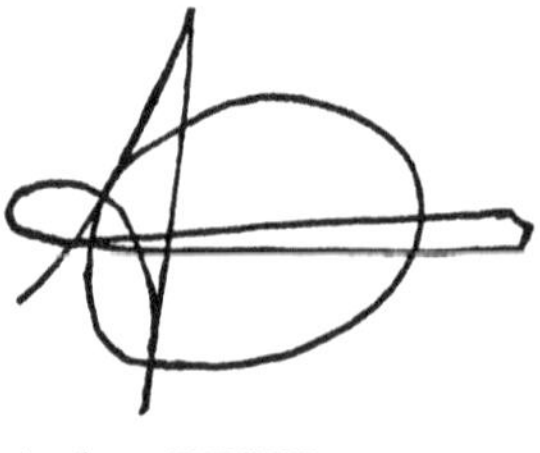

Arthur BESSE

Le Faux, le 15 Juin 1976

---

* L'Université Paris VII et le C.N.R.S. ont rendu cette réunion possible et je les en remercie.

** Je remercie les Editions Springer d'avoir bien voulu accepter de le publier.

Table of Contents

Chapter 0. Introduction

A. Motivation and History

0.1. Suppose that a Riemannian manifold (M, g) is one of the following: $(S^d$, can), $(\mathbb{R}P^d$, can), $(\mathbb{C}P^n$, can), $(\mathbb{H}P^n$, can), $(\mathbb{C}aP^2$, can), namely one of the compact symmetric spaces of rank one, the so-called CROSSes in 3.16—endowed with their canonical Riemannian structure *denoted* by can. Then all geodesics of (M, g) are closed (if you doubt this fact see 3.31). For a mathematician the basic question is then: are these examples the only manifolds all of whose geodesics are closed (up to isometry, of course)? If not, give counter-examples and then try to classify—even describe—all manifolds with that property.

0.2. To be more precise let us introduce for a Riemannian manifold (M, g) the two following assertions:

SC: there exists a positive real number l such that for every unit tangent vector ξ in the unit tangent bundle UM of (M, g), the geodesic γ with initial velocity vector $\dot{\gamma}(0) = \xi$ is

i) a periodic map with *least* period l (i.e. $\gamma(t+l) = \gamma(t)$ for every real number t and this does not hold for any l' in $]0, l[$);

ii) γ is *simple*, i.e. γ is injective on $[0, l[$.

SC^m: there exists a point m in M such that the above property is supposed to hold for every ξ in the unit sphere $U_m M$ at m (assumption on m only).

0.3. Then the ten thousand dollar question is: what can be said about an SC (or an SC^m) Riemannian manifold (M, g)? First it should be clear to the reader that in the SC^m case we cannot hope to get isometric or metric implications on M. For example,

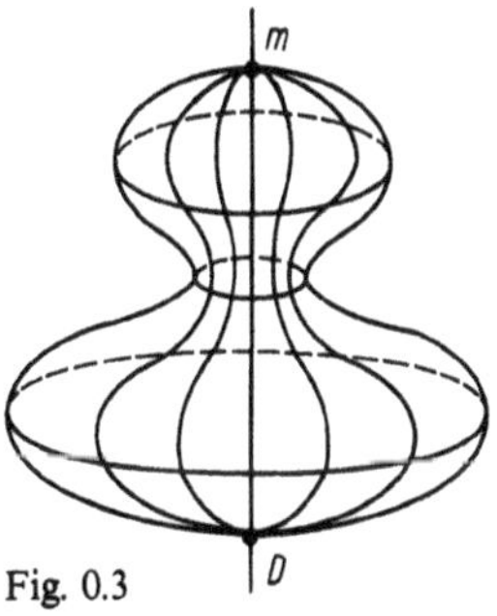

Fig. 0.3

any surface S of revolution in $\mathbb{R}^3$ about some axis D will satisfy the SC^m property if S meets D at some point m; but we can hope to be able to prove that a two-dimensional (M,g) with the SC^m property is necessarily homeomorphic or diffeomorphic to S^2 or $\mathbb{R}P^2$ (see 7.23).

On the contrary, if SC holds for (M,g) we can hope to prove that (M,g) is isometric to one of the CROSSes above.

History of the SC condition

0.4. It seems that the first historical attempt about the SC property is that of the pages 6—9 of Darboux [DX], who gave the explicit Condition 4.11 for the equation of a plane curve to generate by rotation around an axis a Riemannian surface (M,g) with the SC property. But Darboux did not establish the existence of such a globally defined metric on S^2 [of course non-isometric to the standard sphere (S^2, can)!].

0.5. In 4.27 the reader will find a pear constructed by Tannery. Every geodesic is closed and has least period l except the equator which has period l but least period $l/2$. Moreover, it has a singularity at one point. It is of historical interest to check the taste of mathematicians at the end of the century on this example. Tannery's pear is not a smooth manifold but it is algebraic! So this pear did not settle the question: does there exist a non-standard metric on S^2 with the SC property?

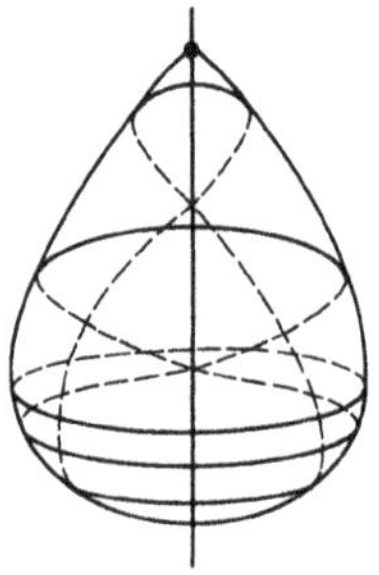

Fig. 0.5

0.6. The question was settled in 1903 by Zoll (see [ZL]) who gave an explicit (real analytic by the way) Riemannian manifold of revolution (S^2,g) with the SC property (see Chapter 4).

0.7. Having such a counter-example we are still far from the complete solution of the problem. A first question is: are there many non-isometric SC Riemannian manifolds (S^2,g)? But also secondly, are there such SC Riemannian structures on the other candidate manifolds $\mathbb{R}P^d$, $S^d (d \geqslant 3)$, $\mathbb{C}P^n (n \geqslant 2)$, $\mathbb{H}P^n (n \geqslant 2)$, $\mathbb{C}\mathrm{a}P^2$?

Believe it or not (and in any case deduce from this fact that the problem is a very hard one despite its simple geometric formulation) nothing appeared in the literature before L. Green's theorem in 1961! Almost all of the main questions are open. The present state of the problem is as follows.

0.8. In 1913 Funk in [FK 1] studied the deformations of (S^2, can) by a one-parameter family $(S^2, g(t))$ with $g(0) = \text{can}$ and $g(t) = \varphi(t)\,\text{can}$, such that $g(t)$ is SC for every t.

What he essentially proved is the non-existence of such deformations under the additional condition that for every t the function $\varphi(t)$ on S^2 is *even* (i.e. invariant under the antipodal map) or, equivalently, that those $(S^2, g(t))$ are Riemannian coverings of a family of *SC* Riemannian metrics $\underline{g}(t)$ on the real projective space $\mathbb{R}P^2$. Otherwise stated: infinitesimally near the canonical one, there is no non-trivial *SC* Riemannian structure on $\mathbb{R}P^2$.

0.9. Moreover, Funk, in the same paper [FK 1], tried to construct a one-parameter family of *SC* metrics (S^2, $\varphi(t)$ can) with initial conditions $\varphi(0)=1$ and $\frac{d\varphi(t)}{dt}(0)=h$ (for any *odd* function h on S^2, i.e. satisfying $h\circ\sigma=-h$ for the antipodal map σ of S^2) as sum of a series. He failed to achieve convergence of his series. The existence of such deformations, for every odd initial derivative, has just been proved (1976) by V. Guillemin in [GU].

0.10. The existence of an *SC* structure itself on $\mathbb{R}P^2$ was studied by Blaschke in 1927 (see first edition of [BE] or [GN 2]), who conjectured that there is no non-trivial one. Truly speaking Blaschke was studying the problem lifted up to S^2 from the preceding one by the canonical covering $S^2 \to \mathbb{R}P^2$, see 0.30. Notice here again how long it took for mathematicians to really think about the abstract manifold $\mathbb{R}P^2$. This is confirmed by the fact that the elliptic geometry was founded long after the hyperbolic one (despite its greater simplicity). The non-existence of such non-trivial *SC* structures was proved by L. Green in 1961 (see [GN 2] or 5.59).

0.11. The existence of a nontrivial *SC* Riemannian metric g on S^d for $d \geqslant 3$ is not too hard to show (once one knows Zoll's example) and was settled by A. Weinstein (unpublished, see 4.E). At the moment of our writing, the problem of the existence of an *SC* Riemannian structure non-isometric to a CROSS on any of the manifolds $\mathbb{R}P^d$ $(d \geqslant 3)$, $\mathbb{C}P^n$ $(n \geqslant 2)$, $\mathbb{H}P^n$ $(n \geqslant 2)$, $\mathbb{C}a\,P^2$ is open. See however Appendix *D*.

0.12. However, in 1972 R. Michel in [ML 2] extended Funk's infinitesimal non-deformation result 0.8 to any dimension d for $\mathbb{R}P^d$. Otherwise stated: there is no hope of finding a non-trivial *SC* structure on $\mathbb{R}P^d$ for $d \geqslant 3$ by only an infinitesimal method.

0.13. The only general condition known in order that a Riemannian manifold (M, g) satisfy the *SC* property is the following result of A. Weinstein (see [WN 3] or 2.21): the total volume $\mathrm{Vol}(g)$ of a d-dimensional Riemannian manifold (M, g), which is a *SC* with period 2π, is an integral multiple of the volume $\beta(d)$ of (S^d, can).

History of the SC^m condition

0.14. As seen in 0.3 the realm of conclusions is now no longer Riemannian geometry but algebraic topology. What are the compact C^∞ manifolds M on which there exists a Riemannian metric g which makes (M, g) into an SC^m manifold? Are there others besides S^d, $\mathbb{R}P^d$, $\mathbb{C}P^n$, $\mathbb{H}P^n$ and $\mathbb{C}aP^2$? The first and basic contribution on the subject is that of R. Bott in 1954 [BT], who proved that such an M should have the same integral cohomology ring $H^*(M; \mathbb{Z})$ as that of a CROSS, see also 7.23.

0.15. But we are still left with many problems since the above cohomology ring condition is far from a condition implying the same diffeomorphism, homeomorphism or homotopy type. For example any exotic sphere would do it. More precisely, L. Bérard Bergery found in 1975 an exotic sphere S^{10} with an SC^m Riemannian structure (see [B.B 1] or Appendix C). Other examples of manifolds with the same cohomology ring as that of a CROSS are the Eells-Kuiper exotic quaternionic projective planes [E-K], but up to the present day no SC^m Riemannian structure is known to exist on them.

0.16 Note. A completely different theme in Riemannian geometry is that of the existence of one, or many, closed geodesic(s) on a Riemannian manifold (M, g). For that subject we refer the reader to the very extensive and complete new book by W. Klingenberg [KG 2].

B. Organization and Contents

Chapter 1. Basic Facts about the Geodesic Flow

0.17. Despite our good intentions we could not write a self-contained book. However, we have put in Chapter 1 a good deal of the Riemannian geometry which is needed for the rest of the work. The unifying object in Chapter 1 is the unit tangent bundle of a Riemannian manifold, equipped with the geodesic flow. That emphasis on the unit tangent bundle is needed because it was the basic escalation of the second half of this century, which permitted Green's theorem, symplectic geometry (and we dare say Fourier integral operators!), etc. ... As a matter of fact L. Green was familiar with ergodic theory.

0.18. Chapter 1 is then quite developed and tries to help the reader become acquainted with the unit tangent bundle and the double tangent bundle. Foundations of Riemannian geometry are developed from the beginning, starting with the search for curves which minimize energy via the calculus of variations. One then gets the geodesic *spray* Z_g via the equation $i_{Z_g} d\alpha_g = -dE_g$ (cf. 1.47).

0.19. The problem here is to define the so-called Levi-Civita connection. We take this opportunity to prove that any spray gives rise to a unique symmetric connector and conversely. The method we adopt to construct the Levi-Civita connection is rather different from the one given in [BR 3], [G-K-M], or [K-N 1]. In particular the usual cyclic permutation trick which gives existence and uniqueness of the Levi-Civita covariant derivative does not play the same role here.

0.20. The end of the chapter is devoted to the Riemannian geometry of the unit tangent bundle, which is needed in Chapters 2 and 5.

Chapter 2. The Manifold of Geodesics

0.21. This chapter is devoted to the study of the manifold one can construct from the set CM of geodesics of a C manifold [i.e., a Riemannian manifold (M, g) with the same requirements SC as in 0.2 except that closed geodesics need not necessarily be simple curves]. The set CM has a natural structure of a C^∞ symplectic manifold, which moreover makes the map $q: UM \to CM$ into a fibration by circles. Here for ξ in UM $q(\xi)$ denotes the closed geodesic whose initial velocity vector at time 0 is precisely ξ. This is a particular case of a construction due to J.-M. Souriau [SU]. This construction is of great importance in mechanics and gives some new openings in order to study non-integrable Hamiltonian systems.

0.22. Chapter 2 systematically exploits the existence of CM and of the bundle $UM \to CM$; in particular, the various Riemannian structures that one can put—in a natural way— on CM and the Riemannian properties of these structures.

0.23. A basic fact is that the canonical 1-form of UM (see 1.58) turns out to be a connection form for the circle bundle $UM \to CM$, whose curvature form is simply the symplectic form of CM (see 2.11). It seems that this fact was exploited for the first time by Reeb in [RB 1] and it culminates in A. Weinstein's theorem 2.21 on the volume of a C manifold. Notice that the historical remark 0.10 can typically be applied here.

Chapter 3. Compact Symmetric Spaces of Rank one from a Geometric Point of View

0.24. We saw in 0.1 that compact symmetric spaces of rank one are our basic examples in the book. Thus they deserve an expository chapter. We think the reader will be glad to have most of the facts concerning them at hand, facts which come from various fields. This explains why almost any book on symmetric spaces is concerned with only one or two aspects. Fields coming into the picture are: homogeneous spaces of Lie groups, Riemannian geometry via the covariant derivative of the curvature tensor and also via the geodesic symmetries around points (hence the name symmetric spaces!), projective geometry and analysis to study the spectrum of the Laplacian.

0.25. We have tried to give a lucid exposition centered on projective spaces, carefully expliciting in the homogeneous space framework the isotropy action (which to us did not seem well described in the literature). Also we have carefully written down the two-fold inheritance: two kinds of inclusions, one from $\mathbb{R} \subset \mathbb{C} \subset \mathbb{H} \subset \mathbb{C}a$ and the other one $\mathbb{K}P^n \subset \mathbb{K}P^{n'}$ $(n \leqslant n')$. Finally we include an explicit classification of all totally geodesic submanifolds of our CROSS'es.

0.26. All of this program is not too hard to carry out for $\mathbb{K} = \mathbb{R}$, $\mathbb{C}$, $\mathbb{H}$ or for the spheres. However we encounter a prime difficulty when defining $\mathbb{C}a\,P^2$. The reason for this is the nonexistence of a Hopf fibration $S^{23} \to \mathbb{C}a\,P^2$ with fiber S^7 which would allow us to define $\mathbb{C}a\,P^2$ as a suitable quotient of S^{23} (as we did for the other $\mathbb{K}P^n$). The basic reason for this non-existence is the fact that the system of Cayley numbers $\mathbb{C}a$ is not associative. This difficult is illustrated by the following two dependent observations: firstly there is no printed text in which $\mathbb{C}aP^2$ is constructed and studied in detail and secondly the number of young (or aged) geometers asking us for references on this wonderful but somewhat frustrating $\mathbb{C}aP^2$ is fairly large.

Chapter 4. Some Examples of C- and P-Manifolds: Zoll's and Tannery's Surfaces

0.27. If Chapter 3 considers the *classical* examples of *SC* manifolds, Chapter 4 studies the 2-dimensional *exotic* examples, namely those of Zoll and of V. Guillemin. Zoll's original construction (see [ZL] or [BR 3]) was almost a miracle: the meridian curve was built up in two pieces Γ', Γ'' whose union is real analytic (believe it or not!).

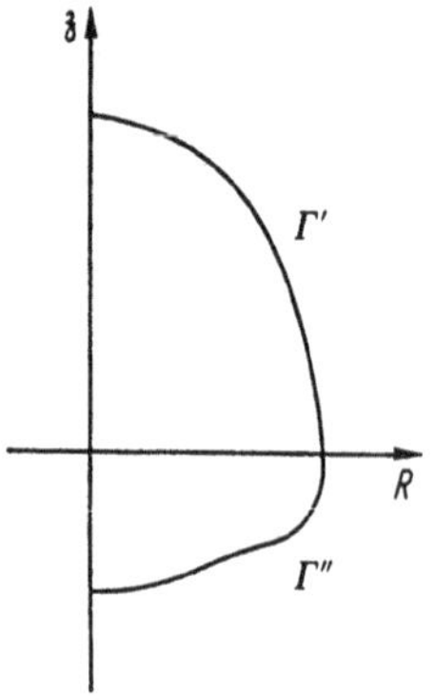

Fig. 0.27

0.28. We give an abstract expression for the *SC* Riemannian metrics g on S^2 which admit an S^1 action by isometries. This expression is very simple and involves an arbitrary odd real-valued function on $[0, \pi]$. It extends Zoll's example to a wider class and also avoids the above tedious construction.

We also give various properties of these Zoll metrics and in particular an explicit $\mathbb{R}^3$-embedding for some of them.

0.29. The chapter ends with Guillemin's result quoted in 0.9, see 4.H. The proof is only sketched, the reader being referred to [GU] because it is fairly technical. We only note that V. Guillemin's result proves, among other facts, that there are many *SC* metrics on S^2 with no isometries but the identity. It is in order to quote here Gambier (1925, see [GR] or 4.C) and Funk (1923, see [FK 2] or 4.B) who studied this problem with some success. We finish with a proof following A. Weinstein [WN 4] that the geodesic flows of two *SC* metrics on S^2 are conjugate by a symplectic diffeomorphism.

Chapter 5. Blaschke Manifolds and Blaschke's Conjecture

0.30. There is a striking difference between the behaviour of geodesics of (S^2, can) and those of a Zoll surface. In (S^2, can) all geodesics starting from a given point m pass at time π the same point m', namely the antipod of m. But in a Zoll surface of revolution the geodesics, starting from a point m which is not on the axis of revolution, have in general a non-trivial envelope Γ:

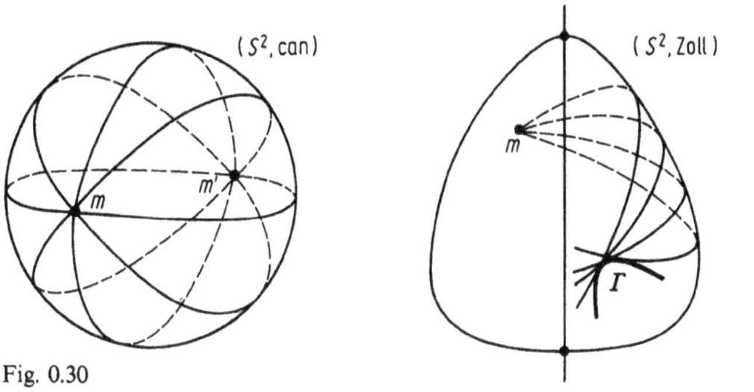

Fig. 0.30

The problem mentioned in 0.10 and studied by Blaschke was the following: suppose that (S^2, g) is a Riemannian manifold such that every geodesic from any point m is at time π at a constant point m' different from m. Is then (S^2, g) necessarily isometric to (S^2, can)? We mentioned in 0.10 that this problem is equivalent to studying SC structures on $\mathbb{R}P^2$.

0.31. More generally it is natural to study Riemannian manifolds (M, g) whose geodesic behaviour is that of a CROSS in the following sense (compare with 3.35): for every pair of points (m, n) in M, such that n belongs to the cut-locus Cut (m) of m, the shortest geodesics (called segments, see 5.12) from m to n have velocity vectors in U_nM which form a whole great sphere of U_nM (see 5.36). The case of only two antipodal points in U_nM is that of $(\mathbb{R}P^2, \text{can})$; that of all possible directions is the generalization of Blaschke's case. For $(\mathbb{C}P^n, \text{can})$ one gets a circle of the Hopf fibration of U_nM:

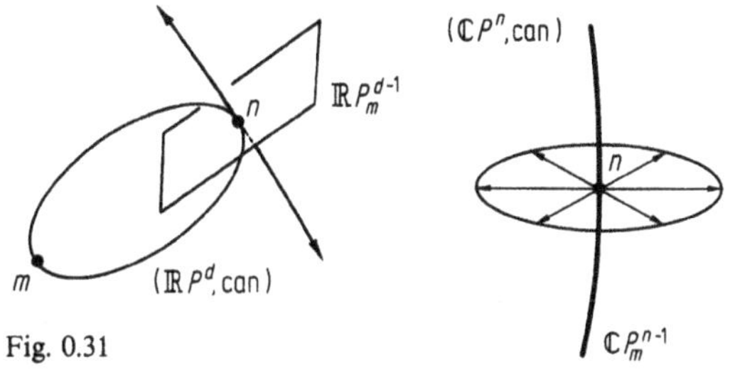

Fig. 0.31

0.32. Chapter 5 is devoted to the study of Riemannian manifolds (M, g) with such a geodesic structure. They are called *Blaschke manifolds*, and *Blaschke manifolds at m* if the above condition is required to hold only for every point n in the cut-locus of some point m in M. They will be characterized by two other equivalent conditions: the first one is that the distance from m to any point n in its cut-locus is constant. The second one is the fact that M can be written as $D \bigcup_a E$, where D is a closed d-dimensional ball, E is a closed disc bundle over a nice compact manifold, and a is an attaching diffeomorphism between the boundary of D and that of E (a boundary which has then to be homeomorphic to S^{d-1}).

0.33. Blaschke's conjecture is that any Blaschke manifold is isometric to a CROSS. L. Green's theorem mentioned in 0.10 proves the conjecture when the dimension is equal to 2. For dimensions $d \geqslant 3$ the conjecture is completely open. However, we shall give Michel's theorem, which is the extension to any dimension d of Funk's result 0.8. See also Appendix D.

Chapter 6. Harmonic Manifolds

0.34. Harmonic manifolds "slept" in this book by the side door. Their definition has nothing to do with closed geodesics. For a Riemannian manifold (M, g), a point m in M and a small enough real number r the geodesic sphere $S(m, r)$ is a nice submanifold. To compute its volume we can work in polar coordinates centered at m. The volume of $S(m, r)$ will be given by the formula $\operatorname{Vol}(S(m, r)) = \int_{U_m M} \delta(\xi, r) d\sigma$ where ξ denotes a unit vector in $U_m M$ and σ the canonical measure on $U_m M$ (see 1.120). Then we say that (M, g) is locally harmonic if the density $\delta(\xi, r)$ depends only on r and not on ξ or on m. One also defines a notion of global harmonicity.

0.35. The problem of finding all locally or globally harmonic manifolds is completely open, except for dimensions less than 5. But a result of Allamigeon implies that compact and simply connected globally harmonic manifolds are Blaschke manifolds (with of course additional properties). That is why we have decided to include harmonic manifolds in this book.

0.36. The beginning of the chapter gives the local and global definitions above, together with a third one: that of strongly harmonic manifolds, which requires compactness and involves the heat equation or, equivalently the spectrum of the Laplacian and its eigenfunctions. Equivalence of these notions is studied in detail.

0.37. Harmonic manifolds necessarily satisfy curvature conditions. The first condition yields the fact that such a manifold should be an Einstein manifold, the second is quadratic in the curvature tensor R, the third is cubic in R and quadratic in its covariant derivative DR. The other conditions are so complicated that nobody has ever worked them out. From the curvature conditions above we prove the Lichnerowicz-Walker theorem: harmonic manifolds of dimension less than or equal to 4 are locally isometric to a ROSS or are flat.

0.38. Allamigeon's theorem is proved essentially in Chapter 5. At the end of Chapter 6 we study strongly harmonic manifolds. They can be embedded as minimal submanifolds in Euclidean spheres and moreover these embedded submanifolds have all their geodesics congruent as curves in the Euclidean space. This last condition is a good support for conjecturing that simply connected globally harmonic manifolds are isometric to CROSSes.

Chapter 7. On the Topology of SC- and P-Manifolds

0.39. Roughly speaking this chapter concerns SC^m manifolds (for a definition see 0.14 and 0.15) and gives the proof of R. Bott's theorem. But in fact, when thinking of weakening the very strong SC^m assumption and also studying links between SC and SC^m conditions, one encounters disguised difficulties. An example might help the reader: consider (S^3, can) and a subgroup G of the special orthogonal group $SO(4)$ which acts on S^3 without fixed point. Then we can construct a quotient Riemannian manifold $(M,g)=(S^3/G, \text{can}/G)$, called a *lens space*. On (M,g) every geodesic is periodic with period 2π as is the case on (S^3, can). But 2π is not the least period of these geodesics, some have a least period equal to $2\pi/k$ for some integer $k>1$.

0.40. *Worse*: if (M,g) is a Riemannian manifold with the property that every geodesic is periodic with a certain period, it is not obvious that there is a *common* period for all these geodesics. This is in fact a very nice result due to Wadsley ([WY 2]) (it is included in Appendix A, by D. Epstein, for the convenience of the reader). Notice that this result, considered as a result about one-dimensional foliations (this is precisely the case for the geodesic flow on UM) is only valid for geodesic foliations; namely our assumption transferred to the unit tangent bundle UM is precisely that UM admits a geodesic foliation with compact leaves (the integral curves of the geodesic flow). The desired conclusion about a common period is equivalent to the boundedness of the length of the leaves. But for a non-geodesic foliation of dimension one on a compact manifold, there is no such result since Thurston has constructed an example of a compact 5-dimensional manifold with a one-dimensional foliation whose leaves are compact but have unbounded lengths. This example will be sketched in Appendix A.

0.41. We hope we have convinced the reader of the existence of snags hidden in the subject. Then he will understand the necessity at the beginning of Chapter 7 of formalizing how a manifold can have all its geodesics closed. After these definitions, we give some non-trivial implications between them and some open problems on the subject. A weakened form of R. Bott's theorem is proved for a class of manifolds all of whose geodesics are closed (namely P-manifolds).

0.42. The chapter ends with the complete classification of homogeneous manifolds all of whose geodesics are closed: up to isometry these are the CROSSes.

Chapter 8. The Spectrum of P-Manifolds

0.43. This chapter is mainly dedicated to a proof of the surprising Duistermaat-Guillemin result ([D-G]) to the effect that P_l-manifolds (i.e. Riemannian manifolds such that every geodesic admits l as *a* period) can be characterized as those compact Riemannian manifolds for which the square root of the spectrum of the Laplace operator Δ is asymptotically an arithmetic progression (for example Zoll's surfaces will necessarily have such a spectrum). The initial value and the ratio of the arithmetic sequence yield exactly the least common period l and modulo 4 the Morse index of our closed geodesics. The chapter ends with a property of the first eigenvalue λ_1 of Δ for an SC structure on S^d.

Appendices

0.44. We have added to the chapters of the book five appendices on related topics. Appendix A, written by David Epstein, presents a proof of Wadsley's result 0.40 and contains a brief description of Thurston's counter-example 0.40.

0.45. Appendix B by Jean-Pierre Bourguignon was motivated by the following question which arises naturally when studying *SC* manifolds: Consider a Riemannian metric on S^2 with the Blaschke property mentioned in 0.30 (i.e. all geodesics from m meet again at time π at a different point m', and not before). If we take the transverse derivative of this condition along a fixed geodesic γ, the general philosophy of Jacobi fields will yield the following consequence: for every t, the Jacobi field Y_t along γ defined by $Y_t(t)=0$ and $Y_t'(0)=1$ will satisfy $Y_t(t+\pi)=0$ and $Y_t(s)\neq 0$ for every s in $]t,t+\pi[$. One way to prove L. Green's theorem (cf. 0.10) would be to prove that the corresponding Jacobi equation $Y''+\sigma Y=0$, where σ is the curvature, is necessarily $Y''+Y=0$, namely that (S^2,g) has constant curvature hence is isometric to (S^2, can). But only quite recently did some analysts give the explicit general form of functions σ such that the equation $Y''+\sigma Y=0$ has only antiperiodic solutions. Appendix B gives this explicit form (which is due to [NN]) and also some considerations about the case of a system (with the higher dimensional Blaschke conjecture in mind).

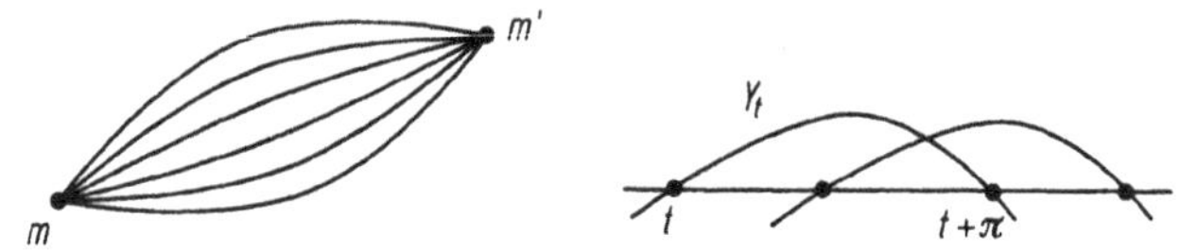

Fig. 0.45

0.46. Appendix C by Lionel Bérard Bergery gives the construction (cf. 0.15 and 0.32) of a Blaschke structure at the center m of a disk D on a compact manifold $M=D\bigcup_a E$. The existence of a Riemannian metric g on $M=D\bigcup_a E$ such that all geodesics from m are not only loops but are *closed* geodesics is then studied. Finally an example of an exotic sphere with such a Riemannian structure is given.

0.47. Appendix D by Marcel Berger contains the proof of Blaschke's conjecture for spheres and real projective spaces. The proof rests eventually on an inequality proved by Jerry L. Kazdan in Appendix E.

C. What is New in this Book?

For the convenience of the reader and in order to be fair we give here a list of what is really new in the book.

It should be clear that these new things are at various levels of difficulty.

Chapter 1. An exposition of the foundations of both Riemannian and symplectic geometry starting from the geodesic flow, an intrinsic version of the relationship between connectors and sprays.

Chapter 2. A systematic treatment of the manifold of geodesics of a C manifold, of its Riemannian geometry and in particular of its geodesics and its curvature.

Chapter 3. A lucid exposition (one must admit not always with detailed proofs) of the Cayley projective plane, with a good set of references; and a presentation of the symmetric spaces of rank one with a detailed exposition of the isotropy group and their two-fold hereditary character (change of dimension and change of field).

Chapter 4. The complete determination of all Zoll metrics on S^2 which are invariant under an S^1 action, and their realization from their abstract expression as equivariant imbeddings in $\mathbb{R}^3$; explicit computation of some of these examples, their cut-loci and conjugate loci.

Chapter 5. A new proof of Michel's theorem, whose original proof was quite difficult and somewhat mysterious.

Chapter 6. The introduction of various definitions of harmonicity on a Riemannian manifold and the relations between these definitions in the differentiable case. In particular the introduction of the notion of strong harmonicity and the fact that a strongly harmonic manifold admits isometric imbeddings in Euclidean spheres S, which yields minimal submanifolds in S all of whose geodesics are congruent curves of the Euclidean space.

Chapter 7. The introduction of various definitions for manifolds "all of whose geodesics are closed" and various implications among these definitions. A proof of a Bott's type theorem for P-manifolds. The complete classification of homogeneous manifolds all of whose geodesics are closed.

Chapter 8. The proof which we give for the Theorem 8.9 (Duistermaat-Guillemin) is slightly different from the original one. The result in Proposition 8.47 is new, at least to our knowledge.

Appendix A. An improved proof of A. Wadsley's theorem (see [WY 2]).

Appendix B. Contains a slightly different method of proving L. Green's theorem (see B.25).

Appendix C. A new proof of A. Weinstein's converse to the Allamigeon-Warner theorem, using Riemannian submersion techniques. An example of an exotic sphere with a Riemannian metric such that for some point every geodesic through that point is periodic.

Appendix D. An extension of L. Green's Theorem to every dimension (see 0.10, 0.33).

Appendix E. An integral inequality used in Appendix D.

D. What are the Main Problems Today?

In various chapters the reader will find many problems, in particular in 5.74—79, 6.16, 6.88, 6.105, 7.66. However, for his convenience we extract here what are, we think, the main ones.

Pb. 1: classify C-structures on S^d ($d=2$ and $d>2$) (do they form a nice subset of Riemannian metrics on S^d?).

Pb. 2: do there exist non-canonical C-structures on $\mathbb{C}P^n$ ($n\geqslant 2$), $\mathbb{H}P^n$ ($n\geqslant 2$), $\mathbb{C}a\, P^2$? If yes, classify them.

Pb. 3: settle Blaschke's conjecture.

Pb. 4: do there exist C-structures with a Weinstein integer different from the standard one?

Pb. 5: what exactly are the manifolds admitting a C-structure or a P-structure?

Pb. 6: what is the right metric on $C^g M$ (the manifold of geodesics)?

Pb. 7: does "P" imply "C" for simply connected manifolds?

Pb. 8: extend R. Michel's result 5.90 to more general deformations on $\mathbb{C}P^n$ and to other CROSSes.

Pb. 9: find all harmonic manifolds (infinitesimally, locally, globally, strongly).

Pb. 10: classify the geodesic flow associated to C_l-metrics from a symplectic point of view (see 8.43 for the case of S^2).

Pb. 11: generalize 8.47 to higher dimension.

Note that Pb. 9 is a subproblem of what is a challenging problem in Riemannian geometry at the moment: find all compact Einstein manifolds.

Chapter 1. Basic Facts about the Geodesic Flow

A. Summary

1.0. This chapter is intended to provide the basic tools in both Riemannian and symplectic geometry, both domains to which the geodesic flow belongs.

It only assumes a basic knowledge of differential geometry such as manifolds, differentiable maps, the tangent functor, exterior differential forms and the exterior differential, vector fields and the Lie derivative. Good references for this material are [AM], [GO 1], [SG], [WR 3].

It does not contain any new result (how could it?): the itinerary at least is not the usual one since after brief preliminaries we start right away with the geodesic vector field and the geodesic flow as they arise from the calculus of variations. Then we come to sprays, connectors and eventually to covariant derivatives.

We have tried to give only intrinsic definitions, but, as often as we can, we give the coordinate expressions.

Some more details: quite often it has appeared necessary to generalize some notions in order to understand them better: this is why we start in Paragraph B by reviewing some generalities on vector bundles such as the role of the dilations and homogeneity properties derived from it, the fiber derivative, the vertical subbundle of the tangent bundle to a vector bundle and pull-back bundles.

Paragraph C reviews the natural symplectic structure on the cotangent bundle deduced from the Liouville differential 1-form.

A thorough study of the double tangent bundle is the aim of Paragraph D (for example we give a definition of it in 1.18 as equivalence classes of one-parameter families of curves, a definition which turns out to be well adapted to the calculus of variations or the geodesic problem as looked at in the book). We then introduce its canonical involution, the notion of a jet. The vertical endomorphism will play a crucial role in the calculus of variations in its Lagrangian formulation.

Paragraph E is devoted to basic definitions concerning Riemannian metrics such as the musical isomorphisms, kinetic energy, energy and length of curves together with their normal parametrization and the unit tangent bundle.

We then turn to the calculus of variations for a Lagrangian L in Paragraph F: we develop the Lagrange equations in a completely intrinsic manner (this involves the notion of vertical differential $\hat{d}$ of L) 1.28 as due (only recently: 1961!) to J. Klein ([KL]). We examine the regular case ($d\hat{d}L$ is then a symplectic differential form) and in particular properties of the vector field X_L, solution of the Lagrange equations, viewed both as a Hamiltonian vector field and a second-order differential equation. We conclude by presenting the usual Hamilton equations on the cotangent bundle.

Modern references for the calculus of variations or notions it involves are [GO 1], [MN] or [SG].

The geodesic vector field Z_g and its flow are eventually presented in Paragraph G. We prove that Z_g is a spray and establish the equations of geodesics in their usual form. We then come to the exponential maps and their basic properties. Liouville's theorem for the geodesic flow is proved. We also examine properties of Z_g restricted to the unit tangent bundle.

We found it necessary to devote the whole of Paragraph H to the notion of a connector (sometimes called connection map). This is one definition among many others of a linear connection (it was introduced in [DK]). It turns out to be well adapted, for example, to manifolds of maps (see [E-M] and [EN 2]). We introduce the horizontal subbundle, the horizontal lift and the local expressions of a connector. We then examine how a connector on the tangent bundle can be adapted to a second order differential equation (Lemma 1.65 seems to be new). The crucial fact is Theorem 1.69 which says that a symmetric connector is equivalent to a spray (cf. [A-P-S]). We define the Levi-Civita connector associated to the geodesic vector field and study the symplectic properties of its horizontal subbundle.

We then come in Paragraph I to the more usual notion of a covariant derivative and how it relates to that of connectors, together with parallel vector fields and parallel transport along a curve. We define the curvature tensor and give the Bianchi identities which it satisfies. We discuss at some length the covariant derivative of differential forms and prove the "Fundamental theorem in Riemannian geometry" (Theorem 1.83).

Paragraph J makes use of almost all notions previouly introduced: we start from one-parameter families of geodesics, establish the Jacobi equation and show how one can use Jacobi fields to study the linearized geodesic flow (Propositions 1.92 and 1.94). We then introduce conjugate points and the conjugate locus.

The tangent bundle TM is made into a Riemannian manifold in paragraph K. We also prove that, in order that the geodesic flow on UM be an isometry, M must have constant sectional curvature 1.

We finish by proving the classical formulas for the first and second variations of the length of curves in Paragraph L. For applications one can consult [MR 1].

Paragraph M presents the canonical measures that one can define on a Riemannian manifold.

B. Generalities on Vector Bundles

1.1. Throughout this book, we will consider C^∞ manifolds. All of the geometric objects we are interested in are implicitely assumed to be smooth.

The *tangent bundle* of a manifold M will be denoted by $p_M: TM \to M$, its *cotangent bundle* by $p_M^*: T^*M \to M$. Generic points will be denoted in M by m, n, p, in TM by u, v, w, in T^*M by λ, μ, ν.

We will usually denote by (x^i) a local coordinate system for M, by $\left(\frac{\partial}{\partial x^i}\right)$ the natural coordinate vector fields and by (dx^i) the natural coordinate 1-forms.

1.2. If $\tau: E \to M$ is a real vector bundle of rank r, then for m in M, E_m is the fibre at m. We will denote by $\mathscr{E}$ the space of C^∞ sections of E and by $\mathring{E} = E - \{0_E\}$, where 0_E is the zero section of E. On E, a special one-parameter family of bundle isomorphisms is defined: the *dilations* a_E^t (for $t \in \mathbb{R}^*$, $a_E^t(v) = tv$). The vector field generating the flow of positive dilations $a_E^{\exp t}$ is called the *Liouville vector field* of E and is denoted by A_E. At v in E, $A_E(v)$ is simply the position vector, i.e. v translated at v in the fiber. Its expression in bundle coordinates (x^i, v^j) at v in E, is

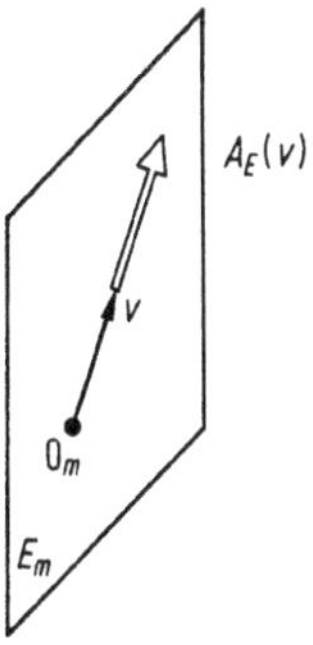

$$A_E(v) = \sum_{j=1}^{r} v^j(v) \frac{\partial}{\partial v^j}.$$

Fig. 1.2

The dilation a_E^{-1} is called the *opposite map* in E.

1.3. We define the *homogeneous bundle* associated with $\tau: E \to M$ to be $\tau_U: UE \to M$, where UE is the quotient of $\mathring{E}$ by the positive dilations; it is a sphere bundle of dimension $r-1$ if E is a vector bundle of rank r. By definition, the vector field A_E spans the kernel of the projection from E to UE.

1.4. A C^∞-function f on $\mathring{E}$ is said to be *homogeneous of degree k* if f satisfies the *Euler identity*

$$A_E \cdot f = kf$$

(where by $A_E \cdot f$ we *denote the derivative of f in the direction* A_E).

A function f on $\mathring{E}$ extends to E if f is polynomial in the v^i's. If f is homogeneous of degree 0, it extends smoothly only if f is constant along the fibers, i.e. if f gives rise to a function $\bar{f}$ on M such that $f = \bar{f} \circ \tau$. The bundle coordinate functions v^i's are homogeneous of degree 1 and the x^i's of degree 0.

1.5. A C^∞ vector field X on $\mathring{E}$ is said to be *homogeneous of degree k* if

$$\mathscr{L}_{A_E} X = (k-1)X$$

(where $\mathscr{L}_{A_E}$ denotes *the Lie derivative in the direction* A_E, i.e. $\left.\frac{d}{dt} a_E^{\exp t}(X)\right|_{t=0}$, where the diffeomorphism $a_E^{\exp t}$ acts naturally by push forward on vector fields).

A vector field X on $\mathring{E}$ can be projected down onto M if and only if X is homogeneous of degree 1. Clearly A_E is homogeneous of degree 1 together with the bundle coordinate vector fields $\left(\frac{\partial}{\partial x^i}\right)$. The vector fields $\left(\frac{\partial}{\partial v^i}\right)$ are homogeneous of degree 0.

1.6. A differential form λ on $\mathring{E}$ is said to be *homogeneous of degree k* if

$$\mathscr{L}_{A_E}\lambda = k\lambda .$$

A differential form λ, smooth on E (and not only on $\mathring{E}$), is the pull-back by τ of a differential form on M if and only if λ is homogeneous of degree 0. Such differential forms are called *basic*. The bundle coordinate 1-forms dx^i's are homogeneous of degree 0; the 1-forms dv^i's are homogeneous of degree 1.

1.7 Remark. Clearly, using the flow $a_E^{\exp t}$ rather than the vector field A_E we could have given similar definitions without any regularity assumption, but this will not be needed later on. Our definition agrees with the classical definition on functions, i.e. f is homogeneous of degree k if for v in E and λ in $\mathbb{R}$, $f(\lambda v) = \lambda^k f(v)$.

1.8. Given a fiber preserving map φ from a vector bundle $\tau_1 : E_1 \to M$ to a vector bundle $\tau_2 : E_2 \to M$, we can define its *fiber derivative* $F\varphi$ which is a fiber preserving map from $\tau_1 : E_1 \to M$ to the vector bundle of linear bundle maps from E_1 to E_2, $L(\tau_1, \tau_2) : L(E_1, E_2) \to M$. It is defined as follows: for each m in M, we define $F\varphi \restriction E_{1,m}$ as the differential of the vector-valued function $\varphi \restriction E_{1,m}$ which maps $E_{1,m}$ into $E_{2,m}$. In the special case where φ is a function on E, we can consider it as a fiber-preserving map from E to the trivial vector bundle $\mathbb{R} \times M$. In this case $F\varphi$ is a fiber-preserving map from E to $E^* = L(E, \mathbb{R})$.

1.9. If we now consider the tangent bundle to E, we can put it into the following commutative diagram

$$\begin{array}{ccc} TE & \xrightarrow{T\tau} & TM \\ {\scriptstyle p_E}\downarrow & & \downarrow{\scriptstyle p_M} \\ E & \xrightarrow{\tau} & M \end{array}$$

which can be thought of in two ways: $(T\tau, \tau)$ is a vector bundle map from $p_E : TE \to E$ onto $p_M : TM \to M$; or (p_E, p_M) is a vector bundle map from $T\tau : TE \to TM$ onto $\tau : E \to M$.

So Ker $T\tau$ is a canonical subbundle of TE, called the *vertical subbundle* of TE and *denoted by* VE; VE can be identified canonically with $E \times_\tau E$ (where $E \times_\tau E$ denotes the fiber product of $\tau : E \to M$ over τ) according to the following remark. Fix m in M, then E_m injects into E. Therefore $TE_m \cong E_m \times E_m$ injects into TE and in fact under this isomorphism, for v in E, $(\text{Ker } T\tau)_v \simeq E_{\tau(v)}$. In this way we *define* a bundle map (ι, τ) of $p_E \restriction VE : VE \to E$ onto $\tau : E \to M$ which is an isomorphism along the fibers.

The Liouville vector field A_E is clearly a *vertical vector field* (i.e. a section of $p_E \restriction VE : VE \to E$).

1.10. If f is a map from N to M and $\tau: E \to M$ a vector bundle over M, then $f^*\tau: f^*E \to N$ denotes the *pull-back bundle* by f. This construction is natural with respect to the tangent functor in the following sense: the two bundles $(Tf)^*T\tau:(Tf)^*TE \to TN$ and $T(f^*\tau): T(f^*E) \to TN$ coincide and we will use the second notation without parentheses. Notice that the total space of the pull-back bundle $\tau^*\tau: \tau^*E \to E$ is nothing but $E \times_\tau E$. Moreover, as a vector bundle over E, $V\tau^*E$ is isomorphic to τ^*VE.

C. The Cotangent Bundle

1.11. To any coordinate system (x^i) on M is canonically associated a coordinate system (x^i, ξ_i) on T^*M, when ξ_i is defined as follows: for ξ in T_m^*M, $x^i(\xi) = x^i(m)$ and $\xi_i(\xi) = \xi\left(\frac{\partial}{\partial x^i}\right)$. A vector Λ in TT^*M is vertical if and only if Λ can be written as

$$\Lambda = \sum_{i=1}^{d} \Lambda_i \frac{\partial}{\partial \xi_i}$$

in natural coordinates.

1.12. There exists on T^*M a canonical differential 1-form α (on a manifold M, $\Omega^k M$ will denote the space of C^∞ sections of $\Lambda^k T^*M$, so α is in $\Omega^1 T^*M$) called the *Liouville form* defined for Λ in TT^*M in the following way

$$\alpha(\Lambda) = p_{T^*M}(\Lambda)\,(Tp_M^*(\Lambda)).$$

In a natural chart, it is clear that $\alpha = \sum_{i=1}^{d} \xi_i dx^i$.

From the local expression, for example, one can see that α is homogeneous of degree 1.

1.13 Proposition. *The Liouville form α is characterized by the following properties:*

i) *it is* semi-basic, *i.e. annihilates the vertical subbundle of the tangent bundle to T^*M;*

ii) *any differential 1-form λ on M is equal to the pull-back of α by λ.*

1.14 *Proof.* First of all notice that in ii), λ is considered both as a 1-form on M and as a map from M to T^*M as the diagram below shows:

$$\begin{array}{ccc} T^*M & \xleftarrow{(T\lambda)^*} & T^*T^*M \\ {\scriptstyle \lambda^*(\alpha)}\uparrow & & \uparrow{\scriptstyle \alpha} \\ M & \xrightarrow{\lambda} & T^*M \end{array}.$$

Let v be in TM. By the definition of a pull-back and of α

$$\lambda^*(\alpha)(v) = \alpha(T\lambda(v)) = \lambda(p_M(v))(Tp_M^* \circ T\lambda(v)).$$

As λ is a section of p_M^*, $p_M^* \circ \lambda = id_M$ and therefore

$$\lambda^*(\alpha)(v) = \lambda(v).$$

The converse follows from the fact that in a natural chart (x^i, ξ_i) for any λ in ΩM and for any i, $\lambda^*(dx^i) = dx^i$, and then from a local computation. □

1.15. Of crucial importance is the fact that the exterior differential 2-form $d\alpha$ on T^*M defines a *symplectic structure*. In general a symplectic form is a non-degenerate closed differential 2-form $\left(d\alpha \text{ is indeed a symplectic form since, in natural coordinates, } d\alpha = \sum_{i=1}^{d} d\xi_i \wedge dx^i\right)$.

Moreover, we see from this formula that natural coordinates are canonical (in the old terminology), i.e. the matrix of $d\alpha$ in this basis is $\begin{pmatrix} 0 & I \\ -I & 0 \end{pmatrix}$. Such a coordinate system will be called a *Darboux system*.

Being non-degenerate, $d\alpha$ defines a volume form $d\alpha \wedge \ldots \wedge d\alpha$ (d times if M is d-dimensional); T^*M is therefore canonically oriented.

The differential 2-form $d\alpha$ is homogeneous of degree 1.

The symplectic structure on T^*M is related with exterior differentiation of differential 1-forms on M by the following formula which can be derived directly from 1.13.ii): let λ be in $\Omega^1 M$, then $\lambda^*(d\alpha) = d\lambda$. In particular, $d\alpha$ restricted to the graph of a closed differential 1-form is zero. Such submanifolds are called *Lagrangian submanifolds*; other examples of Lagrangian submanifolds of T^*M are the fibers of the projection $p_M^*: T^*M \to M$.

In a symplectic vector space any maximal subspace in the family of subspaces on which the symplectic form restricts to zero is called *Lagrangian* (it automatically has half the dimension). The tangent space to a Lagrangian submanifold is clearly a Lagrangian subspace of the ambient tangent space.

D. The Double Tangent Bundle

1.16. The total space TTM of the tangent bundle to TM can be considered in two ways as a vector bundle over TM according to the following diagram:

$$\begin{array}{ccc} TTM & \xrightarrow{Tp_M} & TM \\ {\scriptstyle p_{TM}}\downarrow & & \downarrow{\scriptstyle p_M} \\ TM & \xrightarrow{p_M} & M \end{array} .$$

Then for v in TM if we consider W_1 and W_2 in T_vTM (which is a fiber for the fibration p_{TM}), we can define W_1+W_2. The fiber of the other fibration Tp_M above the zero section of TM is the vertical space VTM. For W_1 and W_2 in TTM such that $Tp_M(W_1)=Tp_M(W_2)$, the sum for this other vector bundle structure is denoted by $W_1 \mathbin{+\!\!\!+} W_2$ to avoid any confusion.

We could also have distinguished the two multiplications by scalars, but we shall not need them later on.

The Liouville vector field on TM is simply *denoted by* A and the *homogeneous tangent bundle* by UM.

1.17. Let (x^i) be a coordinate system. Then (x^i, dx^i) is a natural coordinate system for TM. For the sake of clarity, we replace dx^i (when considered as a function on TM) by X^i. From the coordinate system (x^i, X^i) on TM we deduce a natural coordinate system $(x^i, X^i, d_{TM}x^i, d_{TM}X)$ on TTM which we will denote (again for the sake of clarity!) by $(x^i, X^i, Y^i, \mathscr{X}^i)$.

To make explicit the convention we are making regarding the relative order of the various components, we specify the local expression of p_{TM}

$$p_{TM}(x^i, X^i, Y^i, \mathscr{X}^i)=(x^i, X^i),$$

the local expression of Tp_M being $Tp_M(x^i, X^i, Y^i, \mathscr{X}^i)=(x^i, Y^i)$.

Therefore, a vertical vector in TTM can be written in a local coordinate system as $(x^i, X^i, 0, \mathscr{X}^i)$. The $+$ operation in TTM is then given by

$$(x^i,X^i, Y^i,\mathscr{X}^i)+(x^i,X^i, \bar{Y}^i,\bar{\mathscr{X}}^i)=(x^i,X^i, Y^i+\bar{Y}^i,\mathscr{X}^i+\bar{\mathscr{X}}^i).$$

Lastly, the $\mathbin{+\!\!\!+}$ operation is given by

$$(x^i, X^i, Y^i,\mathscr{X}^i)\mathbin{+\!\!\!+}(x^i,\bar{X}^i, Y^i,\bar{\mathscr{X}}^i)=(x^i,X^i+\bar{X}^i, Y^i,\mathscr{X}^i+\bar{\mathscr{X}}^i).$$

1.18. The double tangent bundle TTM can be identified with equivalence classes of two-parameter families of points in M (or one-parameter families of curves in M) as follows: let c_1 and c_2 be two maps from a neighbourhood of $(0,0)$ in $\mathbb{R}\times\mathbb{R}$ parametrized by (t,s) into M; c_1 is said to be *equivalent* to c_2 if in a chart centered at $m=c_1(0,0)=c_2(0,0)$ the local representatives $\bar{c}_1$ and $\bar{c}_2$ satisfy

1.19
$$\begin{cases} \left.\dfrac{\partial\bar{c}_1}{\partial t}\right|_{(0,0)}=\left.\dfrac{\partial\bar{c}_2}{\partial t}\right|_{(0,0)}, \\ \left.\dfrac{\partial\bar{c}_1}{\partial s}\right|_{(0,0)}=\left.\dfrac{\partial\bar{c}_2}{\partial s}\right|_{(0,0)}, \\ \left.\dfrac{\partial^2\bar{c}_1}{\partial t\partial s}\right|_{(0,0)}=\left.\dfrac{\partial^2\bar{c}_2}{\partial t\partial s}\right|_{(0,0)}. \end{cases}$$

This notion is independent of a chart at m since local representatives $\bar{\bar{c}}_1$ (and $\bar{\bar{c}}_2$) in another chart will be related to $\bar{c}_1$ (and $\bar{c}_2$) by the formulas (Φ denotes the local diffeomorphism relating the two charts)

$$\frac{\partial \bar{\bar{c}}_1}{\partial t}\bigg|_{(0,0)} = T_0\Phi\left(\frac{\partial \bar{c}_1}{\partial t}\bigg|_{(0,0)}\right),$$

$$\frac{\partial \bar{\bar{c}}_1}{\partial s}\bigg|_{(0,0)} = T_0\Phi\left(\frac{\partial \bar{c}_1}{\partial s}\bigg|_{(0,0)}\right),$$

$$\frac{\partial^2 \bar{\bar{c}}_1}{\partial s\partial t}\bigg|_{(0,0)} = \frac{\partial}{\partial s}\left(T_{\bar{c}_1(0,s)}\Phi\left(\frac{\partial \bar{c}_1}{\partial t}\bigg|_{(0,s)}\right)\right)\bigg|_{s=0}$$

$$= T_0T\Phi\left(\frac{\partial \bar{c}_1}{\partial t}\bigg|_{(0,0)}, \frac{\partial \bar{c}_1}{\partial s}\bigg|_{(0,0)}\right) + T_0\Phi\left(\frac{\partial^2 \bar{c}_1}{\partial s\partial t}\bigg|_{(0,0)}\right).$$

The identification is then given in a chart by associating with an equivalence class $[b]$ of a one-parameter family of curves the point $\left(\bar{b}(0,0), \frac{\partial \bar{b}}{\partial t}\bigg|_{(0,0)}, \frac{\partial \bar{b}}{\partial s}\bigg|_{(0,0)}, \frac{\partial^2 b}{\partial s\partial t}\bigg|_{(0,0)}\right)$ of TTM in the chart. For a systematization of this point of view, see [TW].

1.20. The exchange of the two parameters clearly defines an involution on the set of two-parameter families of points. Moreover, this involution is compatible with the equivalence relation (because of commutation of derivatives) and therefore defines an involution j on TTM called the *canonical involution.* In a natural coordinate system, j is given by

$$j(x^i, X^i, Y^i, \mathscr{X}^i) = (x^i, Y^i, X^i, \mathscr{X}^i).$$

The map j exchanges the two vector bundle fibrations p_{TM} and Tp_M of TTM over TM. In particular, we see that the fixed points of j are just the elements of TTM for which the two projections coincide. They are also the 2-jets of curves on M viewed as special two-parameter families of points in M, via the map $(s,t)\mapsto s+t$ from $\mathbb{R}\times\mathbb{R}$ to $\mathbb{R}$; locally, if $(x^i(t))$ denotes the coordinates of $c(t)$, these elements are given by $\left(x^i(0), \frac{dx^i}{dt}\bigg|_0, \frac{dx^i}{dt}\bigg|_0, \frac{d^2x^i}{dt^2}\bigg|_0\right)$. As elements of TTM, they will be referred to as *jets*.

For a curve c in M, $\dot{c}$ will denote the *velocity curve* in TM $\left(\text{given locally by } \left(x^i(t), \frac{dx^i(\tau)}{d\tau}\bigg|_t\right)\right)$ and $\ddot{c}$ the *acceleration curve* in TTM (velocity curve of the velocity curve).

1.21. In the special case of the vector bundle $p_M: TM \to M$, the isomorphism of VTM with $TM \times_{p_M} TM$ described in 1.9 plays a specific role. In a natural coordinate system, with the point $((x^i, X^i), (x^i, Y^i))$ it associates $(x^i, X^i, 0, Y^i)$. As in this case we also have a map from TTM to $TM \times_{p_M} TM$, namely $p_{TM} \times_{p_M} Tp_M$, we get in this way a canonically defined map from TTM to VTM. This map v, called the *vertical endomorphism* of TTM, is given locally by $v(x^i, X^i, Y^i, \mathscr{X}^i) = (x^i, X^i, 0, Y^i)$.

The vertical endomorphism v is a nilpotent fiber map for the fibration p_{TM} (its square is zero).

E. Riemannian Metrics

1.22. The bundle of symmetric 2-forms on M is denoted by O^2T^*M and its space of sections by $\mathscr{S}^2M$. We shall be interested in *Riemannian metrics*, i.e. in elements g of $\mathscr{S}^2M$ which for each m in M induce a scalar product on T_mM. Such a g determines a linear isomorphism between TM and T^*M denoted by $\flat$ (since it lowers the indices of forms) whose inverse is denoted by $\sharp$ (since it raises the indices of forms).

The *norm* induced by the Riemannian metric g is also denoted for v in TM by

$$\|v\|_g = \sqrt{g(v,v)}\,.$$

With g is associated a *kinetic energy function* E_g on TM, given by

$$E_g(v) = 1/2\, g(v,v)$$

for v in TM.

Clearly, $A \cdot E_g = 2E_g$.

In the geodetic problem, E_g is also called the *Lagrangian* (see Paragraph G).

1.23. By using $\flat$ or $\sharp$, we can pull back or map to TM all objects naturally defined on T^*M. For example, we shall call α_g the pull-back $\flat^*(\alpha)$ of the Liouville form. For X in TTM, we have

$$\alpha_g(X) = g(Tp_M(X), p_{TM}(X))\,.$$

In particular, if X is a jet, $\alpha_g(X)$ is non-negative since

$$Tp_M(X) = p_{TM}(X)\,.$$

In natural coordinates (x^i, X^i), if $g = \sum_{i,j=1}^{d} g_{ij}\, dx^i \otimes dx^j$, then for v in TM,

$$\alpha_g(v) = \sum_{i,j=1}^{d} g_{ij}\, X^i(v) dx^j$$

(notice in the formula the abuse of notations for dx^j).

Of course, α_g is also homogeneous of degree 1.

The local expression of the symplectic structure $d\alpha_g$ on TM is then for v in TM,

$$d\alpha_g(v) = \sum_{i,j,k=1}^{d} \frac{\partial g_{kj}}{\partial x^i} X^k(v)\, dx^i \wedge dx^j + \sum_{i,j=1}^{d} g_{ij}\, dX^i \wedge dx^j\,.$$

When the Riemannian metric g is fixed, we can speak of the *energy* of a curve. Let $c:[0,1]\to M$ be a curve, then $\mathbb{E}_g(c)=\int_0^1 E_g(\dot c(t))dt$.

We can also define its *length*, $\mathbb{L}_g(c)=\int_0^1 \sqrt{2E_g(\dot c(t))}\,dt$.

Notice that any curve has a parametrization such that $E_g(\dot c(t))$ is constant in t: such a parametrization is called *normal.*

1.24. We will denote by U^gM the submanifold $E_g^{-1}(1)$ of TM called the *unit tangent bundle.* It is clear that U^gM is diffeomorphic to the homogeneous bundle UM. A vector V in TTM is tangent to U^gM if and only if $V\cdot E_g=0$.

At v in TM, in a natural coordinate system we have

$$dE_g(v)=1/2\sum_{i,j,k=1}^{d}\frac{\partial g_{ij}}{\partial x^k}X^i(v)X^j(v)dx^k+\sum_{i,j=1}^{d} g_{ij}X^i(v)dX^j\,;$$

hence, V must satisfy the equation

$$\sum_{i,j,k=1}^{d}\frac{\partial g_{ij}}{\partial x^k}X^i(V)X^j(V)Y^k(V)+\sum_{i,j=1}^{d} g_{ij}X^i(V)\mathscr{X}^j(V)=0\,.$$

Notice that the Liouville vector field A is everywhere transverse to U^gM.

F. Calculus of Variations

1.25. The main object of study later on will be the geodesic flow associated with a Riemannian metric. We shall first develop the calculus of variations for a general Lagrangian. Let us start by giving some details about the two formulations of the calculus of variations, the Lagrangian and the Hamiltonian, and also on the Legendre transformation which identifies them in the nice cases. We consider a smooth function L on TM and we first construct its *vertical differential* $\hat{d}L$: we define the latter as the composite of the vertical endomorphism υ (see 1.21) and the usual differential dL. Since υ is surjective on vertical vectors and its square is zero, $\hat{d}L$ is a semi-basic differential form on TM.

In a natural chart, it is clear that

$$\hat{d}L=\sum_{i=1}^{d}\frac{\partial L}{\partial X^i}dx^i\,.$$

For example,

$$\hat{d}E_g=\sum_{i,j=1}^{d} g_{ij}X^i dx^j=\alpha_g\,.$$

In particular, we notice that for X in TTM which is represented in a natural chart by $(x^i, X^i, Y^i, \mathscr{X}^i)$,

$$\hat{d}L(X) = \sum_{j=1}^{d} \frac{\partial L}{\partial X^j}(x^i, X^i) Y^j .$$

On the other hand, $A \cdot L(p_{TM}(X)) = \sum_{j=1}^{d} \frac{\partial L}{\partial X^j}(x^i, X^i) X^j$.

Therefore, when X is a jet, we get

1.26 $$\hat{d}L(X) = A \cdot L(p_{TM}(X)) .$$

This point will turn out to be crucial in the next paragraph since it relates a special kind of differential of L to the value of the function $A \cdot L$.

1.27. Let $c:[0,1] \to M$ be a curve with $c(0) = m_0$ and $c(1) = m_1$. We want to express that c is *extremal* for the *integrated Lagrangian*

$$\mathbb{L}(c) = \int_0^1 L(\dot{c}(t))dt$$

among all curves from m_0 to m_1. In particular, it will be necessary for a one-parameter family of curves from m_0 to m_1 $(c_s)_{s \in]-\varepsilon,\varepsilon[}$ such that $c_0 = c$, that the function $s \mapsto \mathbb{L}(c_s)$ be stationary at $s = 0$.

We shall denote by $\tilde{X}$ the elements of TTM along $\dot{c}$ determined by this one-parameter family of curves (see 1.18). We shall refer to $Tc\left(\frac{\partial}{\partial s}\right) = Tp_M(\tilde{X})$ as the *transverse vector field along* c.

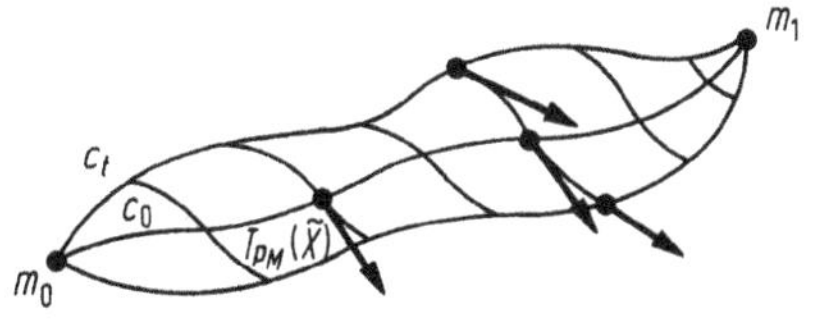

Fig. 1.27

1.28 Theorem. *If a curve* c *from* m_0 *to* m_1 *is extremal for the integrated Lagrangian, its acceleration curve* $\ddot{c}: \mathbb{R} \to TTM$ *satisfies the* Lagrange equations

$$i_{\ddot{c}} d\hat{d}L = d(L - A \cdot L)$$

(recall that both sides are evaluated at $\dot{c}(t)$ *in* TM*).*

1.29 *Proof.* We will suppose that $\tilde{X}$ is not only defined along $\dot{c}$, but is a vector field defined in a neighbourhood of $\dot{c}$.

First of all we evaluate

1.30 $$\frac{d}{ds}\mathbb{L}(c_s)_{|0} = \int_0^1 \frac{d}{ds} L(\dot{c}_s(t))_{|0} dt = \int_0^1 \tilde{X} \cdot L(\dot{c}(t))dt .$$

Since by Formula (1.26) we know that for a jet X in TTM

$$\hat{d}L(X) = A \cdot L(p_{TM}(X)),$$

we have $\int \dot{c}_s^*(\hat{d}L) = \int_0^1 A \cdot L(\dot{c}_s(t))dt$ (here by $\int \dot{c}^*(\hat{d}L)$ we denote the line integral of the 1-form $\hat{d}L$ along the curve $\dot{c}$). Therefore by taking the derivative of the previous equation with respect to s, (1.30) can be replaced by the equation

$$\frac{d}{ds} \mathbb{L}(c_s)_{|0} = \int_0^1 \tilde{X} \cdot L(\dot{c}(t))dt - \int_0^1 \tilde{X} \cdot A \cdot L(\dot{c}(t))dt + \int \dot{c}^*(\mathcal{L}_{\tilde{X}} \hat{d}L).$$

We now come to what should be considered as the "*fundamental relation in the calculus of variations*": the Lie derivative with respect to the vector field $\tilde{X}$, when acting on exterior differential forms, can be expressed in terms of d and interior product by $\tilde{X}$ as follows (see for example [GO 1] p. 93):

1.31 $$\mathcal{L}_{\tilde{X}} = i_{\tilde{X}} d + d i_{\tilde{X}}.$$

Therefore

$$\int \dot{c}^*(\mathcal{L}_{\tilde{X}} \hat{d}L) = \int \dot{c}^*(i_{\tilde{X}} d\hat{d}L) + \int \dot{c}^*(d i_{\tilde{X}} \hat{d}L).$$

As d commutes with pull-backs, the last term on the right-hand side reduces to $i_{\tilde{X}(m_1)}\hat{d}L - i_{\tilde{X}(m_0)}\hat{d}L$, i.e. to zero since $\tilde{X}$ is associated with a one-parameter family of curves through m_0 and m_1.

Stationariness of the integrated Lagrangian is then given by

$$0 = \int_0^1 [\tilde{X} \cdot (L - A \cdot L)(\dot{c}(t)) + d\hat{d}L(\tilde{X}(t), \ddot{c}(t))]\, dt$$

or equivalently by

1.32 $$0 = \int_0^1 (d(L - A \cdot L) - i_{\ddot{c}} d\hat{d}L)(\tilde{X}(t))dt.$$

If we prove that the 1-form $d(L - A \cdot L) - i_{\ddot{c}} d\hat{d}L$ is semi-basic along $\dot{c}$ we are done since the integral will then depend only on $Tp_M(\tilde{X})$ along $\dot{c}$, which is arbitrary ($Tp_M(\tilde{X})$ is the transverse field along c of an arbitrary variation of c).

To check that $d(L - A \cdot L) - i_{\ddot{c}} d\hat{d}L$ is semi-basic, we consider the terms in dX^i in its local expression. From 1.25 and the formula

1.33 $$d\hat{d}L = \sum_{i,j=1}^{d} \frac{\partial^2 L}{\partial X^i \partial X^j} dX^i \wedge dx^j + \frac{\partial^2 L}{\partial x^i \partial X^j} dx^i \wedge dx^j,$$

we get $\frac{\partial L}{\partial X^i}$ from dL, $-\sum_{j=1}^{d} \frac{\partial^2 L}{\partial X^i \partial X^j} X^j(\dot{c}(t)) - \frac{\partial L}{\partial X^i}$ from $-d(A \cdot L)$ and $\sum_{j=1}^{d} \frac{\partial^2 L}{\partial X^i \partial X^j} Y^j(\ddot{c}(t))$ from $-i_{\ddot{c}} d\hat{d}L$, which cancel each other since $\ddot{c}$ is a jet.

Notice also that the local expression of the equation is for $i=1, \ldots, d$,

$$\frac{\partial L}{\partial x^i} - \sum_{j=1}^{d} \frac{\partial^2 L}{\partial X^i \partial X^j} \mathscr{X}^j(\ddot{c}(t)) - \sum_{j=1}^{d} \frac{\partial^2 L}{\partial x^j \partial X^i} Y^j(\ddot{c}(t)) = 0,$$

in which we recognize the usual local expression of the Lagrange equations

$$\frac{d}{dt}\left(\frac{\partial L}{\partial X^i}(x^i(\dot{c}(t)), X^i(\dot{c}(t)))\right) - \frac{\partial L}{\partial x^i}(x^i(\dot{c}(t)), X^i(\dot{c}(t))) = 0. \quad \square$$

1.34. The problem of the calculus of variations determined by the Lagrangian L is called *regular* (or L is said to be regular) whenever $d\hat{d}L$, as a differential 2-form on $\mathring{T}M$, is non-degenerate. In this case $d\hat{d}L$ defines a symplectic structure on TTM. According to (1.33) this is so when in a natural coordinate system the matrix $\left(\frac{\partial^2 L}{\partial X^i \partial X^j}\right)$ has maximal rank, i.e. when $\operatorname{Det}\left(\frac{\partial^2 L}{\partial X^i \partial X^j}\right) \neq 0$. Notice that this condition only involves the variation of L along the fibers. In particular, if L restricted to any fiber is convex, then L is regular. We will see another interpretation of this condition in 1.42.

When L is regular, we can study directly the vector field X_L determined by the equation

1.35 $\qquad i_{X_L} d\hat{d}L = d(L - A \cdot L).$

The quantity $E_L = A \cdot L - L$ is called the *energy* associated with L and $A \cdot L$ is called the *action*. Since $d\hat{d}L$ is an exterior differential 2-form, E is preserved along the integral curves of X_L.

The vector field X_L is an *infinitesimal symplectic transformation* for $d\hat{d}L$ (one also says *Hamiltonian vector field*), since $\mathscr{L}_{X_L} d\hat{d}L = 0$ from (1.35) (this is a general fact for a vector field associated via a symplectic structure with a closed differential 1-form).

1.36. To discuss further properties of X_L, we need a notion which is specific to an iterated tangent bundle.

Again as TTM is fibered in two ways on TM, among vector fields on TM, i.e. sections of $p_{TM}: TTM \to TM$, there are some special ones, namely those which are at the same time sections of $Tp_M: TTM \to TM$ or equivalently, vector fields X on TM such that, for all v in TM, $X(v)$ is a jet.

The ordinary differential equation associated with such an X gives rise to a second-order differential equation since if X is given locally by

$$X = \sum_{i=1}^{d} Y^i(X) \frac{\partial}{\partial x^i} + \mathscr{X}^i(X) \frac{\partial}{\partial X^i},$$

then necessarily $Y^i(X) = X^i$ and an integral curve $v(t) = (x^i(t), X^i(t))$ of X satisfies

1.37
$$\begin{cases} \dfrac{dx^i}{dt} = X^i(t) \\[2mm] \dfrac{dX^i}{dt} = \mathscr{X}^i(X(v(t))) \end{cases}$$

which can be summarized in the system of equations

$$(i=1,\dots,d), \qquad \frac{d^2x^i}{dt^2}=\mathscr{X}^i\left(x^i(t),\frac{dx^i}{dt}(t)\right).$$

Therefore integral curves of X are necessarily velocity curves of curves on M of which X is the acceleration at each point.

Such vector fields are called *second-order differential equations.*

1.38 Proposition. *For a regular Lagrangian L, the vector field X_L, solution of the Lagrange equations* (1.35), *is a second-order differential equation.*

1.39 *Proof.* The local expression of X_L, $X_L=\sum_{i=1}^{d} Y^i(X_L)\frac{\partial}{\partial x^i}+\mathscr{X}^i(X_L)\frac{\partial}{\partial X^i}$ has to satisfy

$$\textbf{1.40}\quad \left\{\begin{aligned} &\sum_{j=1}^{d}\frac{\partial^2 L}{\partial X^i\partial X^j}Y^j(X_L)=\sum_{j=1}^{d}\frac{\partial^2 L}{\partial X^i\partial X^j}X^j\\ &\sum_{j=1}^{d}\frac{\partial^2 L}{\partial X^i\partial X^j}\mathscr{X}^j(X_L)=\frac{\partial L}{\partial x^i}-\sum_{j=1}^{d}\frac{\partial^2 L}{\partial x^j\partial X^i}X^j.\end{aligned}\right. \qquad (i=1,\dots,d)$$

So if we suppose L to be regular, we can invert the first d equations and get $Y^j(X_L)=X^j$ for $j=1,\dots,d$.

Notice that if we add to the right hand side of (1.35) any semi-basic differential form (this corresponds in mechanics to considering external forces), the result of Proposition 1.38 would not be affected. □

1.41. When L is homogeneous of degree 1, the right hand side of (1.35) is zero. So we would have difficulty determining X_L. This is not, however, the case since in this situation L cannot be regular. Indeed as $A\cdot L=L$, for $j=1,\dots,d$, $\sum_{i=1}^{d}\frac{\partial^2 L}{\partial X^i\partial X^j}X^i=0$ and therefore the matrix $\left(\frac{\partial^2 L}{\partial X^i\partial X^j}\right)$ cannot have maximal rank. We could have seen this more conceptually in the following way. If X is a locally defined vector field on TM, which is homogeneous of degree 1, then $v(X)$ is homogeneous of degree 0. Therefore by duality, if L is homogeneous of degree 1, $\hat{d}L=dL\circ v$ is homogeneous of degree 0. But as $\hat{d}L$ is semi-basic,

$$0=\mathscr{L}_A\hat{d}L=i_A d\hat{d}L,$$

which shows that, when L is homogeneous of degree 1, $d\hat{d}L$ cannot be a symplectic structure on TM.

If we suppose that $d\hat{d}L$ is as nondegenerate as possible, i.e. its rank is $2d-2$ $\left(\text{the matrix }\left(\frac{\partial^2 L}{\partial X^i\partial X^j}\right)\text{ is then of rank } d-1\right)$, then the kernel of $d\hat{d}L$ will be 2-dimensional and we know that the Liouville vector field is in this kernel. In this case the solutions will be curves on M defined up to parametrization, since from (1.40) we still get a family of second-order differential equations.

1.42. As defined in 1.8, the fiber derivative F associates to a function L on TM a fiber preserving map FL from TM to T^*M, which is called the *Legendre transformation.*

In a natural coordinate system, for v in TM,

$$FL(v) = \left(x^i(v), \frac{\partial L}{\partial X^i}(x^j(v), X^j(v))\right).$$

In particular, we see that FL is a local diffeomorphism from $\mathring{T}M$ to $\mathring{T}^*M$ if and only if L is regular. Indeed, as FL is fiber preserving, we just have to check its rank along the fibers, but this is just the rank of the matrix $\left(\frac{\partial^2 L}{\partial X^i \partial X^j}\right)$. This gives a new interpretation of the regularity of the Lagrangian L.

1.43. We will use the Legendre transformation to pull back objects defined on T^*M. For example, one can check that

$$FL^*(\alpha) = \hat{d}L.$$

If we suppose FL to be a global diffeomorphism from TM to T^*M (L is then said to be *hyperregular*), we can consider its inverse FL^{-1}. Then the function $E^* = (A \cdot L - L) \circ FL^{-1}$ is called the *Hamiltonian* and we can equivalently transform the Lagrange equations (1.35) into the *Hamilton equations*,

1.44 $$i_{X_{E^*}} d\alpha = -dE^*.$$

In a natural coordinate system (x^i, ξ_i), the integral curves $c(t)$ of X_{E^*} satisfy the equations

$$\begin{cases} \dfrac{dx^i(c(t))}{dt} = \dfrac{\partial E^*}{\partial \xi_i} \\ \dfrac{d\xi_i(c(t))}{dt} = -\dfrac{\partial E^*}{\partial x^i}. \end{cases} \quad (i = 1, \ldots, d)$$

If we start from a Hamiltonian function E^* on T^*M and if we define the Hamiltonian vector field X^* as the solution of the equation (1.44), then we say that $\alpha(X^*)$ is the *action* and $L^* = \alpha(X^*) - E^*$ the *Lagrangian associated with* E^*. The fiber derivative FE^* is a map from T^*M to $T^{**}M$ (which is canonically identified with TM) and we call E^* *regular* if FE^* has maximal rank and *hyperregular* if FE^* is a diffeomorphism from T^*M to TM. If we set $L = L^* \circ (FE^*)^{-1}$, one can check that FE^* coincides with $(FL)^{-1}$ and $E = A \cdot L - L$ with $E^* \circ (FE^*)^{-1}$.

1.45. We have seen that, though equivalent in the nicest cases, the Hamiltonian and Lagrangian formalisms, which take place respectively on T^*M and TM, appeal to the different structures of these two spaces: T^*M is naturally a symplectic manifold and TM has a very special differential calculus.

G. The Geodesic Flow

1.46. We come back to our original situation. We have associated a Lagrangian E_g, the kinetic energy function on TM, with a given Riemannian metric g on M. Since the function E_g is homogeneous of degree 2, E_g as a Lagrangian coincides with the energy associated with it, which justifies the notation E_g. The Legendre transformation in this case is just $\flat$. Hence E_g is hyperregular. In this case $\flat = FE_g$ is a vector bundle isomorphism from TM to T^*M. (Conversely one can prove that for a Lagrangian L, FL is a vector bundle isomorphism only if L is up to a constant the kinetic energy function associated to some non-degenerate symmetric differential 2-form on M, which is not necessarily Riemannian.)

We are interested in curves in M through two given points m_0 and m_1 which have extremal energy: such curves are called the *geodesics* from m_0 to m_1. As E_g is a hyperregular Lagrangian, these curves will be projections on M of integral curves of a vector field (which will be called the *geodesic vector field* and *denoted by* Z_g) which is the solution of the equation

1.47 $$i_{Z_g} d\alpha_g = -dE_g$$

(since we noticed in 1.25 and also in 1.43 that $\alpha_g = \hat{d}E_g$).

Its flow (ζ_g^t) is called the *geodesic flow.*

1.48 Proposition. *The geodesic vector field Z_g satisfies the properties of a* spray, *i.e.:*

i) *Z_g is a second-order differential equation,*

ii) *Z_g is homogeneous of degree 2, a property which has the following consequence: for any v in TM and for small enough real numbers t and t', $\zeta_g^t \circ a^{t'}(v)$ (the point at time t on the integral curve of Z_g with initial point $t'v$) coincides with $a^{t'} \circ \zeta_g^{tt'}(v)$ (the image by the dilation $a^{t'}$ of the point at time tt' on the integral curve of Z_g with initial point v).*

1.49 *Proof.* Part i) is clear from Proposition 1.38. Notice that we can express the fact that Z_g is a second-order differential equation by saying that the two maps Z_g and $j \circ Z_g$ from TM to TTM (where j is the canonical involution of TTM) coincide.

For ii) we establish homogeneity properties of Z_g. The vector field Z_g is defined by the Equation (1.47); Z_g is therefore homogeneous of degree 2 since

$$\mathscr{L}_A i_{Z_g} d\alpha_g = i_{[A,Z_g]} d\alpha_g + i_{Z_g} \mathscr{L}_A d\alpha_g$$

and as α_g is homogeneous of degree 1 together with $d\alpha_g$, and E_g is homogeneous of degree 2 together with dE_g,

$$i_{[A,Z_g]}d\alpha_g + i_{Z_g}d\alpha_g = -2dE_g\,,$$

so that $[A, Z_g] = Z_g$.

From this equality we deduce that the two homomorphisms from $\mathbb{R}$ to vector fields $s \mapsto a^{\exp s}(Z_g)$ (for the notation, see 1.5) and $s \mapsto e^s \cdot Z_g$ coincide. Therefore for each s, the flows of these vector fields also coincide, i.e. $a^{\exp -s} \circ \zeta_g^t \circ a^{\exp s} = \zeta_g^{t\cdot \exp s}$, and so if we replace $\exp s$ by t', we have completed the proof of ii). □

1.50. Notice that if we had been looking for curves from m_0 to m_1 which are extremal for the length, we would have found the same geometric curves. In this case since the length is homogeneous of degree one, the curves are only defined up to a change of parameter, as we pointed out in 1.41. Notice that the curves that we have called geodesics have normal parametrization, i.e., their velocity vectors have constant length.

The geodesics have the following uniqueness property: if v is given in U^gM, then there exists a unique geodesic γ such that $\dot\gamma(0) = v$.

The geodesic vector field has the special property of being at the same time a Hamiltonian vector field and a spray: these two notions belong to different areas of differential geometry: symplectic geometry and connection theory, as we will see later on. Therefore for each of its properties, it will be interesting (but sometimes not so easy) to determine to which area it pertains.

1.51. In a natural coordinate system (x^i, X^i) for TM, the local expression of Z_g, $Z_g = \sum_{i=1}^{d} Y^i(Z_g)\dfrac{\partial}{\partial x^i} + \mathscr{X}^i(Z_g)\dfrac{\partial}{\partial X^i}$, is determined by the Equation (1.40) with $E_g = L$, where $\dfrac{\partial^2 E_g}{\partial X^i \partial X^j} = g_{ij}$, $\dfrac{\partial E_g}{\partial x^i} = 1/2 \sum_{j,k=1}^{d} \dfrac{\partial g_{jk}}{\partial x^i} X^j X^k$ and $\dfrac{\partial^2 E_g}{\partial x^j \partial X^i} = \sum_{k=1}^{d} \dfrac{\partial g_{ik}}{\partial x^j} X^k$.

Therefore, we get $Y^i(Z_g) = X^i$ (which was expected since Z_g is a second-order differential equation) and if (g^{ij}) denotes the inverse matrix of (g_{ij}),

$$\mathscr{X}^k(Z_g) = 1/2 \sum_{i,j,l=1} g^{ki}\left(\frac{\partial g_{jl}}{\partial x^i} - 2\frac{\partial g_{il}}{\partial x^j}\right) X^j X^l\,.$$

The quantity $\Gamma^i_{jk} = 1/2 \sum_{l=1}^{d} g^{il}\left(\dfrac{\partial g_{lk}}{\partial x^j} + \dfrac{\partial g_{jl}}{\partial x^k} - \dfrac{\partial g_{jk}}{\partial x^l}\right)$ is called a *Christoffel symbol* associated with g in this coordinate system. We finally get the following formula at v in TM

$$Z_g(v) = \sum_{i=1}^{d} X^i(v)\frac{\partial}{\partial x^i} - \sum_{i,j,k=1}^{d} \Gamma^i_{jk} X^j(v) X^k(v)\frac{\partial}{\partial X^i}\,.$$

The local equations of geodesics are then given for $i = 1, \ldots, d$ by

$$\frac{d^2x^i}{dt^2} + \sum_{j,k=1}^{d} \Gamma^i_{jk}\frac{dx^j}{dt}\frac{dx^k}{dt} = 0\,.$$

Notice that in a natural coordinate system $\mathscr{X}^k(Z(v))$ is homogeneous of degree two in $X^i(v)$ if and only if Z is a spray.

Indeed, we have $\mathscr{X}^k(\mathscr{L}_A Z(v)) = \sum_{i=1}^{d} \mathscr{X}^i(A(v)) \frac{\partial \mathscr{X}^k(Z(v))}{\partial X^i} - \mathscr{X}^i(Z(v)) \frac{\partial \mathscr{X}^k(A(v))}{\partial X^i}$; for a spray Z from the specific value of $\mathscr{X}^i(A)$ we get

$$\sum_{i=1}^{d} X^i(v) \frac{\partial \mathscr{X}^k(Z(v))}{\partial X^i} = 2\mathscr{X}^k(Z(v))$$

which proves that $\mathscr{X}^k(Z(v))$ (being C^∞ up to the zero section) is a polynomial of degree 2 in $X^i(v)$. In particular, $Z(0_{TM}) = 0_{TTM}$. The case of generalized sprays which are not C^∞ up to the zero section of TM is more complicated since in this case $\mathscr{X}^k(Z)$ can be a homogeneous function which does not have to extend smoothly up to the origin, and which can be more general than polynomials.

1.52. Since it is often simpler to consider M rather than TM, we will present another way of looking at the geodesic flow. We suppose for convenience that the geodesic flow is *complete*, i.e. ζ_g^t is defined for each t on all TM.

What follows including the next proposition is true for the flow associated with a spray.

We define a map $\text{Exp}: TM \to M \times M$, called the *exponential map*: for v in TM

$$\text{Exp}(v) = (p_M(v), p_M \circ \zeta_g^1(v)).$$

We will also need the *pointed exponential maps*: let m be a point in M, then

$$\exp_m = p_M \circ \zeta_g^1 = pr_2 \circ \text{Exp} \restriction T_m M$$

(where pr_2 denotes the projection onto the second factor).

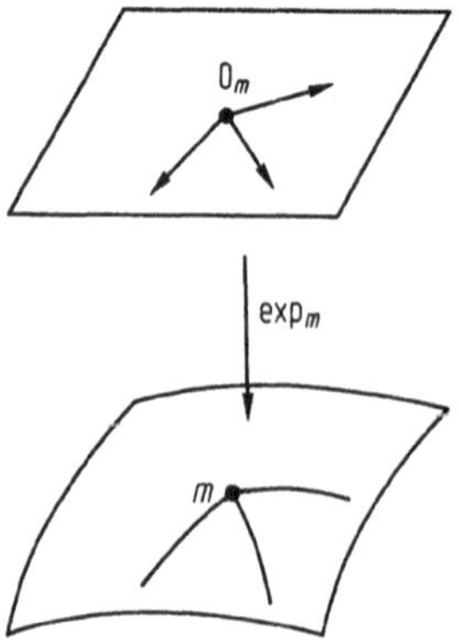

Fig. 1.52

1.53 Proposition. i) *For each m in M, $\exp_m$ is a diffeomorphism of a neighbourhood of 0 in $T_m M$ onto a neighbourhood of m in M.*

ii) *The exponential map* Exp *is a diffeomorphism from a neighbourhood of the zero section of* TM *onto a neighbourhood of the diagonal of* $M \times M$.

1.54 *Proof.* We first consider the differential of $\exp_m$ for m in M at 0. Let X be a vector in $T_0 T_m M$ identified with $T_m M$; we then have

$$T_0 \exp_m(X) = \frac{d}{dt} \exp_m(tX)|_0$$

since $t \mapsto tX$ is a curve whose tangent vector at 0 is X.

But as $\exp_m(tX) = p_M \circ \zeta_g^1(tX)$, from 1.48. ii) we deduce that

$$\exp_m(tX) = p_M \circ \zeta_g^t(X).$$

We then get that $\frac{d}{dt} \exp_m(tX)|_0 = Tp_M \circ Z(X) = X$ from (1.48). This proves i) by the inverse function theorem. With some uniformity argument ii) will be proved if we show that Exp has maximal rank along 0_{TM}.

We just saw that $T_{0_m}\mathrm{Exp}$ is a submersion onto the second factor; as it is clear that Exp is a submersion onto the first factor, $T_{0_m}\mathrm{Exp}$ is surjective from $T_{0_m}TM$ onto $T_m M \times T_m M$ and therefore an isomorphism. □

1.55. In the last Proof 1.54, we showed that $\exp_m$ can be used as a chart centered at m, since $\exp_m$ is a diffeomorphism from a neighbourhood of 0_m in $T_m M$ onto a neighbourhood of m in M. Suppose that we choose a Euclidean coordinate system (x^i) on $T_m M$; we will still denote by (x^i) the coordinate system on M deduced via the exponential map at m. Such a coordinate system is called *normal*. These coordinate systems turn out to be very convenient to perform computations with; they can be characterized by the fact that in such coordinates the functions x^j are equal to $\sum_{i=1}^{d} g_{ij} x^i$ which means that considered in this chart the $\flat$ isomorphism from $T_m M$ to $T_m^* M$ maps A_m into the Liouville vector field of the cotangent bundle (the proof is left as an exercise or see [EPN 2]). In particular in such charts, the Christoffel symbols at m are zero (this can be deduced easily from the previous relation by differentiating at the origin), showing that, in some sense, the Euclidean geometry of $T_m M$ mapped into M by the exponential map at m is osculating to the Riemannian geometry of M in a neighbourhood of m.

1.56 Proposition. i) (Liouville's theorem) Z_g *preserves the symplectic structure* $d\alpha_g$ *on* TTM *and the volume element canonically derived from it.*

ii) *The geodesic vector field* Z_g *is tangent to* $U^g M$ *and if* e^1 *denotes the injection of* $U^g M$ *into* TTM, $Z_g \restriction U_g M$ *preserves the differential form* $e^{1*}(\alpha_g)$.

1.57 *Proof.* Part i) is clear since Z_g is determined by Equation (1.47) and $\alpha_g = \hat{d}E_g$, so that $\mathscr{L}_{Z_g} d\alpha_g = d i_{Z_g} d\alpha_g = 0$.

Part ii) is also straightforward since $U^gM = E_g^{-1}(1)$ and Z_g is the vector field associated by the symplectic structure with the differential of E_g. For the other property, let us compute

$$\mathscr{L}_{Z_g \restriction U_gM} e^{1*}(\alpha_g) = i_{Z_g \restriction U_gM} e^{1*}(d\alpha_g) + d i_{Z_g \restriction U_gM} e^{1*}(\alpha_g).$$

The first term is zero by assumption and the second also, since $i_{Z_g}\alpha_g \restriction U^gM = 1$, Z_g being a jet. □

1.58. Part ii) of Proposition 1.56 will be used extensively in the future. We will generally consider Z_g *as a vector field on* U^gM rather than on TM and the geodesic flow as a one-parameter group of diffeomorphisms of the unit tangent bundle, without changing their names. The restriction of α_g to U^gM will also be denoted by α_g.

The $(2d-1)$ differential form $\alpha_g \wedge (d\alpha_g)^{d-1}$ is a volume form on U^gM: indeed as A is everywhere transverse to U^gM and as $i_A d\alpha_g = \alpha_g$ (α_g is homogeneous of degree 1 and $\alpha_g(A) = 0$), $\alpha_g \wedge (d\alpha_g)^{d-1}$ is non-degenerate.

If we suppose M to be *compact*, then U^gM is also compact and therefore the geodesic flow will automatically be complete as is the flow of a smooth vector field on any compact manifold. This is the case we will be interested in later on in the book, see also 5.14.

H. Connectors

1.59. We have already noticed that the acceleration curve of a given curve cannot be interpreted in terms of vectors on M: it is an object of higher order. We now introduce a notion which will circumvent this difficulty and which will turn out to be equivalent to that of a spray. We present it in the setting of vector bundles.

Let $\tau: E \to M$ be a vector bundle of rank r over M.

A *connector* K for τ is a map from TE to E such that

$\mathscr{K}$i) (K, p_M) is a vector bundle map from $T\tau: TE \to TM$ onto $\tau: E \to M$;

$\mathscr{K}$ii) (K, τ) is a vector bundle map from $p_E: TE \to E$ onto $\tau: E \to M$;

$\mathscr{K}$iii) $K \restriction VE$ coincides fiberwise with the canonical map ι of the vertical subbundle VE to E given in 1.9.

In the special case of the tangent bundle to M, $p_M: TM \to M$, K can have the further property of being invariant under the canonical involution j of TTM (see 1.20), i.e., $K \circ j = K$; we then say that K is *symmetric* [in this case, $\mathscr{K}$ii) is equivalent to $\mathscr{K}$i)].

1.60. From $\mathscr{K}$i) and $\mathscr{K}$ii), Ker K is a subbundle of both $T\tau: TE \to TM$ and $p_E: TE \to E$, Ker K is called the *horizontal subbundle* of TE associated to K and is denoted by HE. The word horizontal is justified by the fact that K being fiberwise an isomorphism of the vertical subbundle onto all TM, HE and VE are transverse to each other which gives rise to a decomposition

$$TE = HE \oplus VE.$$

Therefore, $p_E \restriction HE: HE \to E$ is a vector bundle of rank d over E. Moreover, for each v in E the restriction $T\tau \restriction HE: HE \to TM$ is a linear isomorphism of H_vE onto $T_{\tau(v)}M$. The inverse map *denoted* by h_v will be called the *horizontal lift at* v. This shows that $p_E \restriction HE: HE \to E$ is isomorphic to $\tau^* p_M: \tau^* TM \to E$. Furthermore, $(T\tau \times_\tau K, \tau)$ is a vector bundle map which is an isomorphism on the fibers from $p_E: TE \to E$ onto $\tau \circ \tau^* \tau: \tau^* E \to M$: in other words, the decomposition $H_vE \oplus V_vE$ is mapped isomorphically onto the decomposition $T_{\tau(v)}M \oplus E_{\tau(v)}$.

In the special case of the tangent bundle, if v is an element of T_mM, this isomorphism along the fibres is $Tp_M \times K: H_vTM \oplus V_vTM \to T_mM \oplus T_mM$. In particular, if K is symmetric, the horizontal bundle is invariant under the canonical involution j of TTM. Moreover, from this property and the definition of h_v, we get for w in TM

$$h_v(w) = j(h_w(v)).$$

1.61. Let us take bundle coordinates (x^i, v^i) on E. We denote by (x^i, v^i, Y^i, V^i) the natural coordinates on TE. Let us suppose that the local expression of K is $(k^i(x^j, v^j, Y^j, V^j), \mathscr{K}^i(x^j, v^j, Y^j, V^j))$.

From $\mathscr{K}$i) or $\mathscr{K}$ii) in the definition of K, k^i is necessarily x^i. Since

$$(x^j, v^j, Y^j, V^j) = (x^j, v^j, Y^j, 0) + (x^j, v^j, 0, V^j),$$

from the three axioms of a connector we deduce term by term that $\mathscr{K}^i(x^j, v^j, Y^j, V^j)$ can be written as $\sum_{j=1}^{r} \sum_{k=1}^{d} \gamma^i_{jk} v^j Y^k + V^i$, where γ^i_{jk} is a function on the domain of the coordinate system.

Therefore a vector V in TE represented locally by (x^i, v^i, Y^i, V^i) is horizontal if and only if for $i = 1, \ldots, d$,

$$\sum_{j=1}^{r} \sum_{k=1}^{d} \gamma^i_{jk} v^j Y^k + V^i = 0.$$

Also at $v = (x^i, v^i)$ the horizontal lift to H_vE of the element $Y = (x^i, Y^i)$ of $T_{\tau(v)}M$ is

1.62 $$h_v(Y) = \left(x^i, v^i, Y^i, -\sum_{j=1}^{r} \sum_{k=1}^{d} \gamma^i_{jk} v^j Y^k\right).$$

We notice from (1.62) that in the case of the tangent bundle, with any connector K, we can associate a symmetric connector, namely $1/2\ (K + K \circ j)$. Indeed, the axioms $\mathscr{K}$i) and $\mathscr{K}$ii) are exchanged by composition with j. Axiom $\mathscr{K}$iii) follows from the expression for $\mathscr{K}^i$.

From this formula we also find that in the case of a symmetric connector K defined for the tangent bundle (i.e., locally the functions γ^i_{jk} are symmetric in j and k) for $v = (x^i, v^i)$ and $w = (x^i, w^i)$

$$h_v(w) = \left(x^i, v^i, w^i, -\sum_{j,k=1}^{d} \gamma^i_{jk} v^j w^k\right) = j(h_w(v)).$$

1.63. Let $f:N\to M$ be a map and $\tau:E\to M$ a vector bundle over M. We suppose that a connector $K:TE\to E$ is given. We can define the *pulled-back connector* $Tf^*K:Tf^*E\to f^*E$ by the canonical formula: for (V,X) in Tf^*E, i.e., such that $T\tau(V) = Tf(X)$ (see 1.10), we set

$$Tf^*K(V,X)=(K(V),p_N(X)).$$

Notice that $(K(V),\ p_N(X))$ is an element of f^*E since

$$\tau\circ K(V)=p_M\circ T\tau(V)=p_M\circ Tf(X)=f\circ p_N(X),$$

K being a connector and (V,X) an element of Tf^*E.

This construction will be used when f is an embedding (of a curve for example). Notice that in this case, elements of f^*E can be thought of as *vectors of E along* $f(N)$. But especially when f is not an embedding, for example not injective, the notion is still interesting and we will comment on a variant of it later.

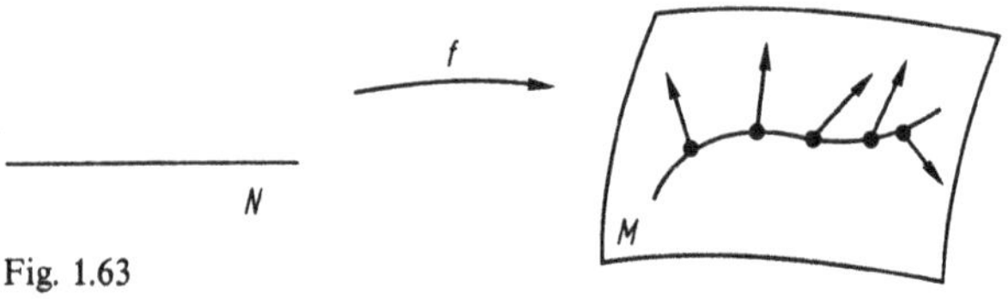

Fig. 1.63

1.64. In the case of the tangent bundle of M, we now come to the problem of determining a connector adapted to a second-order differential equation. First of all we need a lemma which exhibits a decomposition of any element of TTM using a second-order differential equation and the two vector bundle structures of TTM.

Its statement may seem complicated, but its proof will show that the terms of Formula (1.66) are minimal to get a nontrivial expression. On the other hand, in natural coordinates the proof copies the process of polarizing a quadratic form.

1.65 Lemma. *Let X be an element of TTM and Z a second-order differential equation. Then*

1.66 $$[X+Z(p_{TM}(X))]\mathbin{+\!\!\!+}[j(X)+Z(p_{TM}\circ j(X))]=Z(p_{TM}(X)+p_{TM}\circ j(X))+2X_1\,,$$

where X_1 is a vertical vector in TTM (for the notation $\mathbin{+\!\!\!+}$, *see* 1.16).

1.67 *Proof.* First of all note that

$$Tp_M(X+Z(p_{TM}(X)))=Tp_M(j(X)+Z(p_{TM}\circ j(X)))=p_{TM}(X)+Tp_M(X)$$

since (Tp_M,p_M) is a vector bundle morphism from $p_{TM}:TTM\to TM$ onto $p_M:TM\to M$ and Z is a second-order differential equation: we have just proved that the left-hand side of (1.66) is well defined. Moreover its projection onto TM by p_{TM} is $p_{TM}(X)+Tp_M(X)$, since (p_{TM}, p_M) is a vector bundle map from $Tp_M: TTM \to TM$ onto $p_M: TM \to M$. Since Z is a second-order differential equation, $Z(p_{TM}(X)+p_{TM}\circ j(X))$ differs from the left-hand side by a vertical vector. □

1.68. The preceding lemma shows that given a second order differential equation Z, there exists a natural map φ_Z from TTM to VTM. Notice that for X in T_vTM, $X_1=\varphi_Z(X)$ is based at $v+Tp_M(X)$. Therefore, φ_Z cannot be a bundle map for either of the two vector bundle structures of TTM over TM.

From 1.65, it is clear that $\varphi_Z \circ j=\varphi_Z$.

If we now consider the restriction of φ_Z to the vertical subbundle, from the Relation (1.66) we get for X in VTM

$$(X+Z(p_{TM}(X))) \mathbin{+\!\!\!+} (j(X)+Z(0))=Z(p_{TM}(X))+2\varphi_Z(X).$$

As soon as $Z(0_{TM})=0_{TTM}$, we get from an easy direct computation that $X=\varphi_Z(X)$, which shows that φ_Z is the identity on VTM.

It is also interesting to check the effect of φ_Z on Z itself. For v in TM, we get the expression

$$\varphi_Z(Z(v))+Z(2v)=2Z(v) \mathbin{+\!\!\!+} 2Z(v);$$

notice that the right hand side is just $Ta^2(2Z(v))$ (where a^2 is the dilation by 2).

1.69 Theorem. i) *Let Z be a second-order differential equation on M. There exists a connector K for which Z is horizontal if and only if Z is a spray (moreover, if K is symmetric, K is unique).*

ii) *Conversely given any symmetric connector K, there exists a unique spray Z horizontal for K.*

1.70 *Proof.* i) Let K be a connector. We apply K to both sides of Relation (1.66). As K is a bundle map over p_M for both fibrations of TTM over TM and if we suppose Z horizontal for K, we get

$$K(X)+K(j(X))=2K(\varphi_Z(X)).$$

As K on VTM coincides with the fiber isomorphism ι over p_M defined in 1.9, we finally have

$$K(X)+K(j(X))=2\iota(\varphi_Z(X)),$$

which gives the symmetric part of K.

The existence of K will be established if we prove that $\iota\circ\varphi_Z$ has the properties of a connector. We already noticed in 1.68 that φ_Z was the identity on VTM if $Z(0)=0$ and so Axiom $\mathscr{K}$ iii) (see 1.59) will follow as soon as $Z(0)=0$.

From the relation $\varphi_Z \circ j = \varphi_Z$, it is enough to check Axiom $\mathscr{K}\, ii)$ (see 1.59). In a natural coordinate chart, for X in TTM the quantity

$$\begin{aligned}\mathscr{X}^i(X) - 1/2[&\mathscr{X}^i(Z(p_{TM}(X) + p_{TM} \circ j(X)))\\ &- \mathscr{X}^i(Z(p_{TM}(X))) - \mathscr{X}^i(Z(p_{TM} \circ j(X)))]\end{aligned}$$

has to be linear in X with respect to $+$. This is the case if and only if $\mathscr{X}^i(Z(v))$ is quadratic in $X^i(v)$, i.e., if and only if Z is a spray as we saw in 1.51.

The equality $Z(0_{TM}) = 0_{TTM}$ follows also from this assumption.

The last thing to check is that Z is horizontal for $K = \iota \circ \varphi_Z$. This is true since at the end of 1.68 we noticed that

$$\varphi_Z(Z(v)) + Z(2v) = Ta^2(2Z(v)),$$

which is equal to $Z(2v)$ if Z is a spray, proving that $\varphi_Z(Z) = 0$.

ii) Conversely let K be a symmetric connector. We defined in 1.60 the horizontal lift h associated to K: for v in T_mM, h_v maps isomorphically T_mM onto H_vTM. We set

$$Z_K(v) = h_v(v).$$

It is clear that $p_{TM} \circ Z_K = Id_{TM}$. By the definition of h, we also have $Tp_M \circ Z_K = Id_{TM}$. Therefore, Z_K is a second-order differential equation. Moreover by Formula (1.62), $\mathscr{X}^i(Z_K)$ is quadratic in $X^i(Z_K)$, a condition which is known from 1.51 to be equivalent to the fact that Z_K is a spray. Now from i) K is the unique symmetric connector for which Z_K is horizontal. □

1.71. In Theorem 1.69 we established the complete equivalence between symmetric connectors and sprays, the advantage of connectors being that they can be defined on general vector bundles. These notions are used classically in different contexts.

We have some more comments on sprays: first of all, a dynamical system is associated to a spray as to any vector field; on the other hand, it is known that one can consider a vector field on a manifold as a derivation of the algebra of C^∞ functions on this manifold. This point of view for sprays is developed in the next paragraph and gives rise to a special type of derivation. Notice that the relationship between the two points of view is just that for a vector field V and a function f on M, $V \cdot f$ is the ordinary derivative of f along integral curves of V: this property for sprays will be interpreted later in terms of covariant differentiation.

1.72. Let K_g be the unique symmetric connector associated with the geodesic vector field Z_g defined by a Riemannian metric g. The connector K_g is called the *Levi-Civita connector*. Let γ be a geodesic. We saw in 1.46 that $\dot{\gamma}$ is an integral curve of Z_g. Therefore, $\ddot{\gamma}$ is horizontal for K_g (one can also check this directly from the local expressions). One can also say that the relation

$$\ddot{\gamma} = h_{\dot{\gamma}}(\dot{\gamma})$$

characterizes integral curves of Z_g, i.e. the velocity curves of geodesics.

Moreover the splitting of TTM, $TTM = VTM \oplus HTM$, deduced from the Levi-Civita connector K_g, has the further property of being Lagrangian (i.e., at v in TM, V_vTM and H_vTM are Lagrangian subspaces of T_vTM for the symplectic form $d\alpha(v)$).

Indeed in 1.15 we have already noticed that the vertical subbundle VT^*M was Lagrangian. As $\flat$ identifies VT^*M with VTM and pulls back α to α_g, the same is true for VTM.

For HTM this can be seen as follows; pick a natural coordinate system on TTM. From 1.23 for $X_1 = (x^i, X^i, Y_1^i, \mathscr{X}_1^i)$ and $X_2 = (x^i, X^i, Y_2^i, \mathscr{X}_2^i)$ we get

$$d\alpha_g(X_1, X_2) = \sum_{i,j,k=1}^{d} \frac{\partial g_{kj}}{\partial x^i} X^k(Y_1^i Y_2^j - Y_2^i Y_1^j) + \sum_{i,j=1}^{d} g_{ij}(\mathscr{X}_1^i Y_2^j - \mathscr{X}_2^i Y_1^j).$$

From 1.61 and the exact value of the Christoffel symbols Γ_{jk}^i as given in 1.51, we get by a direct check that this formula can be intrinsically written as

$$d\alpha_g(X_1, X_2) = g(K(X_1), Tp_M(X_2)) - g(K(X_2), Tp_M(X_1)).$$

In particular for horizontal vectors X_1 and X_2 we get $d\alpha_g(X_1, X_2) = 0$ proving that HTM is Lagrangian. It would be interesting to have an intrinsic proof of this fact.

This once more illustrates the fact that the geodesic vector field and notions deduced from it play at the same time a symplectic and a purely Riemannian role.

I. Covariant Derivatives

1.73. We now come to present one of the most classical tools used in computations in Riemannian geometry: the Levi-Civita covariant derivative. We already gave the name of Levi-Civita to a connector (which is not usual).

Let $\mathscr{T}M$ be the space of sections of $p_M: TM \to M$ (for $\mathscr{E}$, see 1.2).

A map D from $\mathscr{T}M \bigotimes_{\Omega^0 M} \mathscr{E}$ to $\mathscr{E}$ is called a *covariant derivative* (for V in $\mathscr{T}M$ and W in $\mathscr{E}$, we denote it by $D_V W$) if and only if

$\mathscr{C}\mathscr{D}$i) D is $\Omega^0 M$-linear in its first argument, i.e., for any V in $\mathscr{T}M$, W in $\mathscr{E}$ and f in $\Omega^0 M$, $D_{fV} W = f(D_V W)$ (this means that at m in M, $(D_V W)(m)$ depends only on $V(m)$);

$\mathscr{C}\mathscr{D}$ii) D is an $\Omega^0 M$-derivation in its second argument, i.e., for any V in $\mathscr{T}M$, W in $\mathscr{E}$ and f in $\Omega^0 M$, $D_V(fW) = f(D_V W) + (V \cdot f)W$.

Moreover, in the case of the tangent bundle, we say that D is *torsion-free* or *symmetric* if for any V, W in $\mathscr{T}M$, $D_V W - D_W V = [V, W]$.

1.74 Proposition. *Let K be a connector for the vector bundle $\tau: E \to M$. Then*

i) *the map from $\mathscr{T}M \bigotimes_{\Omega^0 M} \mathscr{E}$ to $\mathscr{E}$ defined by*

1.75 $$D_V W = K(TW(V))$$

is a covariant derivative;

ii) *in the case of the tangent bundle, D is symmetric if and only if K is symmetric.*

1.76 *Proof.* For i) one has to check first that, for V in $\mathcal{T}M$ and W in $\mathcal{E}$, $D_V W$ is $\Omega^0 M$-linear in V, which is clear since $V \mapsto TW(V)$ is so and K is a vector bundle map, and secondly that $D_V W$ is an $\Omega^0 M$-derivation in W.

Let φ be an element of $\Omega^0 M$, V in $\mathcal{T}M$ and W in $\mathcal{E}$, then

$$T(\varphi W)(V) = Ta^{\varphi(p_M(V))} \circ TW(V) + \iota^{-1}((V \cdot \varphi)W).$$

Therefore, K being a connector,

$$D_V(\varphi W) = K(T(\varphi W)(V)) = \varphi(p_M(V)) D_V W + (V \cdot \varphi) W.$$

ii) Notice first that for V, W in $\mathcal{T}M$ and each m in M, $TW(V(m)) - j(TV(W(m)))$ is a vertical vector [since $Tp_M(TW(V(m))) = V(m)$ and also $p_{TM}(TV(W(m))) = V(m)$] which is identified by ι to the Lie bracket $[V, W](m) = \mathcal{L}_V W(m)$.

Therefore, $K(TW(V) - j(TV(W))) = [V, W]$.

We deduce from this relation that if K is symmetric, so is D.

Conversely, if D is symmetric, $K \circ j(TW(V)) = K(TW(V))$ for any V, W in $\mathcal{T}M$. This is enough to prove that K is symmetric by specific choices of V and W. □

1.77. Let $\tau: E \to M$ be a vector bundle over M equipped with a connector K (the associated covariant derivative is denoted by D).

Let f be a map from N to M. The induced covariant derivative associated with the pulled back connector constructed in 1.63 will be denoted by D^f (D^f is a map from $\mathcal{T}N \underset{\Omega^0 N}{\otimes} f^*\mathcal{E}$ to $f^*\mathcal{E}$ where $f^*\mathcal{E}$ is the space of sections of $f^*\tau: f^*E \to N$).

To underline the fact that $D^f_v W$ is defined for v in $T_n N$, we add the following comment: we need more than the value of $T_n f(v)$ in M to compute $D^f_v W(n)$ since there can be many vectors v' such that $T_n f(v) = T_n f(v')$.

If the covariant derivative D is symmetric, then one proves that D^f has the following symmetry property: for V and W in $\mathcal{T}N$,

$$D^f_V Tf(W) - D^f_W Tf(V) = Tf([V, W]).$$

A section W of $f^*\tau: f^*E \to N$ is called *parallel* if $D^f_v W = 0$ for all v in TN (this is equivalent to saying that $TW(v)$ is horizontal, which can be expressed by the relation $TW(v) = h_W(v)$).

In the special case where the map is a curve c of M, a section W of $c^*\tau$ is parallel if and only if W satisfies the differential equations

1.78 $$D^c_{\frac{\partial}{\partial t}} W = 0$$

(often written in an abuse of notation as $D_{\dot c} W = 0$).

In a natural coordinate system, if $W=(x^i(t), w^i(t))$, the local expression of the system is, for $i=1, \ldots, r$

$$\frac{dw^i}{dt}+\sum_{j=1}^{r}\sum_{k=1}^{d}\gamma^i_{jk}(c(t))\,w^j(t)\dot{c}^k(t)=0\,.$$

We have already met such an equation. Indeed, in the case of the tangent bundle, if γ is a geodesic, then $\ddot{\gamma}=T\dot{\gamma}(\dot{\gamma})$ is horizontal, as we saw in 1.51 and 1.72. The equation of geodesics in covariant derivative notation is $D^{\gamma}_{\frac{\partial}{\partial s}}\dot{\gamma}=0$ (often written as $D_{\dot{\gamma}}\dot{\gamma}=0$).

Along a curve the linear system of r differential equations (1.78) is solvable (r denotes the rank of the bundle E). This gives rise to a linear isomorphism $P^c_{t_0t_1}$ from $E_{c(t_0)}$ onto $E_{c(t_1)}$, called the *parallel transport along c* and defined as follows. Let v be an element of $E_{c(t_0)}$. Then $P^c_{t_0t_1}(v)$ is the value at $c(t_1)$ of the unique solution of the differential Equation (1.78) which at $c(t_0)$ has value v, that is P^c is just the resolvant of (1.78). Of course, this map has a natural extension to spaces constructed from E by taking various tensor products.

What we just said can also be formulated as follows: if v is a tangent vector at m and W a section of E defined in a neighborhood of m, then D_vW is the ordinary derivative in E_m of $P^c_{t0}(W(c(t)))$ where c is any curve tangent to v.

1.79. If for V_1, V_2 in $\mathcal{T}M$ and W in $\mathscr{E}$ we consider the quantity

$$R(V_1, V_2)W=D_{[V_1,V_2]}W-D_{V_1}(D_{V_2}W)+D_{V_2}(D_{V_1}W),$$

then R turns out to be Ω^0M-linear in all its arguments, $[V_1, V_2]$ being also the commutator of the derivations of Ω^0M defined by V_1 and V_2. This tensor field R called the *curvature tensor* of the covariant derivative, can therefore be identified with a section of the vector bundle $\Lambda^2p_M^*\otimes\tau^*\otimes\tau:\Lambda^2T^*M\otimes E^*\otimes E\to M$.

The curvature tensor satisfies the following naturality condition with respect to pull backs: if f is a map from N to M and R^f denotes the curvature of the covariant derivative D^f, then for v_1, v_2 in TN and w in f^*E we have

$$R^f(v_1, v_2)w=R(Tf(v_1), Tf(v_2))w\,.$$

In the case of the tangent bundle equipped with a symmetric connection, the curvature tensor, an element of $\Omega^2M\otimes\mathcal{T}^*M\otimes\mathcal{T}M$, satisfies two identities called respectively the *first Bianchi identity*: for all v_1, v_2, v_3 in TM,

$$R(v_1, v_2(v_3+R(v_2, v_3)v_1+R(v_3, v_1)v_2=0$$

and the *second Bianchi identity*: for all v_1, v_2, v_3, w in TM,

$$\tilde{D}_{v_1}R(v_2, v_3)w+\tilde{D}_{v_2}R(v_3, v_1)w+\tilde{D}_{v_3}R(v_1, v_2)w=0$$

(where $\tilde{D}$ is the extension of D as a derivation for the tensor product and the contraction). These two identities can be checked directly from the very definition of R by choosing any extension by vector fields of the v_i's and w.

1.80. Given a Riemannian metric g, a symmetric covariant derivative D is called a *Levi-Civita covariant derivative for* g if for any V, W in $\mathcal{T}M$,

$$V\cdot g(W,W)=2g(D_V W,W).$$

This property is equivalent to saying that the tensorial extension $\tilde{D}$ of D satisfies

$$\tilde{D}g=0.$$

To justify the name Levi-Civita given in 1.72 to the unique symmetric connector K_g for which the geodesic vector field of a Riemannian metric g is horizontal, we have to prove that the symmetric covariant derivative associated to K_g is a Levi-Civita covariant derivative. But before proving this fact we must establish some formulas for the covariant derivative of differential forms. First of all, notice that any differential form can also be considered as a function on TM. Let λ be a differential 1-form. Define for v, w in TM

1.81 $$(\tilde{D}_v\lambda)(w)=v\cdot\lambda(W)-\lambda(D_v W),$$

where W is any vector field extending w (i.e. whose value at $p_M(w)$ is w).

Notice now that $v\cdot\lambda(W)$ can also be written as $TW(v)\cdot\lambda$ (where λ is then considered as a function on TM).

The horizontal part of $TW(v)$ is $h_W(v)$.

Since for any vertical vector V in TTM we have $V\cdot\lambda=\lambda(p_{TM}(V))$, we get

$$\lambda(D_v W)=\iota_W^{-1}(D_v W)\cdot\lambda.$$

But $\iota_W^{-1}(D_v W)$ is precisely the vertical part of $TW(v)$ since

$$K(\iota_W^{-1}(D_v W))=D_v W=K(TW(v)).$$

The Formula (1.81) reduces for all v, w in TM to

1.82 $$(\tilde{D}_v\lambda)(w)=h_w(v)\cdot\lambda,$$

Specialized to the case $v=w$ this gives the important relation

$$(\tilde{D}_v\lambda)(v)=Z(v)\cdot\lambda.$$

This gives the expected link between the spray and the covariant derivative in the case of differential forms (because they can be considered as functions on TM); for v in TM and λ in $\mathcal{T}^*M$, $(\tilde{D}_v\lambda)(v)$ is nothing but the ordinary derivative of λ considered as a function along the velocity curve of the geodesic $t\mapsto\exp_{p_M(v)}tv$. In particular, for the differential of a function f on M, to compute $\tilde{D}df$, called the *hessian* of f, one just has to compute the ordinary second derivative of f along geodesics.

1.83 Theorem (Fundamental theorem in Riemannian geometry). *Let g be a Riemannian metric on M. Then there exists one and only one Levi-Civita covariant derivative D^g associated with g.*

1.84 *Proof.* From the Levi-Civita connector K_g defined in 1.72, we deduce as in Proposition 1.74 a covariant derivative D^g. To prove existence and uniqueness, one just has to check that D^g satisfies the Levi-Civita property and that the Levi-Civita property determines K_g.

It is clear that Formula (1.82) extends to general symmetric differential forms. In particular, for the symmetric differential 2-form g, which when considered as a function on TM is denoted by $2E_g$, we get for v and w in TM

$$(\tilde{D}^g_v g)(w, w) = h_w(v)\cdot(2E_g)\,.$$

We compute the value of the last expression in a natural coordinate system. For $v=(x^i, v^i)$, $w=(x^i, w^i)$ we get

$$h_w(v)\cdot(2E_g) = \sum_{i,j,k=1}^{d} \frac{\partial g_{ij}}{\partial x^k} v^k w^i w^j - 2 \sum_{i,j,k,l=1}^{d} \Gamma^l_{jk} w^j v^k g_{il} w^i\,.$$

Therefore, $(\tilde{D}^g_v g)(w, w)$ will be zero for all v and w if and only if for all i, j, k between 1 and d

$$0 = \frac{\partial g_{ij}}{\partial x^k} - \sum_{l=1}^{d} \left(g_{il}\Gamma^l_{jk} + g_{jl}\Gamma^l_{ik}\right).$$

One can solve for the Γ^i_{jk} by summing three equations of this type with appropriate signs. This reproduces the usual manipulation (which we like to refer to as the "braid" manipulation) proving that a three-tensor which is skewsymmetric in two arguments and symmetric in two others vanishes. This proves at the same time existence and uniqueness. □

1.85. From the Levi-Civita property of D^g, the parallel transport $P^c_{t_0t_1}$ for D^g along a curve $c:[t_0, t_1]\to M$, is an isometry from $(T_{c(t_0)}M, g(c(t_0)))$ onto $(T_{c(t_1)}M, g(c(t_1)))$. In particular, if γ is a geodesic ($\dot\gamma$ is parallel along γ) the parallel transport $P^\gamma_{t_0t_1}$ maps the orthogonal complement of $\dot\gamma(t_0)$ onto the orthogonal complement of $\dot\gamma(t_1)$. Also the length of the image of a vector by parallel transport is constant.

1.86. The curvature tensor associated with the Levi-Civita covariant derivative has some special properties: first notice that by using the ♭ isomorphism associated to g, one can identify the curvature tensor R^g with an element of $\Omega^2 M\otimes(\otimes^2\mathcal{T}^*M)$. Then R^g turns out to be also skew symmetric in its two last arguments: for all v_1, v_2, v_3, v_4 in TM,

$$g(R^g(v_1, v_2)v_3, v_4) = -g(R^g(v_1, v_2)v_4, v_3);$$

Using the first Bianchi identity and the above equality one also gets

$$g(R^g(v_1, v_2)v_3, v_4) = g(R^g(v_3, v_4)v_1, v_2).$$

This follows easily from the Levi-Civita property by direct computation. Therefore, one often considers R^g as an element of $\mathcal{T}^2 \wedge {}^2TM$.

Let π be any two-plane and $\{v_1, v_2\}$ an orthonormal basis of π. We define $\sigma^g(\pi)$, the *sectional curvature* of π, by

$$\sigma^g(\pi) = g(R^g(v_1, v_2)v_1, v_2).$$

One easily checks that σ^g does indeed depend only on π and not on the orthonormal basis $\{v_1, v_2\}$ chosen.

J. Jacobi Fields

1.87. In this paragraph *a Riemannian metric g is fixed.* Therefore, we will drop the subscript g in all objects derived from g.

Our concern here is to study the linearized geodesic flow. There are many ways of doing this. Let us start with the following one. Suppose that we have a map $\Gamma: [0,1] \times]-\varepsilon, \varepsilon[\to M$ such that, for each t in the interval $]-\varepsilon, \varepsilon[$, $\Gamma_t: [0,1] \to M$ is a geodesic with normal parameter s. Such a map is called a *one-parameter family of geodesics*. For any t in $]-\varepsilon, \varepsilon[$, the variation Γ_t gives rise to a vector field $J_t(s) = T_{(s,t)}\Gamma\left(\frac{\partial}{\partial t}\right)$ along Γ_t, the transverse vector field. As Γ_t is a geodesic, $D^{\Gamma}_{\frac{\partial}{\partial s}} \dot{\Gamma}_t = 0$ (as usual, a dot denotes the derivative with respect to the parameter s of the curve Γ_t).

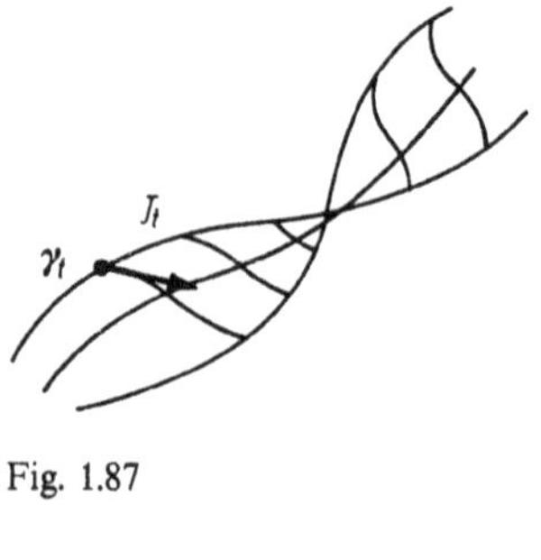

Fig. 1.87

Therefore

$$D^{\Gamma}_{\frac{\partial}{\partial t}} D^{\Gamma}_{\frac{\partial}{\partial s}} \dot{\Gamma}_t = 0.$$

This equation can be transformed, using the remarks at the end of 1.77 and 1.79, into

$$D^{\Gamma}_{\frac{\partial}{\partial s}} D^{\Gamma}_{\frac{\partial}{\partial s}} J_t + R(\dot{\Gamma}_t, J_t)\dot{\Gamma}_t = 0.$$

The curve Γ_t being a geodesic, this equation is a differential equation for the vector field J_t along Γ_t.

1.88. Now let us just consider one geodesic γ.

We will make the convention that covariant derivative along a geodesic is denoted by $'$ (this is consistent with the remark at the end of 1.80 if using contraction with the parallel tensor g we map a vector to a 1-form), so that the previous equation can be written as

1.89 $$J'' + R(\dot\gamma, J)\dot\gamma = 0 .$$

This equation is called the *Jacobi equation*; a vector field J along γ, solution of (1.89), is called a *Jacobi field.* The Jacobi equation being a linear second-order differential equation along γ, the vector space of Jacobi fields is $2d$-dimensional. A Jacobi field is for example determined by two initial conditions at a point $\gamma(s_0)$, its value $J(s_0)$ and the value of its derivative $J'(s_0)$.

Now if we take the scalar product of (1.89) with $\dot\gamma$, we get

$$0 = g(J'', \dot\gamma) = g(J', \dot\gamma)' = g(J, \dot\gamma)''$$

since γ is a geodesic. Therefore, $g(J, \dot\gamma)$ is an affine function of s along γ. Among the Jacobi fields, we can distinguish between the Jacobi fields $(as+b)\dot\gamma$ (where a and b are real numbers) and *normal* Jacobi fields, characterized by the property that, at a point $\gamma(s_0)$, $J(s_0)$ and $J'(s_0)$ are both orthogonal to $\dot\gamma(s_0)$ (and then remain orthogonal to $\dot\gamma$ along γ).

Before giving a converse to the construction 1.87, we make the following remark. A vector field Y along a curve c determines an element of TTM along c, namely $\iota_{\dot c}^{-1}\left(D^c_{\frac{\partial}{\partial s}} Y\right) + h_{\dot c}(Y)$. Now notice that if c_t is a variation of c, the element of TTM obtained from the construction 1.18 is just the above element, since $\iota_{\dot c}^{-1}(D^c_{\frac{\partial}{\partial s}} Y)$ is its vertical part, $h_{\dot c}(Y)$ being its horizontal part. In 1.87, the Jacobi field J arose in this way from Γ. For a Jacobi field J along a geodesic γ, we shall denote by $\mathscr{J}$ the element $\iota_{\dot\gamma}^{-1}(J') + h_{\dot\gamma}(J)$. The next proposition shows that all Jacobi fields arise as in 1.87.

1.90 Proposition. *Any Jacobi field along a geodesic γ is the transverse field of a variation of γ by geodesics.*

1.91 *Proof.* Let J be a Jacobi field along the geodesic γ. Consider the variation $\bar\gamma_t$ of γ defined as follows: associated with J, we have an element of TTM above $\gamma(s_0)$, $\mathscr{J}(s_0) = \iota_{\dot\gamma}^{-1}(J'(s_0)) + h_{\dot\gamma}(J(s_0))$ so that $K(\mathscr{J}) = J'$ and $Tp_M(\mathscr{J}) = J$. Take a one-parameter family of curves, the equivalence class of which is $\mathscr{J}(s_0)$ (see 1.18). It is possible to choose the family so that $\bar\gamma_0$ is precisely γ. We now modify $\bar\gamma_t$ as follows: along the curve $\bar\gamma_t(s_0)$ we replace the curve $s \mapsto \bar\gamma_t(s)$ by the geodesic with initial point $\bar\gamma_t(s_0)$ and with velocity vector $\dot{\bar\gamma}_t(s_0)$ (a more explicit formula for such a construction will be given in 2.14).

We denote by γ_t this new one-parameter family of curves. We still have $\gamma_0 = \bar{\gamma}_0 = \gamma$. Moreover, it is clear that the equivalence classes of γ_t and $\bar{\gamma}_t$ at $(s_0, 0)$ are the same. As γ_t is a one-parameter family of geodesics, from 1.87, we know that its transverse field along γ is a Jacobi field. Moreover, at $(s_0, 0)$ its initial conditions coincide with the initial conditions of J. By uniqueness of the solutions of the Jacobi equation with given initial conditions, these two Jacobi fields are equal along γ. □

1.92 Proposition. *The tangent map to the geodesic flow* (ζ^s) *can be described as follows. Let* X *be an element of* $T_u TM$, *where* u *is in* UM *and let* J *be the Jacobi field along the geodesic* $s \mapsto p_M \circ \zeta^s(u)$ *such that* $\mathscr{J}(0) = X$, *then* $T_u \zeta^s(X) = \mathscr{J}(s)$.

1.93 *Proof.* The claim is a consequence of Proposition 1.90, since $T_u\zeta^s(X)$ is precisely equal to $\frac{d}{dt}\zeta^s(u(t))_{|0}$ (where $u(t)$ is any curve in TM tangent at u for $t=0$ to X), i.e., the transverse field in TTM to a variation of $s \mapsto \zeta^s(u)$ by geodesics. □

1.94 Corollary. *The tangent map to the exponential map at* m *can be described as follows. Let* v, w *be elements of* $\mathring{T}_m M$ *and let* J *be the Jacobi field along the geodesic* $s \mapsto \exp_m(sv)$ *with initial conditions* $J(0) = 0$, $J'(0) = w$, *then* $T_v \exp_m(w) = J(1)$.

1.95 *Proof.* This is clear since $T_{sv}\exp_m(w)$ is simply $\frac{d}{dt}\exp_m(s(v+tw))_{|0}$, which is the transverse field J defined by a one-parameter family of geodesics through m with initial conditions $J(0)=0$, $J'(0)=w$. □

1.96. It follows from Corollary 1.94 and the fact that a Jacobi field along a geodesic γ with initial conditions orthogonal to $\dot{\gamma}$ remains orthogonal to $\dot{\gamma}$ along γ, that the exponential map preserves orthogonality along the geodesics, a property which is known in Riemannian geometry as the *Gauss lemma.*

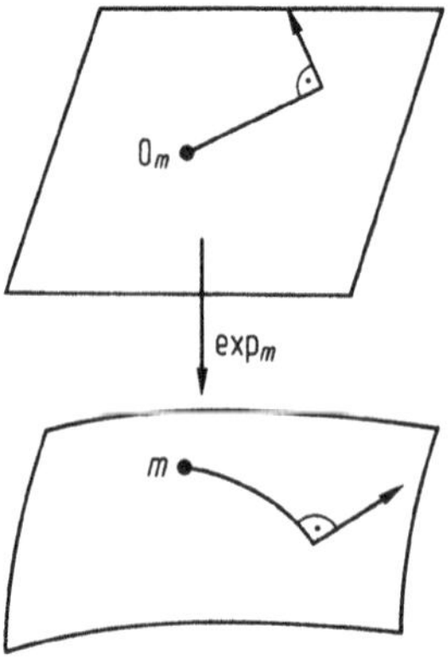

Fig. 1.96

If we fix the interval on which curves are defined, say $[0, 1]$, one can deduce from the Gauss lemma that the *geodesics locally minimize energy.* Consider a point m in M and let $r > 0$ be such that $\exp_m$ restricted to the ball $B(0_m, r)$ is a diffeomorphism (such an r exists from 1.53). It is sufficient to show that any curve c from m to a point p on the sphere $S(m,r)=\{p|p=\exp_m v, \|v\|=r\}$ has an energy $\mathbb{E}_g(c)$ larger than r^2. We consider $\bar{c}=\exp_m^{-1}\circ c$. We write $\bar{c}(t)=\|\bar{c}(t)\|u(t)$ where $u(t)$ has norm 1. It is clear that the radial part of $T_{\bar{c}(t)}\exp_m(\dot{\bar{c}}(t))$ is $\left(\frac{d}{dt}\|\bar{c}(t)\|\right)\dot{\gamma}_t(t)$ (where γ_t is the geodesic $s\mapsto\gamma_t(s)=\exp_m(su(t))$, so that

$$\mathbb{E}_g(c)=\int_0^1\|\dot{c}(t)\|^2dt\geqslant\int_0^1\left(\frac{d}{dt}\|\bar{c}(t)\|\right)^2dt.$$

Therefore by the Schwarz inequality

$$\mathbb{E}_g(c)\geqslant\left(\int_0^1\frac{d}{dt}\|\bar{c}(t)\|dt\right)^2=r^2.$$

Moreover the equality occurs if and only if the curve is radial, i.e. is a geodesic with initial point m and the parametrization is normal.

1.97. From Corollary 1.94, we see that the critical values of $\exp_m$ are precisely the points p for which there exists a geodesic $\gamma: s\mapsto\exp_m(sv)$ along which there is a non zero Jacobi field J with $J(0)=0$ and $J(1)=0$.

Therefore m is also a critical value of $\exp_p$. The points m and p are then said to be *conjugate points* along the geodesic γ. The occurence of conjugate points will be one of the main objects of study in the next chapters. More precisely given a point m, we will be interested in the location of points p conjugate to m along a geodesic with initial point m with the property that between m and p there is no conjugate point (the set of these points p is called the *conjugate locus of* m).

There is another point of view on conjugate points, which is also of interest. Let us fix a point m in M and a geodesic γ starting at m. One can consider the family of bilinear forms I_s called the *index forms* on the space of vector fields X along γ between $m=\gamma(0)$ and $\gamma(s)$ which are zero at m and at $\gamma(s)$:

$$I_s(X,X)=\int_0^s[\|X'(t)\|^2-g(R(\dot{\gamma}(t),X(t))\dot{\gamma}(t),X(t))]dt.$$

Clearly for small values of s, $I_s(X,X)$ is positive definite. Moreover a vector field X is in the nullity of I_s if and only if X is a Jacobi field. One can then say that $p=\gamma(s_0)$ is conjugate to m if and only if I_{s_0} has a non zero nullity. We define the *index* of the points $m=\gamma(0)$ and $p=\gamma(s_0)$ along γ as the nullity of the index form I_{s_0}. We denote it by $Ind_{\gamma(s_0)}$.

We say that $p=\gamma(s_0)$ is the *k-th conjugate point of m along* γ if

$$\sum_{0<s<s_0}Ind_{\gamma(s)}<k\leqslant\sum_{0<s\leqslant s_0}Ind_{\gamma(s)}.$$

Using arguments similar to that of [MR 1] p. 87 or following [ME] p. 235 one can prove

1.98 Proposition: *Let u_0 be a unit tangent vector to M. If the k-th conjugate point of $p_M(u_0)$ along the geodesic $s \mapsto p_M \circ \zeta^s(u_0)$ occurs for the value s_0 of the parameter, then there exist an interval $[s', s'']$ with $s' < s_0 < s''$ and a neighbourhood $\mathscr{U}$ of u_0 in UM such that, for u in $\mathscr{U}$, the k-th conjugate point of $p_M(u)$ along the geodesic $s \mapsto p_M \circ \zeta^s(u)$ occurs for a value s such that $s' < s < s''$.*

1.99. If we pursue the identification of Jacobi fields with elements of TTM as started at the end of 1.88, we can say the following: let J_1, J_2 be Jacobi fields along a geodesic γ with parameter s and $\mathscr{J}_1(s)$, $\mathscr{J}_2(s)$ be the transverse fields in TM associated respectively to J_1 and J_2. From Proposition 1.92 and the fact that the geodesic vector field Z is a Hamiltonian vector field, we see that $d\alpha(\mathscr{J}_1(s), \mathscr{J}_2(s))$ does not depend on s. Since $\mathscr{J}_i(s) = \iota_{\dot\gamma}^{-1}(J_i'(s)) + h_{\dot\gamma}(J_i(s))$ for $i = 1, 2$, from a formula in 1.23 giving $d\alpha_g$ and the fact that $V_{\dot\gamma}TM$ and $H_{\dot\gamma}TM$ are both Lagrangian subspaces of $T_{\dot\gamma}TM$ (see 1.72), we deduce that

$$d\alpha(\mathscr{J}_1(s), \mathscr{J}_2(s)) = g(J_1(s), J_2'(s)) - g(J_2(s), J_1'(s)).$$

The fact that the right hand side is constant in s along the geodesic γ can also be obtained directly from the Jacobi equation (which is of Sturm-Liouville type) since the linear map $J \mapsto R(\dot\gamma, J)\dot\gamma$ is symmetric with respect to g.

K. Riemannian Geometry of the Tangent Bundle

1.100. It will be useful later on to consider also TM as a Riemannian manifold. We would like the Riemannian metric on TM to be nicely related to the given Riemannian metric on M.

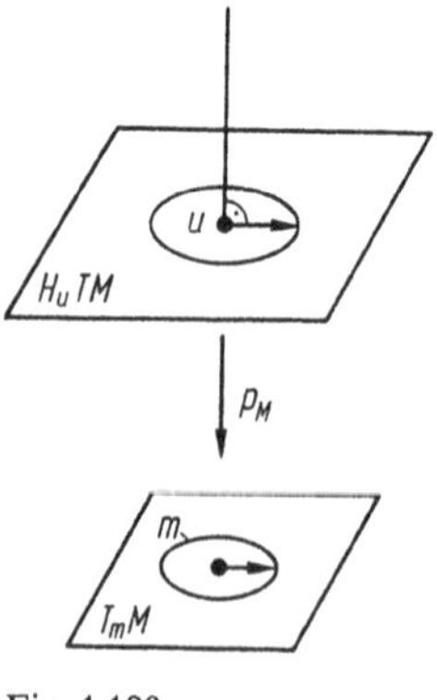

Fig. 1.100

Let us recall (see 1.60) that for any element v in T_mM, $Tp_M \times K$ maps $H_vTM \oplus V_vTM$ isomorphically onto $T_mM \oplus T_mM$.

We *define* a scalar product g_1 on TM by specifying that $Tp_M \times K$ is an isometry of (T_vTM, g_1) with $(T_mM \oplus T_mM, g \times g)$. This means that, for X in T_vTM

1.101 $$g_1(X,X)=g(Tp_M(X), Tp_M(X))+g(K(X), K(X)).$$

In particular with such a definition the horizontal and vertical subspace of TTM are orthogonal. Moreover $p_M:(TM,g_1)\to(M,g)$ appears as a *Riemannian submersion*, the orthogonal space to the kernel of its differential being isometric via Tp_M to the tangent space to the base.

Moreover from the formula for $d\alpha$ in 1.99, g_1 and $d\alpha$ are compatible (i.e. the linear map σ defined for X, Y in TTM by $g_1(X,\sigma Y)=d\alpha(X,Y)$ is an orthogonal transformation and satisfies $\sigma^2=-Id$), and therefore define an almost-Kählerian structure on TM (cf. [LZ 3]).

We denote by D^1 the Levi-Civita connexion of g_1.

The unit tangent bundle UM, which is a submanifold of TM, inherits a Riemannian metric from g_1 which we will also *denote by* g_1.

Notice that the Liouville vector field A is precisely the unit normal vector to UM for the Riemannian metric g_1.

1.102 Proposition. *The fibers of TM are totally geodesic for the Riemannian metric* g_1.

1.103 *Proof.* Let m be a point in M. We shall consider a vertical vector field V on TM and we shall prove that $D^1_V V$ is a vertical vector field. This is sufficient to insure that a curve c in T_mM which is a geodesic for the induced metric on the fiber (i.e. which satisfies $K(D^1_{\dot c}\dot c)=0$) is in fact a geodesic of TM.

To prove that $D^1_V V$ is vertical, we proceed as follows: we first notice that if X is a *basic* vector field (the horizontal lift of a vector field on M), then $\mathscr{L}_X g_1 \restriction VTM=0$. This is due to the fact that the identification of vertical vectors with tangent vectors to M is natural and therefore invariant under the flow of basic vector fields and then that parallel transport along a curve is an isometry.

We now evaluate $X\cdot g_1(V,V)$ in two ways: firstly by using the Levi-Civita property of D^1, we get

$$X\cdot g_1(V,V)=2g_1(D^1_X V,V),$$

and then, by using the previous remark, we also get

$$X\cdot g_1(V,V)=2g_1(\mathscr{L}_X V,V).$$

As D^1 is a symmetric covariant derivative, we deduce from these equalities that

$$g_1(D^1_V X,V)=0.$$

Using again the Levi-Civita property of D^1, we have

$$0=V\cdot g_1(X,V)-g_1(X,D^1_V V).$$

By definition of g_1, X being horizontal and V vertical, we proved that $D^1_V V$ is a vertical vector field. □

1.104 Proposition *The geodesic flow* (ζ^s) *of* (M, g) *acting on the Riemannian manifold* (UM, g_1) *is a one-parameter group of isometries if and only if* (M, g) *has constant sectional curvature* 1.

1.105 *Proof.* From Proposition 1.92, we know that the geodesic flow is a one-parameter group of isometries of (UM, g_1) if and only if $g_1(\mathcal{J}(s), \mathcal{J}(s))$ is independent of the parameter s of a geodesic γ (along which $\mathcal{J}$ is associated to a Jacobi field J). By definition of g_1, this means that

$$\frac{d}{ds}(g(J(s), J(s)) + g(J'(s), J'(s))) = 0.$$

As J is a Jacobi field, this is equivalent to the identity in s

$$(g(J(s), J'(s)) - g(R(\dot{\gamma}, J(s))\dot{\gamma}, J'(s))) = 0.$$

The Jacobi field J being arbitrary, we can for each fixed s consider a Jacobi field such that $J(s) = J'(s) = v$, so that we get

$$g((I - R(\dot{\gamma}, \cdot)\dot{\gamma})v, v) = 0.$$

The linear map $I - R(\dot{\gamma}, \cdot)\dot{\gamma}$ being symmetric with respect to g and the geodesic γ arbitrary, we obtain that for v and w in UM, $R(v, w)v = w$. □

1.106 Proposition. *Let* γ *be a (normally parametrized) geodesic in* M. *The velocity field* $\dot{\gamma}$ *of* γ *is a geodesic of* (TM, g_1).

1.107 *Proof.* We shall use standard arguments in Riemannian submersions. Let γ be a geodesic which minimizes the energy of curves between m_0 and m_1 defined on $[0, 1]$. We pick the point $\dot{\gamma}(m_0)$ in $T_{m_0}M$. The horizontal lift on the curve γ is precisely $\dot{\gamma}$ since γ is a geodesic. Consider any other curve c from $\dot{\gamma}(m_0)$ to $\dot{\gamma}(m_1)$. Since γ minimizes the energy downstairs from m_0 to m_1, we have

$$\mathbb{E}_g(p_M \circ c) \geqslant \mathbb{E}_g(\gamma).$$

On the other hand since $g_1 \geqslant p_M^*(g)$ by construction,

$$\mathbb{E}_{g_1}(c) \geqslant \mathbb{E}_{p_M^*(g)}(c) = \mathbb{E}_g(p_M \circ c);$$

we have equality if and only if c is horizontal, so that $\mathbb{E}_{g_1}(\dot{\gamma}) = \mathbb{E}_g(\gamma)$.
Therefore $\mathbb{E}_{g_1}(c) \geqslant \mathbb{E}_{g_1}(\gamma)$. □

L. Formulas for the First and Second Variations of the Length of Curves

1.108. We are interested here in giving variation formulas for the length of curves in a Riemannian manifold (M, g) by using the formalism of the Levi-Civita covariant derivative D.

We now switch to the length since it is a more geometrical concept, but to get the geodesic vector field and the geodesic flow we were forced to work with the energy, which is a concept borrowed from mechanics.

Let us start with a family of curves $(c_u)_{u\in]-\varepsilon,\varepsilon[}$ with parameter t in $[0, l]$. We are interested in the function $\mathbb{L}(c_u)=\int_0^l \sqrt{g(\dot c_u, \dot c_u)}dt$. Without any loss of generality, we can suppose that c_0 is parametrized by arc length so that l is exactly $\mathbb{L}(c_0)$.

The transverse vector field $Tc\left(\frac{\partial}{\partial u}\right)$ along c_u is denoted by X_u.

We want to evaluate $\frac{d}{du}\mathbb{L}(c_u)_{|0}$ and $\frac{d^2}{du^2}\mathbb{L}(c_u)_{|0}$ in the case where $\frac{d}{du}\mathbb{L}(c_u)_{|0}=0$.

1.109 Proposition (First variation of the length). *Let (c_u) be a family of curves parametrized between* 0 *and l (c_0 is supposed to be parametrized by arc length). Then*

1.110 $$\frac{d}{du}\mathbb{L}(c_u)_{|0}=[g(X_0,\dot c_0)]_0^l-\int_0^l g(D_{\dot c_0}\dot c_0, X_0)dt.$$

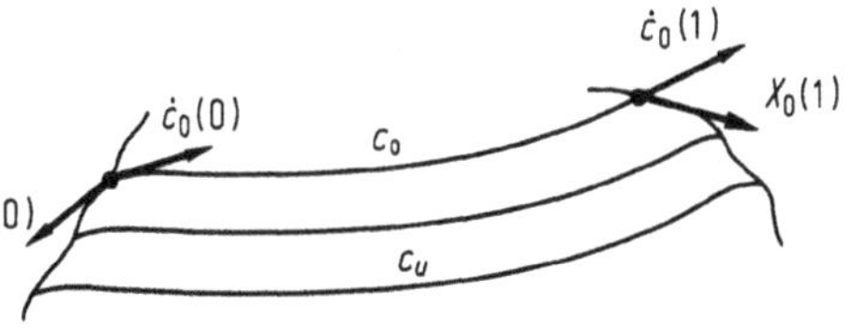

Fig. 1.109

1.111 *Proof.* We compute $\frac{d}{du}\left(\int_0^l \sqrt{g(\dot c_u, \dot c_u)}dt\right)_{|0}$.

From the properties of the Levi-Civita covariant derivative, we get

$$\frac{d}{du}\mathbb{L}(c_u)_{|0}=\int_0^l g(D^c_{\frac{\partial}{\partial u}}\dot c_0, \dot c_0)/\sqrt{g(\dot c_0, \dot c_0)}dt.$$

Since the curve c_0 is parametrized by arc length, the denominator is in fact 1. Since $\frac{\partial}{\partial u}$ and $\frac{\partial}{\partial t}$ commute, by the remark at the end of 1.77 we have

$$D^c_{\frac{\partial}{\partial u}}\dot c_u=D^c_{\frac{\partial}{\partial t}}X_u.$$

Integrating by parts, we get

$$\frac{d}{du}\mathbb{L}(c_u)_{|0}=\int_0^l\left(\frac{\partial}{\partial t}\cdot g(\dot c_0, X_0)\right)dt-\int_0^l g(D_{\dot c_0}\dot c_0, X_0)dt,$$

which immediately gives the announced formula. □

1.112. Two remarks are in order: first of all if c_0 is a geodesic, then only the values of the transverse vector X_0 at the two ends of c_0 matter; secondly, as c_0 is parametrized by arc length, $D_{\dot{c}_0}\dot{c}_0$ is orthogonal to $\dot{c}_0$ all along c_0, so that in the integral of Formula (1.110) only the part of X_0 normal to c_0 matters, which is intuitively clear.

1.113 Proposition (Second variation of the length). *Suppose that $c_0 = \gamma$ is a geodesic and that X_0 is orthogonal to $\dot{\gamma}$ along γ. Then*

1.114 $$\frac{d^2}{du^2}\mathbb{L}(c_u)_{|0} = [g(D_{X_0}X_0,\dot{\gamma}) + g(X_0, D_{\dot{\gamma}}X_0)]_0^l - \int_0^l g(X_0, D_{\dot{\gamma}}D_{\dot{\gamma}}X_0 + R(\dot{\gamma},X_0)\dot{\gamma})dt.$$

1.115 *Proof.* By a direct calculation using the fact that $g(\dot{\gamma},\dot{\gamma}) = 1$, we get

$$\frac{d^2}{du^2}\mathbb{L}(c_u)_{|0} = 1/2\int_0^l \frac{\partial^2}{\partial u^2}\cdot g(\dot{c}_u,\dot{c}_u)_{|0}dt - 1/4\int_0^l\left(\frac{\partial}{\partial u}\cdot g(\dot{c}_u,\dot{c}_u)_{|0}\right)^2 dt.$$

The second term gives no contribution since

$$\frac{\partial}{\partial u}\cdot g(\dot{c}_u,\dot{c}_u)_{|0} = 2g\left(D^c_{\frac{\partial}{\partial u}}\dot{c}_0,\dot{c}_0\right) = 2g\left(D^c_{\frac{\partial}{\partial t}}X_0,\dot{c}_0\right)$$

$$= 2\frac{\partial}{\partial t}\cdot(X_0,\dot{\gamma}) - 2(X_0, D_{\dot{\gamma}}\dot{\gamma}),$$

and each term is zero by assumption. We concentrate on the first term:

$$1/2\frac{\partial^2}{\partial u^2}\cdot g(\dot{c}_u,\dot{c}_u)_{|0} = g\left(D^c_{\frac{\partial}{\partial u}}D^c_{\frac{\partial}{\partial u}}\dot{c}_0,\dot{c}_0\right) + g\left(D^c_{\frac{\partial}{\partial u}}\dot{c}_0, D^c_{\frac{\partial}{\partial u}}\dot{c}_0\right).$$

Applying the relation $D^c_{\frac{\partial}{\partial u}}\dot{c}_0 = D^c_{\frac{\partial}{\partial t}}X_0$ and the definition of curvature, we obtain

$$1/2\frac{\partial^2}{\partial u^2}\cdot g(\dot{c}_u,\dot{c}_u)_{|0} = g(R(\dot{\gamma},X_0)X_0,\dot{\gamma}) + g\left(D^c_{\frac{\partial}{\partial t}}D^c_{\frac{\partial}{\partial u}}X_0,\dot{\gamma}\right) + g\left(D^c_{\frac{\partial}{\partial t}}X_0, D^c_{\frac{\partial}{\partial t}}X_0\right).$$

So applying again the remarks at the end of 1.77 and 1.79, we get

$$1/2\frac{\partial^2}{\partial u^2}\cdot g(\dot{c}_u,\dot{c}_u)_{|0}$$

$$= g(R(\dot{\gamma},X_0)X_0,\dot{\gamma}) - g\left(X_0, D^c_{\frac{\partial}{\partial t}}D^c_{\frac{\partial}{\partial t}}X_0\right)$$

$$+ \frac{\partial}{\partial t}\cdot g\left(D^c_{\frac{\partial}{\partial u}}X_0,\dot{\gamma}\right) + \frac{\partial}{\partial t}\cdot g\left(X_0, D^c_{\frac{\partial}{\partial t}}X_0\right),$$

where we have used the fact that γ is a geodesic.

If we now integrate from 0 to l, we get Formula (1.114). □

1.116. Notice that the assumption in Proposition 1.113 is stronger than is actually necessary. To obtain the formula it would have been sufficient to suppose that $\frac{d}{du}\mathbb{L}(c_u)_{|0}=0$, but it is usually under the assumptions in Proposition 1.113 (or even stronger ones) that the formula is applied.

The boundary terms are of different natures. The first one is the analog of the boundary term in the first variation formula involving second order information on the variation, and the second involves quadratically the transverse vectorfield X.

It is not surprising that the integral part contains exactly the quantity which when set equal to zero gives the Jacobi equation. In particular, for a variation of γ through geodesics, only the boundary terms remain. In the second boundary term we recognize after an integration by parts the index form (see 1.97).

M. Canonical Measures of Riemannian Manifolds

1.117. We will denote by μ_g the *canonical* measure of a Riemannian manifold (M,g) (see [B-G-M], p. 10—11).

In particular, when M is compact the total measure of (M,g) for μ_g is called the *volume* of (M,g) and is denoted alternatively by

1.118 $$\operatorname{Vol}(g)=\operatorname{Vol}(M,g)=\int_M d\mu_g\,.$$

In the special case of the standard sphere (S^d, can) we set

1.119 $$\beta(d)=\operatorname{Vol}(S^d,\mathrm{can}),$$

whose explicit value can be found in [B-G-M], p. 12, and elsewhere. We also set

1.120 $$\sigma=\mu_{\mathrm{can}}\,.$$

Now when we consider the unit tangent bundle of (M,g) endowed with its canonical Riemannian metric g_1 defined by Formula (1.101), we use the abbreviation

1.121 $$\mu_1=\mu_{g_1}\,.$$

In this special case and because UM is oriented, the canonical measure μ_1 can be defined as the absolute value of the *canonical volume form* $\eta_1=\eta_{g_1}$ (see [B-G-M], p. 13—14 if necessary). Here it turns out that this volume form is proportional to the volume form of UM introduced in 1.58. More precisely, one proves (see [BR 3], p. 166—167) straightforwardly that

1.122 $$\eta_1=\frac{1}{(d-1)!}\,\alpha_g\wedge(d\alpha_g)^{d-1}\,.$$

Another way of writing μ_1 is the following: we desintegrate μ_1 along the fibers U_mM for m in M. If we still *denote* by σ the *canonical measure of the unit sphere* U_mM in the Euclidean space (T_mM, g_m), we can write

1.123 $$\int f d\mu_1 = \int f d\sigma \otimes d\mu_g$$

for every f in $C^\infty(UM)$. In particular, Fubini's theorem implies that

1.124 $$\mathrm{Vol}(UM, g_1) = \beta(d-1)\,\mathrm{Vol}(M, g).$$

1.125. Notice that Liouville's theorem 1.56 and Formula (1.122) imply that the geodesic flow of UM leaves the measure μ_1 invariant.

Chapter 2. The Manifold of Geodesics

A. Summary

We are interested in the study of Riemannian manifolds (M, g) (called C_l-manifolds) whose geodesics are periodic and have the same length l. We define the manifold of geodesics $C^g M$ for a C_l-manifold and we relate its tangent spaces to normal Jacobi fields. The existence of a nondegenerate closed two-form on $C^g M$ is the most striking fact. This form endows the manifold with a symplectic structure. Using the fact that the unit tangent bundle of M is fibered in two ways over M and $C^g M$, we prove A. Weinstein's theorem.

Then we discuss some Riemannian metrics which can be naturally defined on $C^g M$, especially the metrics $\bar{g}_0$ and $\bar{g}_1$ which are respectively of Sobolev type H^0 and H^1. We study in detail the geodesics of $\bar{g}_0$ together with its connection and curvature.

B. The Manifold of Geodesics

Let us first give some of the definitions of Chapter 7, which we will need in the present chapter.

2.1 Definition. Let (M, g) be a Riemannian manifold. A curve γ is a *periodic geodesic* with period l (where l is a non-zero real number) if and only if:

i) γ is a geodesic;

ii) γ is periodic as a map from $\mathbb{R}^+$ to M (parametrized by arc length) with least period l. The number l is the *length* of the periodic geodesic.

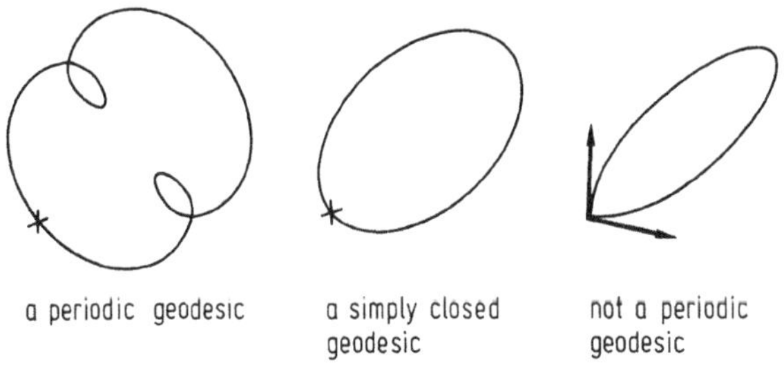

Fig. 2.2

2.3 Definition. A manifold M is a C_l-*manifold* if there exists a Riemannian metric g on M such that all geodesics are periodic geodesics with the same length l. We say that g is a C_l-*metric.*

For every point v in the unit tangent bundle of M the integral curves $s \mapsto \zeta^s(v)$ of the geodesic flow Z_g (see 1.56 for the definition) on the unit tangent bundle U^gM are periodic with least period l and $\zeta^s(v) = v$ if and only if s is an integral multiple of l.

In the same way, a manifold is called an SC_l-*manifold* if all of its geodesics are simply closed with the same length l.

2.4 Remark. Until now all known C_l-manifolds have been SC_l-manifolds and we may wonder whether every C_l-manifolds is in fact an SC_l-manifold (see also 7.G, in particular 7.69 (C').

In the following chapters we will define the only known C_l-manifolds, namely: the compact symmetric spaces of rank one (the CROSSes) with their canonical metric (see Chapter 3) and Zoll metrics (see Chapter 4).

2.5. Let (M, g) be a C_l-manifold. The vector field Z_g generates a free action of $S^1 \cong \mathbb{R}/l\mathbb{Z}$ on the unit tangent bundle U^gM and the quotient space U^gM/S^1 is a $(2d-2)$ dimensional manifold (*the manifold of oriented geodesics*) which we *denote by* C^gM (we will omit the subscript g and the adjective oriented when no confusion is possible).

Clearly the projection $q: UM \to CM$ is a principal bundle with structural group S^1.

2.6. Using the opposite map a^{-1} on UM (see 1.2) we can consider the free action of $\mathbb{Z}_2 \times S^1$ on UM and get another manifold (*the manifold of non-oriented geodesics*) of which CM is a two-sheeted covering.

2.7 Remark. If we only require the geodesic flow to be periodic with least period l (i.e., the periodic geodesics do not necessarily have the same length), then the action of S^1 is not free and we cannot define the manifold of geodesics (for an example, see the lens spaces 0.39).

2.8 Examples. Consider (S^d, can), the d-dimensional sphere in $\mathbb{R}^{d+1}$ endowed with the metric induced by the Euclidean metric of $\mathbb{R}^{d+1}$. A point in the unit tangent bundle of (S^d, can) may be thought of as a pair of orthonormal vectors in $\mathbb{R}^{d+1}$ which determine a two-dimensional oriented subspace P_2 in $\mathbb{R}^{d+1}$. The Euclidean symmetry which keeps P_2 fixed induces an isometry of (S^d, can) whose fixed point set $(P_2 \cap S^d)$ is a totally geodesic submanifold of S^d. Conversely, every geodesic in S^d lies in an oriented two-dimensional subspace (by uniqueness). The manifold (S^d, can) is a $C_{2\pi}$-manifold and oriented geodesics are intersections of the two-dimensional subspaces of $\mathbb{R}^{d+1}$ with the sphere. Then $C^{\text{can}} S^d$ is the *Grassmann manifold of oriented two-planes* in $\mathbb{R}^{d+1}$ (denoted by $G^+_{2,d+1}$).

From a different standpoint we can say that US^d is the homogeneous manifold $SO(d+1)/SO(d-1)$ and the action of S^1 [considered as $SO(2)$] leads to the quotient manifold

$$G^+_{2,d+1} \simeq SO(d+1)/SO(d-1) \times SO(2).$$

If we remark that the geodesics of $(\mathbb{R}P^d, \mathrm{can})$ (the real projective space) are projections of the geodesics of (S^d, can), then we see that $G^+_{2,d+1}$ is the manifold of geodesics of a $C_{2\pi}$-manifold and also of a C_π-manifold.

Let us give a more specific example in the case $d=3$. The Lie group S^3 of unit quaternions is parallelizable and $US^3 \simeq S^3 \times S^2$. We can give an interpretation of the action of S^1 as follows:

2.9 Proposition. *The manifold of geodesics of* (S^3, can) *is diffeomorphic to* $S^2 \times S^2$ *(the product of two 2-dimensional spheres).*

Proof. Following [ST] we prove that $G^+_{2,4}$ is diffeomorphic to $S^2 \times S^2$.

Let us consider $\Lambda^2\mathbb{R}^4$, the space of skew symmetric two-tensors (or bivectors) of $\mathbb{R}^4$ endowed with the scalar product deduced from the canonical scalar product in $\mathbb{R}^4$ and denoted by (,). The Hodge map $*$ generally defined from $\Lambda^r E$ onto $\Lambda^{d-r}E$ is a linear involution of $\Lambda^2\mathbb{R}^4$. Hence $\Lambda^2\mathbb{R}^4$ splits into two orthogonal subspaces E_1 and E_{-1} associated to the eigenvalues $+1$ and -1 of $*$.

A bivector in $\Lambda^2\mathbb{R}^4$ is called *decomposable* if it can be written as the exterior product of two elements of $\mathbb{R}^4$. It is easily verified that a bivector ξ in $\Lambda^2\mathbb{R}^4$ is decomposable if and only if ξ and $*\xi$ are orthogonal for the scalar product (,) (use the normal form of bivectors as given for example in [S-G] p. 16). We claim that the set of decomposable bivectors of norm 1 in $\Lambda^2\mathbb{R}^4$ is diffeomorphic to $S^2 \times S^2$. Indeed, let W be a unit decomposable bivector in $\Lambda^2\mathbb{R}^4$ and let

$$\xi = \frac{W + *W}{2} \quad \text{and} \quad \bar{\xi} = \frac{W - *W}{2}.$$

The pair $(\xi, \bar{\xi})$ belongs to $E_1 \times E_{-1}$ and $\|\xi\|^2 = \|\bar{\xi}\|^2 = \frac{1}{2}$ ($\|\ \|$ is the norm associated with the scalar product on $\Lambda^2\mathbb{R}^4$). Conversely, if $(\xi, \bar{\xi})$ is a pair of bivectors in $S^2_{1/\sqrt{2}} \times S^2_{1/\sqrt{2}}$ (the spheres in E_1, E_{-1} of radius $1/\sqrt{2}$), then $\xi + \bar{\xi}$ is a decomposable bivector of norm 1 since

$$(\xi + \bar{\xi}, *(\xi + \bar{\xi})) = (\xi + \bar{\xi}, \xi - \bar{\xi}) = \|\xi\|^2 - \|\bar{\xi}\|^2 = 0.$$

Moreover, it is clear that $G^+_{2,4}$ is diffeomorphic to the set of decomposable bivectors of norm 1 in $\Lambda^2\mathbb{R}^4$. Indeed, given P in $G^+_{2,4}$ and (e_1, e_2) a direct orthonormal frame of P we can associate with P the decomposable bivector of norm 1 $W = e_1 \wedge e_2$ and conversely. □

2.10. Using the homotopy exact sequence we prove that the manifold of geodesics of any C_l-metric on S^2 is diffeomorphic to S^2. As a matter of fact, $US^2 \simeq SO(3)$, and we can write

$$\begin{array}{ccccccc} \cdots \to & \pi_1(S^1) & \to & \pi_1(SO(3)) & \to & \pi_1(CS^2) \to 0. \\ & \wr\!\wr & & \wr\!\wr & & \\ & \mathbb{Z} & & \mathbb{Z}_2 & & \end{array}$$

Hence, $\pi_1(CS^2)$ is zero or $\mathbb{Z}_2$. But as we shall see later on, CS^2 is an orientable manifold, so that $\pi_1(CS^2) = 0$ and CS^2 is the sphere S^2.

Let us now go back to the general situation and consider the restriction to UM of the 1-form α on TM defined in 1.23. We still denote this restriction by α.

2.11 Proposition. *The 1-form α is a connection form for the principal bundle $q: UM \to CM$ and $d\alpha$ is the curvature form of this connection.*

For the definitions of a connection form and its curvature form on a principal bundle see [K.N-1] p. 63—64 and p. 75—79.

Proof. Define a distribution Q on TUM. For u in UM, we let $Q_u = \{X \in T_u Um \mid \alpha(X) = 0\}$. We have to show that Q is a horizontal distribution whose associated connection form is α.

i) $T_u UM = \mathbb{R} \cdot Z_g(u) \oplus Q_u$.
This is obvious if you recall that

$$T_u UM = \{X \in T_u TM \mid g_1(A_u, X) = 0\}$$

(where A is the Liouville vector field on TM and g_1 is defined in 1.101), and that the vertical tangent space to the fiber of the projection q is generated by the geodesic vector field Z_g which satisfies $\alpha(Z_g) = 1$ (see 1.57).

ii) The 1-form α is invariant under the action of the one-parameter group (ζ^s) generated by Z_g since $\mathscr{L}_{Z_g}\alpha = 0$ on UM (see 1.56). Hence, Q is invariant under the action of S^1 (i.e., for u in UM and s in $\mathbb{R}$, $Q_{\zeta^s(u)} = T\zeta^s(Q_u)$).

iii) The differentiability of the map $u \mapsto Q_u$ is clear. Hence, Q is a horizontal distribution on TUM. The fact that $\alpha(Z_g) = 1$ implies that α is the connection form of the S^1-principal bundle $q: UM \to CM$.

From 1.56 we have $d\alpha(Z_g, .) = 0$, so that the differential two form $d\alpha$ is horizontal and consequently is the curvature form of the connection. □

2.12 Remark. From the invariance of $d\alpha$ under the action of S^1 ($\mathscr{L}_Z d\alpha = 0$) we conclude that $d\alpha$ is basic so that we can find a closed two-form ω on CM such that $q^*(\omega) = d\alpha$.

Let γ be a point in CM and let u be a point in $q^{-1}(\gamma)$. Then $T_u q$ defines an isomorphism from Q_u onto $T_\gamma CM$ and we have the following characterization of the tangent space of CM at γ.

2.13 Proposition. *The tangent space at a point γ in CM is naturally isomorphic (via the horizontal lifting) to the space of normal Jacobi fields along the geodesic γ in M.*

Proof. The two spaces are $(2d-2)$ dimensional and we construct an injective map φ in the following way. Let $\bar{X}$ be a vector in $T_\gamma CM$ and let X_u be the horizontal lifting of X at a point u in $q^{-1}(\gamma)$. Then we claim that $s \mapsto Tp_M(X_{\zeta^s(u)})$ is a normal Jacobi field along the geodesic $s \mapsto p_M(\zeta^s(u))$. By 2.11 and 1.23 we have

$$\begin{aligned} 0 = \alpha(X_{\zeta^s(u)}) &= g(Tp_M(X_{\zeta^s(u)}), p_M(X_{\zeta^s(u)})) \\ &= g(Tp_M(X_{\zeta^s(u)}), \zeta^s(u)) \end{aligned}$$

and normality is verified.

Following the notations introduced in 1.91 we get

$$Tp_M(X_{\zeta^s(u)}) = Tp_M \circ T\zeta^s(X_u) = Tp_M(\mathscr{J}(s)) = J(s),$$

where $J(s)$ is precisely the normal Jacobi field along the geodesic $s \mapsto \gamma(s)$ with initial conditions

$$J(0) = Tp_M(X_u), \qquad J'(0) = K(X_u)$$

(K is the connection map defined in 1.72).

The map $\varphi : \bar{X} \to J$ is linear and independent of the choice of u in $q^{-1}(\gamma)$.

Suppose that $\varphi(\bar{X}) = 0$, so that $J(0)$ and $J'(0)$ are equal to zero. Then $Tp_M(X_u) = 0$ and this implies that X_u is a vertical vector for the fibration $p_M : TM \to M$. Now $K(X_u) = 0$ so that X_u is a horizontal vector for the same fibration. But by 1.60 the subspaces of vertical and horizontal vectors intersect at $\{0\}$ so that $X_u = 0$ and φ is injective.

We will often use this identification and denote a tangent vector in CM indifferently by $\bar{X}$ or J.

2.14. We now exhibit a local chart for CM. Let γ be a point in CM and let $s \mapsto \gamma(s)$ be a normal parametrization of the geodesic γ in M. Fix a point $m = \gamma(0)$ on γ and denote by $N_m\gamma$ the normal space to $\dot{\gamma}(0)$ in T_mM. The space $N_m\gamma$ is isometric to $\mathbb{R}^{d-1}$ whenever we choose an orthonormal basis $(e_2, e_3, \ldots, e_d)$ in $N_m\gamma$. Define a map φ from $\mathbb{R}^{2d-2} \simeq \mathbb{R}^{d-1} \times \mathbb{R}^{d-1}$ into UM as follows: for (u, v) in $N_m\gamma \times N_m\gamma$,

$$\varphi(u, v) = \frac{d}{ds} \exp_{\exp_m(u)} s(P_u)\left(\frac{\dot{\gamma}(0) + v}{\sqrt{1 + \|v\|^2}}\right)\Bigg|_0$$

where P_u is the parallel transport along the geodesic $t \mapsto \exp_m tu$ and $\|v\|$ is the norm of the vector v in $N_m\gamma$ ($\|v\|^2 = g_m(v, v)$). Our map φ is a differentiable map from $\mathbb{R}^{2d-2}$ into CM and $\bar{\varphi} = q \circ \varphi$ is a map from $\mathbb{R}^{2d-2}$ into CM such that $\bar{\varphi}(0) = q \circ \varphi(0) = \gamma$.

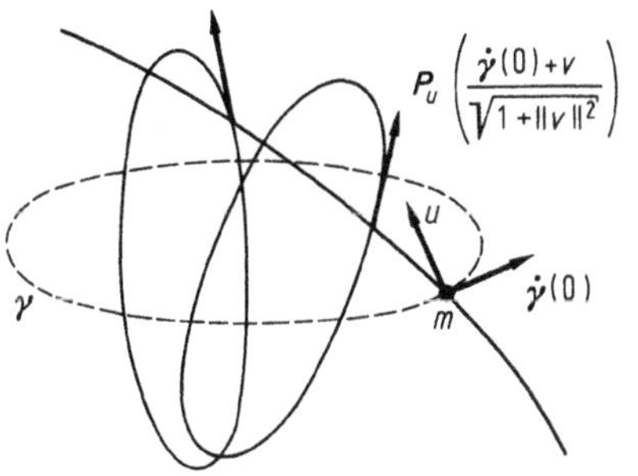

Fig. 2.15

We claim that $\bar{\varphi}$ has maximal rank at 0. If $(u_i)_{i=2,\ldots,d}$, and $(v_i)_{i=2,\ldots,d}$ are coordinates relative to the e_i's $(i = 2, \ldots, d)$ on $\mathbb{R}^{d-1} \times \mathbb{R}^{d-1}$, then $J_i = T\bar{\varphi}\left(\frac{\partial}{\partial u_i}\right)_0$ and $K_i = T\bar{\varphi}\left(\frac{\partial}{\partial v_i}\right)_0$ are normal Jacobi fields along the geodesic γ with initial conditions $(J_i(0) = e_i, J_i'(0) = 0)_{i=2,\ldots,d}$ and $(K_i(0) = 0, K_i'(0) = e_i)_{i=2,\ldots,d}$. This follows from the fact that $(s, v_i) \mapsto \exp_m s\left(\frac{\dot\gamma(0)+v_i e_i}{\sqrt{1+v_i^2}}\right)$ and $(s, u_i) \mapsto \exp_{\exp u_i e_i} s(P_u(\dot\gamma)(0)))$ are variations of the geodesic γ by geodesics (see [B-G-M] p. 89). From the definition of normal Jacobi fields (see 1.89) $(J_i \cup K_i)_{i=2,\ldots,d}$ is a basis of the $(2d-2)$ dimensional vector space of normal Jacobi fields along γ (i.e. of the tangent space $T_\gamma CM$). Then $T_0\bar\varphi$ is an isomorphism of $\mathbb{R}^{2d-2}$ onto $T_\gamma CM$ and $\bar\varphi$ is locally invertible. □

C. The Manifold of Geodesics as a Symplectic Manifold

2.16 Proposition. *The manifold CM with the naturally defined two form ω is a symplectic manifold. For a point γ in CM and two tangent vectors (J_1, J_2) at γ (identified with normal Jacobi fields along γ) we have*

2.17 $$\omega(J_1, J_2) = g(J_1, J_2') - g(J_2, J_1').$$

Proof. The closed two form ω defined in 2.12 is nondegenerate. Indeed, otherwise $d\alpha$ would be degenerate (in fact if there exists X in Q_u such that $d\alpha(X)=0$ on Q_u then $d\alpha(X, Z)=0$ and $d\alpha(X, A_u)=\alpha(X)=0$ implies that $d\alpha(X, \cdot)=0$ on $T_u TM$, a contradiction).

From the very definition of ω and the formula given in 1.99 we get Formula (2.17) (the right hand side of the equality is surely constant along the geodesic γ!). □

Remark. From the formula (2.17) it is obvious that ω is nondegenerate on CM: for any initial conditions $(J_1(0), J_1'(0))$, that is, for any non zero tangent vector at γ, we can find a tangent vector J_2 (i.e., initial conditions $(J_2(0), J_2'(0))$ such that $\omega(J_1, J_2) \neq 0$.

2.18 Corollary. *The manifold of geodesics is an oriented manifold and ω^{d-1} is a natural volume form.*

Remark. In the case where d is an even integer, the manifold of nonoriented geodesics (defined in 2.6) is a nonorientable manifold.

2.19. Using (2.17) we can describe some Lagrangian submanifolds in CM. Let γ be a point in CM and m a point on γ in M. The submanifold generated by geodesics through m and normal to $\dot\gamma$ at m is a $(d-1)$-dimensional Lagrangian submanifold. Indeed, if (J_1, J_2) are two tangent vectors to the submanifold (i.e. two normal Jacobi fields along a geodesic which goes through m), then from (2.17) we deduce that

$$\omega(J_1, J_2) = g(J_1(m), J_2'(m)) - g(J_2(m), J_2'(m)) = 0,$$

since $J_1(m)$ and $J_2(m)$ are zero (they come from variations which keep m fixed).

2.20. Notice that the assumption we have made on the geodesic flow enables us to construct a global space of motions in the sense of Souriau [SU] p. 105. The symplectic structure on CM is the natural symplectic structure of such a space.

2.21 Theorem (A. Weinstein [WN 3]). *If (M,g) is a d-dimensional C_l-manifold then the ratio* $\dfrac{\mathrm{Vol}(M,g)}{\mathrm{Vol}(S^d,\mathrm{can})}\left(\dfrac{2\pi}{l}\right)^d$ *is an integer $i(M,g)$ called* Weinstein's integer.

In particular, if (M,g) is a $C_{2\pi}$-manifold the volume of (M,g) is an integral multiple of the volume of $(S^d$, can).

Proof. The basic fact is that the closed two form $\frac{\omega}{l}$ on the oriented manifold C^gM is via de Rham isomorphism the Euler class $e(q)$ of the principal bundle $q: U^gM \to C^gM$. This follows from the facts that $q^*\omega = d\alpha$ and that $d\alpha$ is the curvature form of a connection on the bundle (for more details on the relations between characteristic classes and curvature see [MI-ST] p. 311 or [K-N 2] p. 314 or the very explicit [G-H-V], p. 321).

The de Rham cohomology class $\left[\left(\frac{\omega}{l}\right)^{d-1}\right] = \left[\frac{\omega}{l}\right]^{d-1}$ of $\left(\frac{\omega}{l}\right)^{d-1}$ is the image of the integral class $[e(q)]^{d-1}$ under the coefficient homomorphism $H^{2d-2}(C^gM,\mathbb{Z}) \to H^{2d-2}(C^gM,\mathbb{R})$ and the value of the class $[e(q)]^{d-1}$ on the fundamental $(2d-2)$ cocycle C^gM is an *integer* $j(M,g)$. The integer $j(M,g)$ is a topological invariant of the fibration $q: U^gM \to C^gM$.

Using Riemannian methods we can compute the integer

$$j(M,g) = \int_{C^gM} \left(\frac{\omega}{l}\right)^{d-1} = \frac{1}{l^{d-1}} \int_{C^gM} \omega^{d-1}.$$

Let us consider the unit Riemannian tangent bundle endowed with the metric g_1 defined in 1.101 and let us compute its volume (See [BR 3] p. 166 for the expression of the volume of (U^gM, g_1) or 1.122).

$$\mathrm{Vol}(U^gM,g_1) = \frac{1}{(d-1)!}\int_{U^gM} \alpha \wedge (d\alpha)^{d-1} = \frac{1}{(d-1)!}\int_{U^gM} \alpha \wedge q^*\omega^{d-1}$$
$$= \frac{1}{(d-1)!}\int_{\gamma\in C^gM} \left(\int_{q^{-1}(\gamma)} \alpha\right) \omega^{d-1};$$

but $\int_{q^{-1}(\gamma)} \alpha = l$ and then

$$\mathrm{Vol}(U^gM,g_1) = \frac{l^d}{(d-1)!} j(M,g).$$

On the other hand by Formula (1.124) we have

$$\mathrm{Vol}(U^gM,g_1) = \mathrm{Vol}(M,g)\cdot \mathrm{Vol}(S^{d-1},\mathrm{can}).$$

Using the formula which gives the volume of the standard sphere in $\mathbb{R}^d$ (see [BR 3] p. 166—169)

$$(d-1)!\,\mathrm{Vol}(S^{d-1},\mathrm{can}) = 2(2\pi)^d/\mathrm{Vol}(S^d,\mathrm{can}).$$

Hence

$$j(M,g)=\frac{(d-1)!}{l^d}\,\mathrm{Vol}(U^gM,g_1)=2\left(\frac{2\pi}{l}\right)^d\frac{\mathrm{Vol}(M,g)}{\mathrm{Vol}(S^d,\mathrm{can})}. \quad \square$$

Now we just have to prove that $j(M,g)$ is an even integer. This is done in the following

2.22 Proposition. *Let $\xi:P\to B$ be a principal bundle with fiber $SO(2)$ and structural group $O(2)$ and let $\beta:\xi\to\xi$ be an involutive mapping of $O(2)$-bundles such that the induced map on B has no fixed point. Moreover, suppose that B is an oriented manifold of dimension $2p$. Then the class $[e(\xi)]^p$ is an even multiple of the generator of $H^{2p}(B,\mathbb{Z})\simeq\mathbb{Z}$.*

Proof. Consider $p\xi$, the $SO(2p)$-bundle obtained by taking the p-fold Whitney sum of ξ with itself. The involution β induces an involution $p\beta$ of the $O(2p)$ bundle $p\xi$, and if $\bar{B}$ denotes the quotient manifold of B by β we can define an $O(2p)$-bundle $\overline{p\xi}$ over $\bar{B}$ such that $\Pi^*\overline{p\xi}=p\xi$, where Π is the projection of B into $\bar{B}$:

$$\begin{array}{ccc} \Pi^*\overline{p\xi}=p\xi & \longrightarrow & \overline{p\xi} \\ \downarrow & & \downarrow \\ B & \xrightarrow{\ \Pi\ } & \bar{B}\,. \end{array}$$

The Whitney classes of $p\xi$ and $\overline{p\xi}$ are related as follows:

$$w_{2p}(p\xi)=\Pi^* w_{2p}(\overline{p\xi})\,.$$

But as Π is a double covering which induces the zero map from $H^{2p}(\bar{B},\mathbb{Z}_2)$ to $H^{2p}(B,\mathbb{Z}_2)$, we have $w_{2p}(p\xi)=0$. Moreover, we know that $w_{2p}(p\xi)$ is the reduction modulo 2 of $[e(p\xi)]=[e(\xi)]^p$, so that $[e(\xi)]^p$ is even (see [MI-ST] p. 99).

Consider the opposite map a^{-1} on U^gM. By definition of the spray Z_g, a^{-1} is an $O(2)$-bundle mapping over the $(2d-2)$-dimensional oriented manifold C^gM, and the induced map on C^gM has no fixed point (in this case B is the manifold of non-oriented geodesics). Then $[e(q)]^{d-1}$ is an even multiple of the generator of $H^{2d-2}(C^gM,\mathbb{Z})$ and $j(M,g)=\langle[e(q)]^{d-1},[C^gM]\rangle$ is an even integer. $\square$

2.23. By non-trivial computations, which can be found in [BR 3] p. 204—210, we can list Weinstein's integers for all known C_l-manifolds

$$i(S^d,\mathrm{can})=1 \qquad i(\mathbb{R}P^d,\mathrm{can})=2^{d-1}$$

$$i(\mathbb{C}P^n,\mathrm{can})=\binom{2n-1}{n-1} \qquad i(\mathbb{H}P^n,\mathrm{can})=\frac{1}{2n+1}\binom{4n-1}{2n-1}$$

$$i(\mathbb{C}aP^2,\mathrm{can})=39\,.$$

As Weinstein's integer is a continuous function of the metric, it remains constant under continuous deformations of the metric. Hence, Weinstein's integer is equal to one for Zoll surfaces. For Zoll surfaces of revolution, this fact will follow from 4.16 by a straightforward computation.

Note: It would be interesting to prove that $i(M, g)$ is independent of the metric g when M is an SC-manifold. Only one result in this direction can be given.

2.24 Proposition (cf. [WN 3]). *For an even-dimensional sphere S^{2p}, $j(S^{2p}, g)$ is equal to 2 (i.e., Weinstein's integer is 1).*

Proof. Let us view $U^g S^{2p}$ as an S^1-oriented bundle over $C^g S^{2p}$ and let us write down the Gysin sequence of the bundle $q: U^g S^{2p} \to C^g S^{2p}$ (see [MI-ST] p. 143).

$$\cdots \to H^{i+1}(US^{2p}) \to H^i(CS^{2p}) \xrightarrow{\cup[e(q)]} H^{i+2}(CS^{2p})$$
$$\to H^{i+2}(US^{2p}) \to H^{i+1}(CS^{2p}) \to \cdots$$

Remark that for every Riemannian metric g on S^{2p} the unit bundle $U^g S^{2p}$ is homotopic to the unit bundle $U^{can} S^{2p}$; indeed the set of Riemannian metrics on S^{2p} is connected. But $U^{can} S^{2p}$ is simply the Stiefel manifold $V_{2p+1,2}$. Its integral cohomology ring is completely known (see for example [BL 4] p. 147) namely:

$$H^i(US^{2p}) = 0, \quad \text{for} \quad i \neq (0, 2p, 4p-1),$$

and

$$H^0(US^{2p}) = \mathbb{Z}, \qquad H^{2p}(US^{2p}) = \mathbb{Z}_2, \qquad H^{4p-1}(US^{2p}) = \mathbb{Z}.$$

Then looking at the Gysin sequence, we find that $H^i(CS^{2p})$ is isomorphic to $H^{i+2}(CS^{2p})$ for $i \neq 2p-2$ and $i < (2p-2)$; then by duality for $i > 2p-2$. The map $H^{2p-2}(CS^{2p}) \to H^{2p}(CS^{2p})$ is the multiplication by 2; finally the map $H^0(CS^{2p}) \xrightarrow{\cup[e(q)]^{d-1}} H^{2p}(CS^{2p})$ gives the result. □

2.24 a) Remark. Proposition 2.24 has been recently extended to odd-dimensional spheres by C. T. C. Yang.

2.24 b) Remark. Recently A. Weinstein ([WN 5]) generalized 2.21 to P-manifolds. Let M be a d-dimensional P_l-manifold which admits some exceptional closed geodesics with period l/k ($k \in \{2, 3, \ldots, 4\}$). Let m be the least common multiple of such k's. Then $m^{d-1}(\mathrm{Vol}(M)/\mathrm{Vol}(S^d))(2\pi/l)^{d-1}$ is an integer. The proof essentially uses the same method as the proof of 2.21. However, here CM is not a smooth manifold, but only a symplectic V-manifold ([SAT]).

D. The Manifold of Geodesics as a Riemannian Manifold

2.25. Recall that the manifold $C^g M$ [the manifold of geodesics of a C_l-manifold (M, g)] is the quotient of the Riemannian unit tangent bundle $(U^g M, g_1)$ under the action of the compact Lie group S^1. There is a natural way of defining a Riemannian metric on C^gM. Namely we average the Riemannian metric g_1 on U^gM under the action of S^1, i.e., we define g_1^Z as:

$$g_1^Z = \frac{1}{l} \int_{S^1} \zeta^s(g_1)ds \quad \text{(see 1.101 for the definition of } g_1).$$

Then g_1^Z is a Riemannian metric on U^gM and S^1 is a group of isometries of (U^gM, g_1^Z). The quotient space C^gM has a natural metric $\bar{g}_1$. More precisely, we have the following (using notations of 2.13 and 1.99)

2.26 Definition. Let $(\bar{X}_1, \bar{X}_2)$ be two tangent vectors at γ in CM (identified with two normal Jacobi fields along the geodesic γ in M). Then $\bar{g}_1$ defined by

$$\bar{g}_1(\bar{X}_1, \bar{X}_2) = \frac{1}{l} \int_{S^1} \zeta^s g_1(\mathscr{J}_1(s), \mathscr{J}_2(s))$$

$$= \int_\gamma g(J_1, J_2) + \int_\gamma g(J_1', J_2')$$

is a Riemannian metric on CM.

2.27. In our specific case, we can define another *Riemannian metric* $\bar{g}_0$ on CM:

for $(\bar{X}_1, \bar{X}_2)$ in $T_\gamma CM$ we let

$$\bar{g}_0(\bar{X}_1, \bar{X}_2) = \int_\gamma g(J_1, J_2).$$

The metrics $\bar{g}_0$ and $\bar{g}_1$ are respectively of Sobolev-type H^0 and H^1 in the sense that they involve the L^2-norm of the field (for $\bar{g}_0$) and the L^2-norm of the field and of its derivative (for $\bar{g}_1$) at each point.

2.28 Remarks. The projection $q:(UM, g_1) \to (CM, \bar{g}_1)$ is a Riemannian submersion if and only if S^1 acts by isometries on the horizontal subspaces of the fibration. Now, S^1 acts by isometries on the tangent space to the fiber (the norm of the vector field Z_g is equal to 1). Hence, we get a Riemannian submersion if and only if S^1 is a group of isometries for (U^gM, g_1). By (1.104), this is equivalent to the fact that (M, g) has constant curvature equal to one (i.e., (M, g) is $(S^d$, can) or $(\mathbb{R}P^d$, can)).

In the general case, we can only notice that the decomposition of $T_uUM = \mathbb{R} \cdot Z_g(u) \oplus Q_u$ (see 2.11) is g_1-orthogonal (and then, g_1^Z-orthogonal). It follows from the expression of the 1-form α, that we have

$$\alpha(X) = g_1(Z_g, X) \quad \text{for any tangent vector } X \text{ to } UM.$$

2.29. Let us consider the case of $(S^d$, can) or $(\mathbb{R}P^d$, can). The two metrics $\bar{g}_0$ and $\bar{g}_1$ are the same (up to a constant factor). This obviously follows from the particular form of the Jacobi fields (see [BR 3] p. 199).

Moreover, $SO(d+1)$ acts by isometries on the unit tangent bundle (US^d, g_1). If φ belongs to $SO(d+1)$, then for X, Y in TUS^d we get (using Definition 1.101 and denoting p_{S^d} by p)

$$\begin{aligned}(T\varphi^*(g_1))(X,Y) &= g_1(T\varphi(X), T\varphi(Y)) \\ &= g(Tp\circ T\varphi(X), Tp\circ T\varphi(Y)) + g(K\circ T\varphi(X), K\circ T\varphi(Y)).\end{aligned}$$

The group $SO(d+1)$ acts by isometries on S^d. If $\tilde{\varphi}$ is the restriction of the action φ on US^d to S^d then we have

$$\begin{aligned}Tp\circ T\varphi &= T\tilde{\varphi}\circ Tp \\ K\circ T\varphi &= T\tilde{\varphi}\circ K.\end{aligned}$$

As a result of the linear action of $SO(d+1)$ we have that if $u=(p(u), v)$ with $p(u)$ in $\mathbb{R}^{d+1}$, v in $\mathbb{R}^{d+1}$,

$$\varphi(u) = (\tilde{\varphi}(u), \tilde{\varphi}(v)) \quad \text{so that} \quad p\circ\varphi(u) = \tilde{\varphi}\circ p(u)$$

which implies that $Tp\circ T\varphi = T\tilde{\varphi}\circ Tp$.

The second equality can be also proved by using local coordinates.

Then we have

$$\begin{aligned}T\varphi(g_1)(X,Y) &= g(T\tilde{\varphi}\circ Tp(X), T\tilde{\varphi}\circ Tp(Y)) + g(T\tilde{\varphi}\circ K(X), T\tilde{\varphi}\circ K(Y)) \\ &= g(Tp(X), Tp(Y)) + g(K(X), K(Y)) = g_1(X,Y). \quad \square\end{aligned}$$

As we have already noticed, S^1 acts by isometries on (UM, g_1) so that $SO(d+1)$ acts by isometries on (UM, g_1^Z) and g_1 is then the canonical Riemannian symmetric metric defined on the Grassmann manifold $G^+_{2,d+1}$.

2.30. By an elementary computation we can show that the manifold of geodesics of (S^3, can) endowed with the Riemannian metric $\bar{g}_0$ (or $\bar{g}_1$) is isometric to the product of two spheres of dimension two and radius $1/\sqrt{2}$ endowed with the product metric.

Going back now to the general situation, we remark that the existence of the closed nondegenerate differential two form ω on $(CM, \bar{g}_0)$ insures the existence of an almost complex structure on $(CM, \bar{g}_0)$.

2.31 Proposition. *There exists an almost complex structure on CM, that is, a field of linear maps of the tangent bundle $\gamma\to F_\gamma$ such that $F_\gamma^2 = -Id_{T_\gamma CM}$. Furthermore, we can define a metric $\bar{g}_\omega$ on CM which is Hermitian for F and such that for $\bar{X}$, $\bar{Y}$ in $T_\gamma CM$*

$$\omega(\bar{X}, \bar{Y}) = \bar{g}_\omega(\bar{X}, F_\gamma \bar{Y}).$$

Proof. The nondegenerate differential two form ω determines a field of invertible skewsymmetric linear maps of $T_\gamma CM$, $\gamma\mapsto A_\gamma$ such that for $\bar{X}$, $\bar{Y}$ in $T_\gamma CM$ $\omega(\bar{X}, \bar{Y}) = \bar{g}_0(\bar{X}, A_\gamma(\bar{Y}))$.

Then we apply Lemma 2.32 to A_γ and we get the result. $\square$

2.32 Lemma. *Let (E, g_0) be a Euclidean vector space and let A be an invertible skewsymmetric linear map of E. Then there exists an orthogonal linear map F such that $F^2 = -Id_E$. Furthermore, we can define a scalar product $(\ ,\ )$ on E for which F is orthogonal [i.e. satisfies $(F(X), F(Y)) = (X, Y)$] and such that for X, Y in E the following equality holds:*

$$g_0(X, A(Y)) = (X, F(Y)).$$

Proof. By elementary linear algebra (see [CY] p. 14) every invertible linear map can be uniquely written as

$$A = FS = SF,$$

where the linear map F is g_0-orthogonal (i.e., $F^{-1} = F^*$, where * denotes the adjoint map), and S is symmetric, positive (i.e., $S = S^*$ and its eigenvalues are positive).

Using the fact that A is skewsymmetric ($A^* = -A$) we get

$$A^* = S^* F^* = SF^{-1} = -A = -SF \quad \text{so that} \quad F^2 = -Id_E.$$

Now if we set $(X, Y) = g_0(X, S(Y))$ for (X, Y) in E we define a new scalar product in E for which

$$(X, F(Y)) = g_0(X, SF(Y)) = g_0(X, A(Y)),$$

so that

$$(F(X), F(Y)) = g_0(F(X), SF(Y)) = g_0(F(X), FS(Y)) = g_0(X, S(Y)) = (X, Y).$$

Hence, the new scalar product is Hermitian for F. □

2.33. Let us consider the special case of the locally symmetric spaces (see Chapter 3 for definitions). We claim that they are characterized by one of the following properties:

(S) for each geodesic, the derivative along the geodesic of a normal Jacobi field is a normal Jacobi field.

(S′) the Levi-Civita covariant derivative of the curvature tensor R satisfies

$$DR(X; X, \cdot)X = 0 \quad \text{for any } X \text{ in } TM.$$

By taking a derivative of (1.89) it is clear that S and S′ are equivalent.
It is obvious that a locally symmetric space (for which $DR = 0$) satisfies S′. The converse is a consequence of the following lemma.

2.34 Lemma. *If a Riemannian manifold satisfies (S′), then along any geodesic γ the symmetric linear map $R(\dot\gamma, \cdot)\,\dot\gamma$ of $T_\gamma M$ has parallel eigenvectors and its eigenvalues are constant.*

Proof. If $E(s)$ is a parallel vector field along the geodesic $s \mapsto \gamma(s)$, then $R(\dot\gamma(s), E(s))\dot\gamma(s)$ is clearly a parallel vector field. Let $(E_i)_{i=2,\dots,d}$ be $(d-1)$ orthogonal eigenvectors relative to eigenvalues $(\lambda_i)_{i=2,\dots,d}$ of $R(\dot{\gamma}_{(s_0)}, \cdot)\dot{\gamma}(s_0)$ at a point $\gamma(s_0)$ of γ. If $\{E_i(s)\}_{i=2,\dots,d}$ are the parallel vector fields along γ with initial conditions $E_i(s_0)=E_i$, then $\lambda_i E_i(s)$ and $R(\dot{\gamma}(s), E_i(s))\dot{\gamma}(s)$ are parallel vector fields with the same initial conditions, so that they are equal along γ. □

2.35 Proposition. *A Riemannian manifold (M, g) which satisfies (S) or (S′) is locally symmetric (i.e. $DR=0$).*

Proof. We first notice that if $Y(s)=\sum_2^d f_i(s)E_i$ is a normal Jacobi field along the geodesic $s\mapsto\gamma(s)$ in M (we use notations of Lemma 2.34), then the functions $f_i(s)$ satisfy the equations

$$f_i''(s)+\lambda_i f_i(s)=0 \qquad i=2,\dots,d\,.$$

Fix a point $\gamma(s_0)$ on γ. The norm of the normal Jacobi field $J(s)$ along γ such that $J(s_0)=0$ and $J'(s_0)=X_0$, where X_0 is any normal tangent vector in $T_{\gamma(s_0)}M$, only depends on the distance between $\gamma(s_0)$ and $\gamma(s)$ (i.e., $|s-s_0|$).

Let m_0 be a point in M, and let σ be the geodesic symmetry at m_0 defined in a normal neighbourhood V of m_0, where the exponential map $\exp_{m_0}$ is a diffeomorphism. For m in V, let γ be the unique geodesic from m_0 to m with $\gamma(0)=m_0$ and $\gamma(s)=m$.

Then $\sigma(m)=\exp_{m_0}(-\exp_{m_0}^{-1}(m))=\gamma(-s)$.

From 1.94, for every tangent vector X_m in T_mM we have

$$\|\sigma^*(X_m)\|=\|-J(-s)\|=\|J(s)\|=\|X_m\|\,,$$

where we realize X_m as the value at s of a normal Jacobi field along γ such that $J(0)=0$, $J'(0)=X_0$ for some X_0 in $T_{m_0}M$. The map σ is an isometry and then (M, g) is symmetric. □

2.36. Let (M, g) be a C_l-symmetric manifold, that is, a symmetric space of rank one. Then the eigenvalues $(\lambda_i)_{i=2,\dots,d}$ are positive and $\sqrt{\lambda_i}\,l=0 \pmod{2\pi}$ (the Jacobi fields being transverse vector fields to variations by geodesics, in a C_l-manifold they have to be periodic with period l). Then we can write down a basis of normal Jacobi fields along any geodesic γ, namely for $i=2,\dots,d$, $J_i(s)=\sin(\sqrt{\lambda_i}s)E_i$, $K_i(s)=\cos(\sqrt{\lambda_i}s)E_i$. The collection $(J_i\cup K_i)_{i=2,\dots,d}$ is a $\bar{g}_0$- (and $\bar{g}_1$-)orthogonal basis of the tangent space $T_\gamma CM$. By a direct computation, we find the expression of the endomorphism F_γ (defined in 2.31) in this basis, namely

$$F_\gamma(J_i)=K_i, \quad F_\gamma(K_i)=-J_i, \quad i=(2,\dots,d).$$

Notice that the three metrics $\bar{g}_0$, $\bar{g}_1$, $\bar{g}_\omega$ are different. We have

$$\bar{g}_1(J_i, J_i)=(1+\sqrt{\lambda_i})\bar{g}_0(J_i, J_i),$$
$$\bar{g}_\omega(J_i, J_i)=\sqrt{\lambda_i}\,\bar{g}_0(J_i, J_i).$$

They coincide for the spheres and the real projective spaces for which all the λ_i's are equal.

2.37. Let us now come to some properties of the geodesics in (CM, g_0). Let $t \mapsto \Gamma(t)$ be a curve in CM and $(t,u) \to \bar{\Gamma}(t,u) = \bar{\gamma}_u(t)$ a variation of the curve $\Gamma(t)$ defined on $I \times]-\varepsilon, +\varepsilon[$. By horizontal lift and projection down to M we get a two-parameter family of geodesics in M. This defines a map Γ from $S^1 \times I \times]-\varepsilon, +\varepsilon[$ into M

$$(s,t,u) \mapsto \Gamma(s,t,u) = \gamma_u(t)(s)$$

such that, for a fixed pair (t,u), $s \mapsto \gamma_u(t)(s)$ is a periodic geodesic of length l and if $T\Gamma\left(\frac{\partial}{\partial t}\right) = J_u(t)(s)$ and $T\Gamma\left(\frac{\partial}{\partial u}\right) = H_u(t)(s)$ denote the vectors tangent to the curves $t \mapsto \gamma_u(t)(s)$ and $u \mapsto \gamma_u(t)(s)$, then J and H are normal Jacobi fields along each geodesic of the family.

Consider the energy function $\bar{\mathbb{E}}_{\bar{g}_0}(u)$ of the curve $t \mapsto \bar{\gamma}_u(t)$ in $(CM, \bar{g}_0)$. We denote by $\dot{\bar{\gamma}}_u(t)$ its tangent vector).

$$\bar{\mathbb{E}}_{\bar{g}_0}(u) = \int_I \bar{g}_0(\dot{\bar{\gamma}}_u(t), \dot{\bar{\gamma}}_u(t))dt = \int_I \left(\int_{\gamma_u} g(J_u(t), J_u(t))\right) dt\,.$$

The curve $\bar{\Gamma}$ is a geodesic if and only if $\bar{\Gamma}$ is an extremum for $\bar{\mathbb{E}}_{\bar{g}_0}$. Following the notations of 1.17 we have $\left(\text{since } D^{\Gamma}_{\frac{\partial}{\partial u}} J = D^{\Gamma}_{\frac{\partial}{\partial t}} H\right)$,

$$\frac{1}{2}\frac{d\bar{\mathbb{E}}_{\bar{g}_0}(u)}{du}\bigg|_0 = \int_{I \times S^1} g\left(D^{\Gamma}_{\frac{\partial}{\partial u}} J, J\right) dt ds = \int_{I \times S^1} g\left(D^{\Gamma}_{\frac{\partial}{\partial t}} H, J\right) dt ds\,.$$

Suppose that the variation $\bar{\Gamma}_u(t)$ keeps the end points fixed. Then integrating by parts we get

2.38

$$\frac{1}{2}\frac{d\bar{\mathbb{E}}_{\bar{g}_0}(u)}{du}\bigg|_0 = \int_{I \times S^1} (J \cdot g(H,J)) dt ds - \int_{I \times S^1} g\left(H, D^{\Gamma}_{\frac{\partial}{\partial t}} J\right) dt ds$$
$$= \int_{I \times S^1} g\left(H, D^{\Gamma}_{\frac{\partial}{\partial t}} J\right) dt ds\,.$$

If $\mathscr{T}_\gamma M$ denotes the normed space of vector fields along a geodesic γ in M (for the H^0 norm), then we may define a linear map $\mathscr{J}^\gamma$ on $\mathscr{T}_\gamma M$ which is the H^0-orthogonal projection onto the finite dimensional subspace of $\mathscr{T}_\gamma M$ consisting of normal Jacobi fields along γ. We can characterize the geodesics of $(CM, \bar{g}_0)$ as follows:

2.39 Proposition. *The curve $\bar{\Gamma}$ in $(CM, \bar{g}_0)$ is a geodesic if and only if $\mathscr{J}^{\gamma_t}\left(D^{\Gamma}_{\frac{\partial}{\partial t}} J\right) = 0$ along the geodesic $s \mapsto \gamma(s)$ in M.*

Proof. The condition is sufficient. It implies that $\frac{d\bar{\mathbb{E}}_{\bar{g}_0}(u)}{du}\Big|_0 = 0$, so that $\bar{\Gamma}$ is a geodesic.

The condition is necessary. Indeed, if $\bar{\Gamma}$ is a geodesic, then $\frac{1}{2}\frac{d\bar{\mathbb{E}}_{\bar{g}_0}(u)}{du}\Big|_0 = 0$ for any variation $\bar{\Gamma}_u$ of $\bar{\Gamma}$, and a classical argument of calculus of variations applied to (2.38) gives $\int_{\gamma_t} g\left(H, D^{\Gamma}_{\frac{\partial}{\partial t}} J\right) ds = 0$ for any normal Jacobi field H along the geodesic $s \to \gamma_t(s)$.

Consider the subset N of M, $N=\{\gamma_t(s)|t\in I, s\in S^1\}$. If from N we remove the set of points where the vector $J_t(s)$ is 0 (this set has measure 0), then we get a two-dimensional immersed manifold N^* of M which can be called a "ruled manifold". If $k(t,s)$ is the mean curvature vector of N^*, we can write the condition of Proposition 2.37 in a more geometrical form.

2.40 Proposition. *The curve $\bar{\Gamma}$ is a geodesic in $(CM, \bar{g}_0)$ if and only if, for any normal Jacobi field H along the curve $s\mapsto\gamma_t(s)$ in N,*

$$\int_{\gamma_t} \|J\|^2 g(H,k)ds + \int_{\gamma_t} (J\cdot \mathrm{Log}\|J\|)g(H,J)ds = 0.$$

Proof. The vectors $E(t,s)=\dfrac{J_0(t)(s)}{\|J_0(t)(s)\|}$ and $\dot{\gamma}_t(s)$ form an orthonormal basis of $T_{\gamma_t(s)}N^*$ so that

$$k(t,s)=(D_E E)^{\perp}+D_{\dot{\gamma}_t(s)}\dot{\gamma}_t(s),$$

where $(\ .\ )^{\perp}$ denotes the orthogonal projection onto the normal space to the submanifold N^*. As $D_{\dot{\gamma}_t}\dot{\gamma}_t=0$, we have

$$k(t,s)=(D_E E)^{\perp}=D_E\cdot E-(D_E E,\dot{\gamma}_t)\dot{\gamma}_t.$$

We evaluate $D^{\Gamma}_{\frac{\partial}{\partial t}} J = D_J J = \|J\| D_E(\|J\|E)$ so that

$$D_J J=\|J\|^2 D_E E+(J\cdot\|J\|)E=\|J\|^2 D_E E+(J\cdot \mathrm{Log}\|J\|)J.$$

For every normal Jacobi field H along γ_t, $g(H,\dot{\gamma}_t)=0$ we then obtain

$$\int_{\gamma_t} g(H,\|J\|^2 k)ds+\int_{\gamma_t} g(H,(J\cdot \mathrm{Log}\|J\|)J)ds=0$$

which is the expected formula. □

2.41. Our aim is to estimate the curvature of the Riemannian metric $\bar{g}_0$ on CM. First we must find out what $\bar{D}^0$, the Levi-Civita connection of $\bar{g}_0$, is.

Before giving a definition of the Levi-Civita connection, we must give some more details about lifting vector fields. Let $\bar{J}_1$ be a locally defined vector field on CM. We call $\mathscr{J}_1$ its horizontal lift in UM (i.e., at u in UM $\mathscr{J}_1(u)$ is the only horizontal vector such that $Tq(\mathscr{J}_1(u))=\bar{J}_1(q(u))$). We will use $J_1=Tp_M(\mathscr{J}_1)$ (a vector field along p_M) as the basic object associated to $\bar{J}_1$ (see 1.91).

If we pick a point m in M, $J_1(m)$ may be multivalued since $J_1(m)=Tp_M(\mathscr{J}_1(u))$ for any u in U_mM. But if we consider a geodesic γ we can specify the value of J_1 at $m=\gamma(s_0)$, namely $J_1(\gamma(s_0))$. Then J_1 is a Jacobi field along γ.

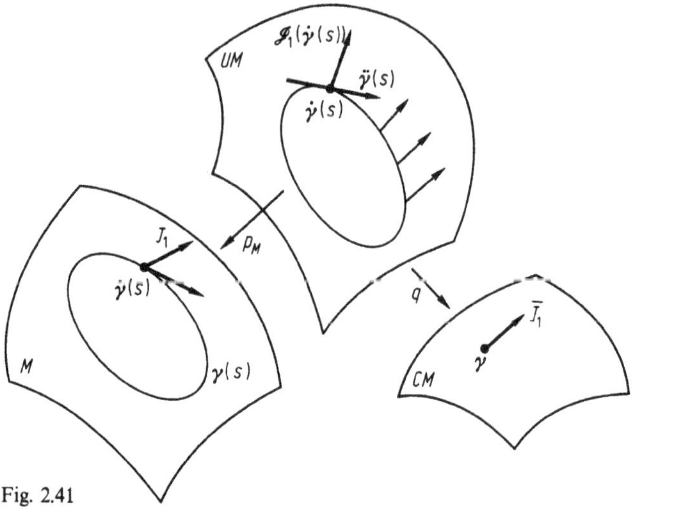

Fig. 2.41

2.42 Proposition. *Let $\bar J_2$ be a vector field on CM defined in a neighbourhood of a geodesic γ and $\bar J_1$ a tangent vector at γ. Then $\bar D^0_{\bar J_1}\bar J_2 = \mathscr{J}^\gamma(D^{p_M}_{\mathscr{J}_1} J_2)$ (where $\mathscr{J}^\gamma$ is the H^0-orthogonal projection onto the space of normal Jacobi fields in the space of vector fields along γ), i.e., for any other tangent vector $\bar J_3$ at γ*

$$\bar g_0(D^0_{\bar J_1}\bar J_2, \bar J_3) = \int_\gamma g(D^{p_M}_{\mathscr{J}_1} J_2, J_3). \tag{2.42}$$

2.43 Remark. With the notation of 1.77, one has

$$D^{p_M}_{\mathscr{J}_1} J_2 = Tp_M(D^1_{\mathscr{J}_1}\mathscr{J}_2).$$

Proof. The first thing to check is that $D^{p_M}_{\mathscr{J}_1} J_2$ is a well-defined vector field along γ. We can certainly take a curve tangent to $\bar J_1$, that is, a variation of γ by geodesics. As we already noticed, along a geodesic, J_2 is well defined. Therefore, $D^{p_M}_{\mathscr{J}_2} J_1$ is also well defined since the covariant derivative of vector fields along the map p_M is well defined (see 1.78).

One has to verify that $(J_1, J_2) \mapsto \mathscr{J}^\gamma(D^{p_M}_{\mathscr{J}_1} J_2)$ verifies the axioms of a covariant derivative. First of all it is clear that $\mathscr{J}(D^{p_M}_{\mathscr{J}_1} J_2)$ depends only of the value of $\mathscr{J}_1$ along γ. On the other hand, when we multiply J_2 by a function f on CM (i.e., constant along each geodesic) we get

$$\mathscr{J}^\gamma(D^{p_M}_{\mathscr{J}_1}(f\circ q)J_2) = \mathscr{J}^\gamma(\mathscr{J}_1(f\circ q))J_2 + (f\circ q)D^{p_M}_{\mathscr{J}_1} J_2).$$

As the functions f and $\mathscr{J}_1\cdot(f\circ q)(=(\bar J_1\cdot f)\circ q)$ are constant along γ and as J_2 is a Jacobi field, we have

$$\mathscr{J}^\gamma(D^{p_M}_{\mathscr{J}_1}(f\circ q)J_2) = (\bar J_1\cdot f)\bar J_2 + (f\circ q)\mathscr{J}^\gamma(D^{p_M}_{\mathscr{J}_1} J_2).$$

To verify the Levi-Civita property, we will use Formula 2.42. One has to check that

$$\bar J_1\cdot\bar g_0(\bar J_2, \bar J_3) = \bar g_0(\mathscr{J}^\gamma(D^{p_M}_{\mathscr{J}_1} J_2), \bar J_3) + \bar g_0(\bar J_2, \mathscr{J}^\gamma(D^{p_M}_{\mathscr{J}_1} J_3)).$$

The left-hand side is simply $\frac{d}{dt}\left(\int_{\gamma_t} g(J_2, J_3)\right)\Big|_0$, where $g(J_2, J_3)$ is considered as a function along γ_t (supposed to be an integral curve of $\bar{J}_1$). We can also say that it is

$$\int_\gamma \mathscr{J}_1 \cdot (g(T_{p_M}(\mathscr{J}_2), T_{p_M}(\mathscr{J}_3)) \circ p_M).$$

But the pulled back connection D^{p_M} satisfies $D^{p_M} p_M^*(g)=0$, so that

$$\int_\gamma \mathscr{J}_1 \cdot (g(J_2, J_3) \circ p_M) = \int_\gamma [g(D^{p_M}_{\mathscr{J}_1} J_2, J_3) + g(J_2, D^{p_M}_{\mathscr{J}_1} J_3)].$$

As J_2 and J_3 are Jacobi fields along γ, the right-hand side of the above formula can also be written as

$$\int_\gamma [g(\mathscr{J}^\gamma(D^{p_M}_{\mathscr{J}_1} J_2), J_3) + g(J_2, \mathscr{J}^\gamma(D^{p_M}_{\mathscr{J}_1} J_3))].$$

This ends the proof. □

2.44. Before going any further we need some remarks about the Lie bracket of vector fields in CM. If $\bar{J}_1$ and $\bar{J}_2$ belong to $\mathscr{T}CM$, we let $\mathscr{J}_1$ and $\mathscr{J}_2$ be their horizontal lifts. Then $[\mathscr{J}_1, \mathscr{J}_2]$ is the sum of the horizontal lift of $[\bar{J}_1, \bar{J}_2]$ and of the vertical vector $l\omega(\bar{J}_1, \bar{J}_2)Z$. We will denote by $[J_1, J_2]$ the normal Jacobi field along the geodesic γ associated to $[\bar{J}_1, \bar{J}_2](\gamma)$. Therefore, along γ

$$T_{p_M}([\mathscr{J}_1, \mathscr{J}_2]) = [J_1, J_2] + l\,\omega(\bar{J}_1, \bar{J}_2)\dot{\gamma}.$$

Notice that if $\bar{g}_0(\bar{J}_1, \bar{J}_2)=0$, then $\omega(\bar{J}_1, \bar{J}_2)$ depends only on the plane spanned by $\bar{J}_1$ and $\bar{J}_2$ (since $\cos^2+\sin^2=1$).

2.45. Notice that the formula established in Proposition 2.37 provides another way of looking at geodesics of the metric $\bar{g}_0$ in CM. Their equation is nothing but $\mathscr{J}^\gamma(D^{p_M}_{\dot{\gamma}} \dot{\Gamma})=0$, where $\dot{\gamma}$ is the velocity field of the horizontal lift of the curve Γ in CM.

2.46 Proposition. *Let (M, g) be a C_l-manifold. Let $\bar{P}$ be a tangent plane at a geodesic γ with $\bar{g}_0$-orthonormal basis $\{\bar{J}_1, \bar{J}_2\}$. Then*

$$\begin{aligned}\bar{\sigma}_0(\bar{P}) = &\int_\gamma g(R(J_1, J_2)J_1, J_2) + \frac{l^2}{2}(\omega(\bar{J}_1, \bar{J}_2))^2 \\ &+ \int_\gamma g(\mathscr{J}^{\gamma,\perp}(D^{p_M}_{\mathscr{J}_2} J_2), \mathscr{J}^{\gamma,\perp}(D^{p_M}_{\mathscr{J}_1} J_1)) \\ &- \int_\gamma g(\mathscr{J}^{\gamma,\perp}(D^{p_M}_{\mathscr{J}_1} J_2), \mathscr{J}^{\gamma,\perp}(D^{p_M}_{\mathscr{J}_2} J_1))\end{aligned}$$

(where we have extended $\bar{J}_1$ and $\bar{J}_2$ in a neighborhood of γ).

Proof. We recall (see 1.86) that

$$\begin{aligned}\bar\sigma_0(\bar P)&=\bar g_0(\bar R_0(\bar J_1,\bar J_2)\bar J_1,\bar J_2)\\&=\bar g_0(\bar D^0_{\bar J_2}\bar D^0_{\bar J_1}\bar J_1-\bar D^0_{\bar J_1}\bar D^0_{\bar J_2}\bar J_1,\bar J_2)+\bar g_0(\bar D^0_{[\bar J_1,\bar J_2]}\bar J_1,\bar J_2)\end{aligned}$$

are extensions of $\bar J_1$ and $\bar J_2$ in a neighborhood of γ.

Therefore we can now transform the expression for $\bar\sigma_0(\bar P)$ as follows:

$$\begin{aligned}\bar\sigma_0(\bar P)=&\bar J_2\cdot g_0(\bar D^0_{\bar J_1}\bar J_1,\bar J_2)-\bar g_0(\bar D^0_{\bar J_1}\bar J_1,\bar D^0_{\bar J_2}\bar J_2)-J_1\cdot\bar g_0(\bar D^0_{\bar J_2}\bar J_1,\bar J_2)\\&+\bar g_0(\bar D^0_{\bar J_1}\bar J_2,\bar D^0_{\bar J_2}\bar J_1)+\bar g_0(\bar D^0_{[\bar J_1,\bar J_2]}\bar J_1,\bar J_2).\end{aligned}$$

If we use Formula (2.42) and what we noticed in 2.42, then we have

$$\bar J_2\cdot\bar g_0(\bar D^0_{\bar J_1}\bar J_1,\bar J_2)=\int_\gamma[g(D^{PM}_{\mathscr J_2}D^{PM}_{\mathscr J_1}J_1,J_2)+g(D^{PM}_{\mathscr J_1}J_1,D^{PM}_{\mathscr J_2}J_2)],$$

$$\bar J_1\cdot\bar g_0(\bar D^0_{\bar J_2}\bar J_1,\bar J_2)=\int_\gamma[g(D^{PM}_{\mathscr J_1}D^{PM}_{\mathscr J_2}J_1,J_2)+(D^{PM}_{\mathscr J_2}J_1,D^{PM}_{\mathscr J_1}J_2)]$$

and also

$$\bar g_0(\bar D^0_{[\bar J_1,\bar J_2]}\bar J_1,\bar J_2)=\int_\gamma(g(D^{PM}_{[\mathscr J_1,\mathscr J_2]}J_1,J_2)-l\omega(\bar J_1,\bar J_2)g(D^{PM}_Z J_1,J_2)).$$

Notice that $D^{PM}_Z J_1=D_{\dot\gamma}J_1$.

We now collect the various terms and get

$$\begin{aligned}\bar\sigma_0(\bar P)=&\int_\gamma g(D^{PM}_{\mathscr J_2}D^{PM}_{\mathscr J_1}J_1-D^{PM}_{\mathscr J_1}D^{PM}_{\mathscr J_2}J_1+D^{PM}_{[\mathscr J_1,\mathscr J_2]}J_1,J_2)\\&+\int_\gamma[g(D^{PM}_{\mathscr J_1}J_1,D^{PM}_{\mathscr J_2}J_2)-g(\mathscr J^\gamma(D^{PM}_{\mathscr J_1}J_1),\mathscr J^\gamma(D^{PM}_{\mathscr J_2}J_2))]\\&-\int_\gamma[g(D^{PM}_{\mathscr J_1}J_2,D^{PM}_{\mathscr J_2}J_1)-g(\mathscr J^\gamma(D^{PM}_{\mathscr J_1}J_2),\mathscr J^\gamma(D^{PM}_{\mathscr J_2}J_1))]\\&-l\omega(\bar J_1,\bar J_2)\int_\gamma g(D_{\dot\gamma}J_1,J_2).\end{aligned}$$

If we introduce $\mathscr J^{\gamma,\perp}$, the H^0-orthogonal projection complementary to $\mathscr J^\gamma$, and if we notice that

$$\int_\gamma g(D_{\dot\gamma}J_1,J_2)=-\int_\gamma g(J_1,D_{\dot\gamma}J_2)=-l\omega(\bar J_1,\bar J_2)\frac{l}{2},$$

then we obtain the expected formula. □

Chapter 3. Compact Symmetric Spaces of Rank One from a Geometric Point of View

A. Introduction

3.0. Since Elie Cartan's work several studies have been published concerning the properties of compact symmetric spaces of rank one.

In the present volume these spaces (we call them CROSSes) play a privileged role. Together with Zoll surfaces they are the only known examples of C_π-manifolds. Their geometrical structure is so rich that one can characterize them in several other ways: They are the symmetric spaces with strictly positive curvature. They are the compact two-point homogeneous spaces (see [WG 2], [HN 1] p. 355). They are the Euclidean spheres, the projective spaces $\mathbb{K}P^n$ (here $\mathbb{K}$ means either the field $\mathbb{R}$ of real numbers, or the field $\mathbb{C}$ of complex numbers, or the noncommutative field $\mathbb{H}$ of quaternions) and the Cayley plane $\mathbb{C}aP^2$ which is exceptional for several reasons.

As a matter of fact, Chapters 5, 6, 7 are respectively devoted to characterization of projective spaces $\mathbb{K}P^n$ in a geometrical, "harmonic", or topological way. Less rigid, the spheres play a separate role as one can see in Chapter 4.

In this chapter, the Euclidean geometry of the sphere and some standard results in geometry and topology are assumed; references for them are given, when necessary.

We start with the very definition of projective spaces. Then we characterize them as symmetric spaces having in mind the double inheritance of their structure. In this way we hope to contribute towards a better visualization and a better handling of these spaces. As a consequence we collect the properties of geodesics and antipodal submanifolds.

A paragraph is devoted to the study of the $\mathbb{Z}$-cohomology ring of $\mathbb{K}P^n$ with the help of its fundamental skew-form: this is not generally feasible, but it is another confirmation of the precious nature of these objects.

Concerning the Cayley plane we shall contemplate very briefly the correspondance between Arguesian (resp. non-Arguesian) plane geometries and associative (resp. alternative) division algebras. Then we shall describe the manifold structure of $\mathbb{C}aP^2$ and merely mention the geometrical approach of J. Tits and the group-theoretical study of H. Freudhental to obtain the exceptional CROSS $F_4/Spin9$ of E. Cartan.

B. The Projective Spaces as Base Spaces of the Hopf Fibrations

3.1. Let $\mathbb{K}$ be either the field $\mathbb{R}$ of real numbers, the field $\mathbb{C}$ of complex numbers or the division algebra $\mathbb{H}$ of quaternions. We set $a=\dim_{\mathbb{R}}\mathbb{K}$. The product space $\mathbb{K}^{n+1}$ is endowed with its right vector space structure,

$$x\cdot\lambda=(x_1,\dots,x_{n+1})\cdot\lambda=(x_1\cdot\lambda,\dots,x_{n+1}\cdot\lambda)\,,$$

its Hermitian product

$$\langle x,y\rangle=\sum_{i=1}^{i=n}\bar{x}_i\cdot y_i\,,$$

and its real scalar product

$$\langle x,y\rangle_{\mathbb{R}}=\mathrm{Re}\langle x,y\rangle\,.$$

3.2 Definition of $\mathbb{K}P^n$. The projective space $\mathbb{K}P^n$ is the orbit space for the right action of the group $\mathbb{K}^*=\mathbb{K}-\{0\}$ on $\mathbb{K}^{n+1}-\{0\}$ (that is, $x\sim y$ if and only if there exists a λ in $\mathbb{K}^*$ such that $x=y\cdot\lambda$).

Let us denote by $\pi(x)$ the orbit of x. Such a quotient space cannot be defined so simply for the Cayley division algebra $\mathbb{C}\mathrm{a}$, because $\mathbb{C}\mathrm{a}$ is only alternative and not associative (see the different articles of the *M.A.A*'s book [AT]).

3.3. The space $\mathbb{K}P^n$ is a C^∞, na-dimensional real manifold. For $1\leqslant i\leqslant n+1$, we define

$$A_i=\{x=(x^1,\dots,x^{n+1})|x^i\neq 0\}\,.$$

The sets $\pi(A_i)=B_i$ are open sets of the quotient topology on $\mathbb{K}P^n$, and the functions

$$B_i\xrightarrow{\varphi_i}\mathbb{K}^n\simeq\mathbb{R}^{na}$$

with

$$\pi(x)\mapsto\left(\frac{x^1}{x^i},\dots,\frac{x^{i-1}}{x^i},\frac{x^{i+1}}{x^i},\dots,\frac{x^{n+1}}{x^i}\right)$$

are the canonical charts for $\mathbb{K}P^n$.

For $\mathbb{C}P^n$ the structure of a real manifold is clearly obtained from a complex analytic one. The set $(\mathbb{K}^{n+1}-\{0\},\ \mathbb{K}P^n,\ \pi,\ \mathbb{K}^*)$ has the structure of a *principal bundle*. The local trivializations are given by

$$A_i=\pi^{-1}(B_i)\xrightarrow{\theta_i}B_i\times\mathbb{K}^*\quad\text{with}\quad x\mapsto(\pi(x),x^i)$$

and the transition functions by

$$B_i \cap B_j \to \mathbb{K}^* \quad \text{with} \quad \pi(x) \mapsto \frac{x^i}{x^j}.$$

We *denote* by $S\mathbb{K}^{n+1}$ (instead of S^{na+a-1}) the unit sphere in $\mathbb{K}^{n+1}$ defined by the equation $\langle x, x \rangle = 1$. Its dimension is $na+a-1$. Then $S\mathbb{K} = S^{a-1}$ is a subgroup of $\mathbb{K}^*$ and one verifes that

3.4 $\qquad S\mathbb{K} \to S\mathbb{K}^{n+1} \to \mathbb{K}P^n$

is a *principal subbundle* of the previous one.

3.5. The total space of the tangent bundle of $S\mathbb{K}^{n+1}$ is classically identified as follows:

$$TS\mathbb{K}^{n+1} = \{(x, u) | x \in S\mathbb{K}^{n+1}, u \in \mathbb{K}^{n+1}, \langle x, u \rangle_{\mathbb{R}} = 0\}.$$

The space tangent to the fiber at x in (3.4) is identified with $V_x S\mathbb{K}^{n+1} = \{(x, x\lambda) | \lambda \in \mathbb{K}, \operatorname{Re} \lambda = 0\}$. We also consider $H_x S\mathbb{K}^{n+1} = \{(x, u) | u \in \mathbb{K}^{n+1}, \langle x, u \rangle = 0\}$ and then $T_x S\mathbb{K}^{n+1} = H_x S\mathbb{K}^{n+1} \oplus V_x S\mathbb{K}^{n+1}$. The decomposition is an orthogonal one for $\langle\ ,\ \rangle$. The $S\mathbb{K}$-action on $TS\mathbb{K}^{n+1}$ in (3.4) is given by

$$((x, u); \lambda) \mapsto (x\lambda, u\lambda),$$

so that the distribution $H_x S\mathbb{K}^{n+1}$ is invariant (since $\langle x\lambda, u\lambda \rangle = \bar{\lambda} \langle x, u \rangle \lambda$) and it defines a connection on the bundle (3.4).

It follows that the space tangent to $\mathbb{K}P^n$ at $\pi(x)$ is isomorphic to the set of classes

$$\{(x\lambda, u\lambda) | \langle x, u \rangle = 0, \lambda \in S\mathbb{K}\} \simeq T_{\pi(x)} \mathbb{K}P^n$$

(for more details concerning the properties above, see for example [K-N 1] p. 65 and [K-N 2] p. 134).

We denote by $\pi(x, u)$ such a tangent vector at $\pi(x)$.

3.6. The projective space $\mathbb{K}P^n$ is endowed with a natural Riemannian structure. First notice that for u, v in $\mathbb{K}^{n+1}$ and λ in $S\mathbb{K}$

$$\langle u, v \rangle_{\mathbb{R}} = \langle u\lambda, v\lambda \rangle_{\mathbb{R}}.$$

In fact,

$$\begin{aligned} 2\langle u\lambda, v\lambda \rangle_{\mathbb{R}} &= \langle u\lambda, v\lambda \rangle + \langle v\lambda, u\lambda \rangle \\ &= \bar{\lambda} \langle u, v \rangle \lambda + \bar{\lambda} \langle v, u \rangle \lambda \\ &= 2\bar{\lambda} \langle u, v \rangle_{\mathbb{R}} \cdot \lambda \\ &= 2\langle u, v \rangle_{\mathbb{R}}. \end{aligned}$$

Then consider the expression

$$g(\pi(x,u),\ \pi(x,v)) = \langle u,v\rangle_{\mathbb{R}}\,.$$

It defines g as a natural metric on $\mathbb{K}P^n$, so that (3.4) is a Riemannian submersion with totally geodesic fibers. Notice that, for $\mathbb{K}=\mathbb{C}$, $\langle u,v\rangle$ is invariant by the S^1 action; but for $\mathbb{K}=\mathbb{H}$, $\langle u,v\rangle$ is not invariant under S^3, because of the lack of commutativity.

3.7. Notions of $\mathbb{K}$-dependence and $\mathbb{K}$-independence. On $T_{\pi(x)}\mathbb{C}P^n$ one defines a $\mathbb{C}$-vector space structure as follows: for λ in $\mathbb{C}$,

$$\pi(x,u)\cdot\lambda = \pi(x,u\cdot\lambda)\,.$$

Indeed, thanks to the commutativity

$$\pi(x\mu,u\mu)\cdot\lambda = \pi(x\mu,u\mu\lambda) = \pi(x\mu,u\lambda\mu) = \pi(x,u\lambda)\,.$$

This multiplication law does not make sense on $\mathbb{H}P^n$. However, we have the following *definition*:

3.8. Two non-zero vectors at $\pi(x)$ in $\mathbb{K}P^n$, $\pi(x,u)$ and $\pi(x,v)$, are $\mathbb{K}$-*dependent* if there exists a λ in $\mathbb{K}^*$ such that $u=v\cdot\lambda$.

This relation is reflexive, symmetric, transitive and invariant under $S\mathbb{K}$ since

$$u\mu = v\lambda\mu = v\mu(\mu^{-1}\lambda\mu)\,.$$

For $X=\pi(x,u)$, $u\neq 0$, we define

$$X\cdot\mathbb{K} = \{\pi(x,v)\ \mid\ \pi(x,u)\ \text{and}\ \pi(x,v)\ \ \mathbb{K}\text{-dependent}\}\cup\{\pi(x,0)\}\,.$$

One sees that $X\cdot\mathbb{K}$ is an $\mathbb{R}$-subspace of $T_{\pi(x)}\mathbb{K}P^n$ and that $\dim X\cdot\mathbb{K}=a$.

3.9. *We now construct an adapted basis of* $T_{\pi(x)}\mathbb{K}P^n$. *Let* $X=\pi(x,u)$ *and* $Y=\pi(x,v)$. *The following properties are equivalent:*

a) $\langle u,v\rangle = 0$;

b) X *is orthogonal to* $Y\cdot\mathbb{K}$ *with respect to* g;

c) $X\cdot\mathbb{K}$ *is orthogonal to* $Y\cdot\mathbb{K}$ *with respect to* g.

Proof. Part b) implies Part a). For β in $\mathbb{K}$ we set $Y'=\pi(x,v\beta)$. Then

$$g(X,Y') = \mathrm{Re}\langle u,v\beta\rangle = \mathrm{Re}\langle u,v\rangle\beta = 0\,,$$

and taking

$$\beta = \overline{\langle u,v\rangle},\quad \text{one gets}\quad |\langle u,v\rangle|^2 = 0\,.$$

Part a) implies Part c). For α in $\mathbb{K}$, we set $X'=\pi(x,u\alpha)$ and

$$g(X',Y') = \mathrm{Re}\langle u\alpha,v\beta\rangle = \mathrm{Re}\,\bar{\alpha}\langle u,v\rangle\beta = 0\,. \qquad \square$$

Hence, if $E=X_1\cdot\mathbb{K}\oplus\ldots\oplus X_i\cdot\mathbb{K}\subset T_{\pi(x)}\mathbb{K}P^n$ $(1\leqslant i\leqslant n)$ is an orthogonal decomposition with respect to g, taking X_{i+1} orthogonal to E, we get the orthogonal sum $E\oplus X_{i+1}\cdot\mathbb{K}$. Thus, by iteration

3.10 $$T_{\pi(x)}\mathbb{K}P^n=X_1\cdot\mathbb{K}\oplus\ldots\oplus X_n\cdot\mathbb{K}\,.$$

3.11 Definition. If $(X_{j,1},\ldots,X_{j,a})$ is an orthonormal basis of $X_j\cdot\mathbb{K}$, then the $a\cdot n$ vectors $(X_{j,s})$, $1\leqslant s\leqslant a$, $1\leqslant j\leqslant n$, form an *adapted basis* of $T_{\pi(x)}\mathbb{K}P^n$.

3.12. The projective spaces $\mathbb{K}P^1$ are called *projective lines*. For $n=1$, one easily shows by considering the charts (B_1,φ_1), (B_2,φ_2) on $\mathbb{K}P^1$ given in 3.3, that this projective line is diffeomorphic to a standard sphere S^a. More precisely, the fibrations

$$\{-1,+1\}\to S^1\to\mathbb{R}P^1\simeq S^1\,,$$
$$S^1\to S^3\to\mathbb{C}P^1\simeq S^2\,,$$
$$S^3\to S^7\to\mathbb{H}P^1\simeq S^4$$

are the classical Hopf fibrations of a sphere by spheres over a sphere. With respect to g, S^2 and S^4 have constant sectional curvature equal to 4 (see for example [OL] for a calculation using Riemannian submersion techniques; another proof of this result will be given in 3.23).

In G we give another presentation of the *projective planes* $\mathbb{K}P^2$ and some characterizations of the Cayley plane $\mathbb{C}\mathrm{a}P^2$. In the latter case one also has the fibration ([SD], p. 108—109)

$$S^7\to S^{15}\to\mathbb{C}\mathrm{a}P^1\simeq S^8\,.$$

C. The Projective Spaces as Symmetric Spaces

3.13. At first we briefly recall some basic facts due to E. Cartan about symmetric and locally symmetric spaces (see [HN 1] p. 162 ff., [K-N 2] p. 222 ff. for a complete study and also [MR 1] p. 109 ff).

Let M be a Riemannian manifold and N a symmetric neighborhood of 0 in T_mM for some m in M such that $\exp_m\restriction N:N\to V_m$ is a diffeomorphism. We define $s:N\to N$ by $s(X)=-X$ and set $s_m=(\exp_m\restriction N)\circ s\circ(\exp_m\restriction N)^{-1}$. This map is called the *geodesic symmetry* with respect to m on V_m.

The manifold M is called *locally symmetric* if for each m there exist N and V_m such that the geodesic symmetry is an isometry. These spaces are characterized by the following integrability condition:

3.14 Theorem. *A Riemannian manifold M is locally symmetric if and only if the curvature tensor R (resp. the sectional curvature) is invariant under parallel transport (or equivalently, if $DR=0$).*

3.15 Definition. A connected Riemannian manifold M is a *symmetric space* if for each m in M there exists an involutive isometry $s_m: M \to M$ such that (on $T_m M$)

$$s_m \circ \exp_m = \exp_m \circ s.$$

Clearly if M is symmetric it is locally symmetric and complete.

3.16. The *rank* of a symmetric space M is defined as the maximal dimension of the flat submanifolds (that is, those with zero curvature) which are totally geodesic in M.

Clearly a C_L-manifold cannot contain any totally geodesic flat torus. Let us *denote by* ROSSes the symmetric spaces of *rank one* and *by* CROSSes the compact ones. The noncompact ROSSes are the various hyperbolic spaces corresponding to $\mathbb{R}$, $\mathbb{C}$, $\mathbb{H}$, $\mathbb{C}a$ and in a dual manner the CROSSes are the various $\mathbb{K}P^n$'s and the Euclidean spheres (see [HN 1], p. 355, [CL], [WF 1]).

In fact, symmetric spaces can be identified in a canonical way with particular homogeneous Riemannian spaces, via the transitive action of their isometry group. This property was used by E. Cartan to classify them. (See [CN 1]).

Below we give this identification in the case of $\mathbb{K}P^n$ and we prove that the rank is one.

3.17. We denote by $U(n+1, \mathbb{K})$ the subgroup of $GL(n+1, \mathbb{K})$ which leaves the Hermitian product $\langle\ ,\ \rangle$ invariant, i.e.,

$$\forall A \in U(n+1, \mathbb{K}), \qquad \langle A(x), A(y)\rangle = \langle x, y\rangle.$$

Then we set

$$U(n+1\ \mathbb{R}) = O(n+1), \quad \text{the } \textit{Euclidean orthogonal group};$$

$$U(n+1, \mathbb{C}) = U(n+1), \quad \text{the } \textit{unitary group}; \text{ and}$$

$$U(n+1, \mathbb{H}) = Sp(n+1), \quad \text{the } \textit{symplectic group} \text{ (see [CY] p. 18).}$$

For A in $U(n+1, \mathbb{K})$, x in $\mathbb{K}^{n+1}$ and λ in $\mathbb{K}$ one has

$$A(x \cdot \lambda) = A(x) \cdot \lambda.$$

Hence, $U(n+1, \mathbb{K})$, which acts transitively on $S\mathbb{K}^{n+1}$, also acts transitively on $\mathbb{K}P^n$. Let $(e_1, \ldots, e_{n+1})$ be the canonical basis of $\mathbb{K}^{n+1}$ and let H be the subgroup of $U(n+1, \mathbb{K})$ keeping the point $\pi(e_{n+1})$ fixed.

3.18. Consider the map $p: U(n+1, \mathbb{K}) \to \mathbb{K}P^n$ defined by $p(A) = \pi A(e_{n+1})$ for A in $U(n+1, \mathbb{K})$. This map is C^∞ and one knows (see [SD] p. 105, [HN 1] p. 114 for example) that the map

3.19 $$\varphi: U(n+1, \mathbb{K})/H \to \mathbb{K}P^n$$

given by $\varphi(A \cdot H) = \pi A(e_{n+1})$ is a diffeomorphism.

For A in H one has $A(e_{n+1})=e_{n+1}\cdot\lambda$ for some λ in $U(1,\mathbb{K})=S\mathbb{K}=S^{a-1}$. Hence, A has the following form: $A=\begin{pmatrix} B & 0 \\ 0 & \lambda \end{pmatrix}$ with B in $U(n,\mathbb{K})$, so that H is isomorphic to (and denoted by)

$$H=U(n,\mathbb{K})\times U(1,\mathbb{K}).$$

Hence one has the resulting diffeomorphisms:

$$O(n+1)/O(n)\times\{-1,1\}\overset{\varphi}{\simeq}\mathbb{R}P^n,$$
$$U(n+1)/U(n)\times U(1)\overset{\varphi}{\simeq}\mathbb{C}P^n,$$
$$Sp(n+1)/Sp(n)\times Sp(1)\overset{\varphi}{\simeq}\mathbb{H}P^n.$$

3.20. Therefore, the projective spaces inherit a *symmetric structure*.

One defines an involution θ on $U(n+1,\mathbb{K})$ by setting

$$\theta(A)=SAS^{-1} \quad \text{with} \quad S=\begin{pmatrix} I_n & 0 \\ 0 & -1 \end{pmatrix}.$$

Then $\theta(A)=A$ if and only if $A\in H$. It follows that $(U(n+1,\mathbb{K}), H, \theta)$ is a symmetric pair (see [HN 1] p. 173 or [K-N 2] p. 225). The canonical decomposition of the Lie algebra $\mathfrak{u}(n+1,\mathbb{K})$ is given by

3.21 $\qquad \mathfrak{u}(n+1,\mathbb{K})=\mathfrak{h}\oplus\mathfrak{m} \quad \text{with} \quad \mathfrak{h}=(\mathfrak{u}(n,\mathbb{K})\oplus\mathfrak{u}(1,\mathbb{K})).$

In fact, b is an element of $\mathfrak{u}(n+1,\mathbb{K})$ if and only if $b+{}^t\bar{b}=0$ and

$$\mathfrak{m}=\left\{\left(\begin{array}{c|c} 0 & \xi \\ \hline -{}^t\bar{\xi} & 0 \end{array}\right)\middle|\ \xi\in\mathbb{K}^n\right\}.$$

In this way $\mathfrak{m}$ is naturally identified with $\mathbb{K}^n$. With this identification the *adjoint action* of H on $\mathfrak{m}$ has the form

$$ad\begin{pmatrix} B & 0 \\ 0 & \lambda \end{pmatrix}(\xi)=B\xi\cdot\bar{\lambda}.$$

3.22. Let us now come to the Riemannian structure of $U(n+1,\mathbb{K})/U(n,\mathbb{K})\cdot U(1,\mathbb{K})$. For ξ,η in $\mathfrak{m}\simeq\mathbb{K}^n$, the real scalar product $\langle\xi,\eta\rangle_{\mathbb{R}}$ is adH-invariant. In fact, by (3.21), 3.5 and the definition of $U(n,\mathbb{K})$ we have

$$\langle B\xi\cdot\bar{\lambda}, B\eta\cdot\bar{\lambda}\rangle_{\mathbb{R}}=\langle B\xi, B\eta\rangle_{\mathbb{R}}=\langle\xi,\eta\rangle_{\mathbb{R}}.$$

Hence, $U(n+1,\mathbb{K})/H$ with the pushed forward metric is a Riemannian symmetric space (see [K-N 2] pp. 200—256).

3.23. We now show that the diffeomorphism φ is an isometry. Consider the following diagrams

$$\begin{array}{ccc} U(n+1,\mathbb{K}) & \xrightarrow{q} & S\mathbb{K}^{n+1} \\ {\scriptstyle j}\downarrow & \searrow^{p} & \downarrow{\scriptstyle \pi} \\ U(n+1,\mathbb{K})/H & \xrightarrow[\varphi]{} & \mathbb{K}P^n \end{array}$$

and

$$\begin{array}{ccc} \mathfrak{h}\oplus\mathfrak{m} & \xrightarrow{Tq} & T_{e_{n+1}}S\mathbb{K}^{n+1} \\ {\scriptstyle Tj}\downarrow & \searrow^{Tp} & \downarrow{\scriptstyle T\pi} \\ T_H(U(n+1,\mathbb{K})/H) & \xrightarrow[T\varphi]{} & T_{\pi(e_{n+1})}\mathbb{K}P^n\,. \end{array}$$

Recall that $Tj\restriction\mathfrak{m}$ is an isomorphism and by construction is an isometry. Thus

$$T\varphi = T\pi\circ Tq\circ(Tj\restriction\mathfrak{m})^{-1},$$

and for ξ in $\mathfrak{m}$, we have

$$\begin{aligned} Tq(\xi) &= Tq\left(\frac{\partial}{\partial t}(\exp t\xi)\Big|_0\right) = \frac{\partial}{\partial t}(q(\exp t\xi))\Big|_0 \\ &= \frac{\partial}{\partial t}(\exp t\xi(e_{n+1}))\Big|_0 = \left(e_{n+1}, \begin{pmatrix}\xi\\0\end{pmatrix}\right). \end{aligned}$$

With the notation (3.5) of tangent vectors of $\mathbb{K}P^n$, $T\pi\circ Tq(\xi) = \pi\left(e_{n+1}, \begin{pmatrix}\xi\\0\end{pmatrix}\right)$, and $T\varphi$ is then an isometry from 3.22. Moreover, since each metric is invariant under the actions of $U(n+1,\mathbb{K})$ on $TU(n+1,\mathbb{K})/H$ and $T\mathbb{K}P^n$ (which commute with $T\varphi$), the result follows. □

D. The Hereditary Properties of Projective Spaces

3.24. We describe here the totally geodesic submanifolds in $\mathbb{K}P^n$ for $\mathbb{K}=\mathbb{R}$, $\mathbb{C}$, or $\mathbb{H}$ (see also [WF 2] p. 451). Each of those submanifolds is associated with a two-fold hereditary inclusion of $\mathbb{K}'^{n'+1}$ into $\mathbb{K}^{n+1}$ with $\mathbb{K}'$, $\mathbb{K}$ either $\mathbb{R}$, $\mathbb{C}$, or $\mathbb{H}$, $\mathbb{K}'\subset\mathbb{K}$ and $n'\leqslant n$. Moreover, each one is a CROSS by the very definition of the rank 3.16. Let a' denote $\dim_{\mathbb{R}}\mathbb{K}'$.

3.25 Theorem. *Let* $(X_{1,0}, \ldots, X_{1,a'-1}, \ldots, X_{n',0}, \ldots, X_{n',a'-1})$ *be* $n'a'$ *vectors extracted from an adapted basis of* $\mathbb{K}P^n$ *at* $p=\pi(x)$ *and let* E *be the subspace which they span. More precisely, if* $\mathbb{K}=\mathbb{H}$ *and* $\mathbb{K}'=\mathbb{C}$, *suppose that*

$$X_{1,0}=\pi(x, u_0), \quad X_{1,1}=\pi(x, u_0\alpha), \ldots, X_{n',0}=\pi(x, u_{n'}), \quad X_{n',1}=\pi(x, u_{n'}\alpha)$$

for some α *in* $\mathbb{H}$ *with* $\operatorname{Re}\alpha=0$. *Then* $\exp_p E$ *is a totally geodesic submanifold of* $\mathbb{K}P^n$.

Conversely, any totally geodesic submanifold is obtained in this way for some n, n', a, a'.

Before giving the proof let us point out a consequence which is convenient for calculations and which gives some idea of what projective spaces look like.

3.26 Corollary. *Let* $(X_{r,s})$, $0 \leqslant s \leqslant a-1$, $1 \leqslant r \leqslant n$, *be an adapted basis of* $T_p\mathbb{K}P^n$. *Then*

a) $\exp(\mathbb{R}X_{r,s} \oplus \mathbb{R}X_{r,t})$, $t \neq s$, $a>1$, *is a totally geodesic 2 sphere isometric to* $\mathbb{C}P^1$ *(with constant curvature 4),*

b) $\exp(\mathbb{R}X_{r,s} \oplus \mathbb{R}X_{k,t})$, $r \neq k$, *is a totally geodesic surface isometric to* $\mathbb{R}P^2$ *(with constant curvature 1).*

Proof of 3.25. Consider the inclusions $\mathbb{R} \subset \mathbb{C} \subset \mathbb{H}$ defined by the corresponding bases 1, $(1, i)$ and $(1, i, j, k)$.

This defines an inclusion for $n' \leqslant n$

$$\mathbb{K}'^{n'} \times \mathbb{K}^{n-n'} \times \mathbb{K}' \to \mathbb{K}^{n'} \times \mathbb{K}^{n-n'} \times \mathbb{K}.$$

The subgroup of $U(n+1, \mathbb{K})$ which leaves the direct sum of the first and the third factor globally invariant and $\mathbb{K}^{n-n'}$ pointwise invariant is isomorphic to $U(n'+1, \mathbb{K}')$. Its subgroup which keeps $\pi(e_{n+1})$ fixed (when acting on $\mathbb{K}P^n$) is isomorphic to $H'=U(n', \mathbb{K}') \times U(1, \mathbb{K}')$.

Clearly, the involution θ in $U(n+1, \mathbb{K})$ (3.20) leaves $U(n'+1, \mathbb{K}')$ invariant. Let θ' be its restriction to this subgroup. It follows that $(U(n'+1, \mathbb{K}'), H', \theta')$ is a symmetric subspace of $U(n+1, \mathbb{K})/H$. The diagram

$$\begin{array}{ccc} U(n+1, \mathbb{K})/H & \xrightarrow{\ \varphi\ } & \mathbb{K}P^n \\ \uparrow{\scriptstyle i_2} & & \uparrow{\scriptstyle i_2} \\ U'(n'+1, \mathbb{K}')/H' & \xrightarrow{\ \varphi'\ } & \mathbb{K}'P^{n'} \end{array}$$

(where i_1, i_2 are natural injections) defines $\mathbb{K}'P^{n'}$ as a totally geodesic submanifold of $\mathbb{K}P^n$ (see [K-N 2] p. 234). Using the fact that $T\varphi \circ Tj = T\pi \circ Tq$ in 3.23, we see that $T_{\pi(e_{n+1})}\mathbb{K}'P^{n'}$ admits the following adapted basis:

$$(\pi(e_{n+1}, e_{r,s}))(1 \leqslant r \leqslant n', 0 \leqslant s \leqslant a'-1)$$

with

$$e_{r,0}=e_r, \quad e_{r,1}=e_r\cdot i, \quad e_{r,2}=e_r\cdot j \quad \text{and} \quad e_{r,3}=e_r\cdot k.$$

Now let $(X_{r,s})$, $1\leqslant r\leqslant n'$, $0\leqslant s\leqslant a'-1$, be some vectors (picked up from an adapted basis, as indicated in Theorem 3.25) at $p=\pi(x)$.

Put $X_{r,0}=\pi(x, u_r)=\pi(xt, u_r t)$ with t in $S\mathbb{K}$. By the assumption $\langle u_r, u_q\rangle=\delta_{rq}$ there exists an element A in $U(n+1, \mathbb{K})$ such that ([CY] p. 21):

3.27 $\qquad u_1 t=A(e_1),\dots, u_{n'}t=A(e_{n'}),\ xt=A(e_{n+1})$.

Moreover, since A acts on $T\mathbb{K}P^n$ as an isometry (by

$$A(\pi(e_{n+1}, v))=\pi(A(e_{n+1}), A(v))),$$

it suffices to prove that A maps $\mathbb{K}'P^{n'}$ onto $\exp_p E$ and hence that $\{A(e_{r,s})|0\leqslant s\leqslant a'-1, 1\leqslant r\leqslant n'\}$ spans E.

If $a'=1$, i.e., $\mathbb{K}'=\mathbb{R}$, this is clear. If $\mathbb{K}'=\mathbb{K}$, one sees that for fixed r $\{A(e_{r,s})|0\leqslant s\leqslant a'-1\}$ spans $X_{r,0}\mathbb{K}$. If $\mathbb{K}'=\mathbb{C}$, $\mathbb{K}=\mathbb{H}$ we set $X_{r,0}=\pi(x, u_0)$ and $X_{r,1}=\pi(x, u_0\alpha)$ with $1\leqslant r\leqslant n'$ and $|\alpha|=1$, $\operatorname{Re}\alpha=0$. There exists a t in $S^3=S\mathbb{H}$, such that $\alpha=tit^{-1}$ ([CY] p. 38).

Let A in $Sp(n+1)$ be as in 3.27. Then

$$A(e_r\cdot i)=u_r\cdot ti=u_r\alpha t \quad \text{for} \quad 1\leqslant r\leqslant n'$$

[that is, $A(\pi(e_{n+1}, e_{r,s}))=\pi(xt, A(e_{r,s}))=X_{r,s}$ with $1\leqslant r\leqslant n'$, $0\leqslant s\leqslant 1$]. □

Conversely, let N be a totally geodesic submanifold in $\mathbb{K}P^n$. Up to isometry we can suppose that $\pi(e_{n+1})$ is in N and that $\pi(e_{n+1}, e_r)$ belongs to $T_{\pi(e_{n+1})}N=E$ if and only if $1\leqslant j\leqslant n'$. If $\dim_{\mathbb{R}}E=n'$, the first part of the proof shows that N is isometric to $\mathbb{R}P^{n'}$. Otherwise, one augments this set of vectors so that it becomes an orthonormal basis of E which is necessarily of the following type:

$$e_1, e_1\cdot\theta_{1,1},\dots, e_1\cdot\theta_{1,s_1},\dots, e_{n'}, e_{n'}\cdot\theta_{n',1},\dots, e_{n'}\cdot\theta_{n',s_{n'}}$$

with $\operatorname{Re}\theta_{lm}=0$.

It is clear that $s_l\leqslant a-1$ for $1\leqslant l\leqslant n'$. In a natural way N is a symmetric space G'/H' ([K-N 2] p. 235) for some subgroup G' of G. The action of ad H' on E (3.21) implies in a straightforward way that $s_1=\dots=s_{n'}=a'-1$. If $a'=4$, then $N=\mathbb{H}P^{n'}$. If $a'=2$ and $\mathbb{K}=\mathbb{H}$, it is necessary that $\theta_{1,1}=\dots=\pm\theta_{n',1}$. This condition is automatically satisfied if $\mathbb{K}=\mathbb{C}$. □

3.28 *Proof of Corollary 3.26.* Notice that $\mathbb{C}P^1$ and $\mathbb{R}P^2$ are 2-dimensional symmetric spaces. The curvature, being invariant under parallel transport, must be constant. Moreover, the curvature of a symmetric space (with the notation of (3.21)) is given by (see [HN 1] p. 180)

$$R(X, Y)Z=[[X\,Y], Z] \quad \text{with} \quad X, Y, Z\in\mathfrak{m}.$$

An elementary calculation then shows that

$$[[e_1, e_2], e_1] = e_2 \quad \text{in} \quad \mathfrak{m} \subset \mathfrak{so}(3)$$

and

$$[[e_1, e_1 \cdot i], e_1] = 4e_1 i \quad \text{in} \quad \mathfrak{m} \subset \mathfrak{u}(2).$$

3.29. It also follows from 3.25 (with $n' = n = 1$ and $\mathbb{K}' = \mathbb{C}$, $\mathbb{K} = \mathbb{H}$) and 3.26, that $\mathbb{H}P^1$ is isometric to a 4-dimensional sphere of constant curvature 4. □

3.30 Theorem. *Let (X, Y) be an orthonormal basis of a tangent plane E to $\mathbb{K}P^n$. The corresponding sectional curvature is given by*

$$\sigma(E) = g(R(X, Y)X, Y) = 1 + 3 \cdot \|pr_{Y \cdot \mathbb{K}} X\|^2.$$

Proof. From (3.10) one has $X = X_1 + X_2$, with X_1 in $Y \cdot \mathbb{K}$ and X_2 orthogonal to $Y \cdot \mathbb{K}$. Then

$$R(X, Y)X = R(X_1, Y)X_1 + R(X_2, Y)X_2 + R(X_1, Y)X_2 + R(X_2, Y)X_1,$$

$$R(X_1, Y)X_1 = 4Y,$$

$$R(X_2, Y)X_2 = Y$$

and $R(X_1, Y)X_2 = R(X_2, Y)X_1 = 0$ (which are verified for example by using the formula recalled in 3.28). Now

$$g(R(X, Y)X, Y) = 4g(X_1, X_1) + g(X_2, X_2) = 1 + 3\|X_1\|^2. \qquad \square$$

Notice that $\sigma(E)$ belongs to the interval $[1, 4]$, the minimal (resp. maximal) value in this interval corresponding to the case where E is tangent to some $\mathbb{R}P^2$ (resp. $\mathbb{C}P^1$)

E. The Geodesics of Projective Spaces

3.31 Proposition. *The projective spaces $\mathbb{K}P^n$ are C_π-manifolds.*

Proof. By Theorem 3.25 with $\mathbb{K}' = \mathbb{R}$ and $n' = 1$ each geodesic is isometric to $\mathbb{R}P^1$. Hence, each geodesic is simply closed and has length π. □

Alternatively, let us express the equation of the geodesic γ with initial conditions $\gamma(0) = p = \pi(x)$, $\dot{\gamma}(0) = X = \pi(x, u)$, $\|X\| = 1$.

3.32 Proposition. *The geodesic γ has the equation*

$$\gamma(s) = \pi(x \cos s + u \sin s) = \exp_p sX.$$

Proof. We look at the diagram 3.23 for $x=e_{n+1}$. The geodesic θ, with initial condition $Tj(u)$ (with u in m) in $U(n+1,\mathbb{K})/H$, is expressed by $\theta(s)=j(\exp su)$ (see [HN 1] p. 177). The geodesic $\varphi\circ\theta=\gamma$ is such that $X=Tp(u)$. Hence,

$$\gamma(s)=\pi\circ q(\exp su)=\pi(\exp su(e_{n+1}))=\pi(x\cos s+u\sin s)\,. \quad \square$$

3.33. We look at the intersection of geodesics issuing from p. Let $X_i=\pi(x_i, u_i)$, $\|X_i\|=1$ ($i=1,2$, $X_1\neq\pm X_2$), be the initial conditions for γ_1 and γ_2 issuing from p. Using the formula 3.32 one easily verifies that

$$\gamma_1(\mathbb{R})\cap\gamma_2(\mathbb{R})=\{p\} \text{ if } X_1 \text{ and } X_2 \text{ are } \mathbb{K}\text{-independent}$$

and that $\gamma_1(\mathbb{R})\cap\gamma_2(\mathbb{R})=\{p, A(p)\}$, if X_1 and X_2 are $\mathbb{K}$-dependent [here $A(p)$ denotes the point antipodal to p on the $\mathbb{K}$-line defined by X_1 or X_2].

3.34. We now compute the *Jacobi fields.*

Proposition. *Let γ be the geodesic with initial conditions $\gamma(0)=p$, $\dot\gamma(0)=X_1$ ($\|X_1\|=1$). Let (X_i) be an adapted basis ($1\leqslant i\leqslant an$) at p, such that X_i belongs to $X_1\cdot\mathbb{K}$ for $1\leqslant i\leqslant a$ and denote by $X_i(s)$ the vector fields along γ, which are extensions of X_i by parallel transport. The space of Jacobi fields along γ admits the following basis:*

$$J_1(s)=s\cdot X_1(s); \qquad K_1(s)=X_1(s)=\dot\gamma(s);$$

$$J_l(s)=\sin 2s\cdot X_l(s); \qquad K_l(s)=\cos 2s\cdot X_l(s)\,, \qquad 2\leqslant l\leqslant a;$$

$$J_l(s)=\sin sX_l(s); \qquad K_l(s)=\cos s\cdot X_l(s)\,, \qquad a+1\leqslant l\leqslant na\,.$$

Hint: For example, one might use Corollary 3.26 and look at the solutions of $Y''=0$ and $Y''+\sigma Y=0$ in a surface a constant curvature σ.

3.35. We now study the cut locus of a point (see 5.20 for definitions).

Proposition. *For each point p in $\mathbb{K}P^n$ one has:*

a) $\mathrm{Cut}(p)=\{q|d(p,q)=\frac{\pi}{2}\}$ *(this is a totally geodesic submanifold isometric to $\mathbb{K}P^{n-1}$ called the* antipodal submanifold of *p);*

b) $\mathrm{Conj}(p)=\{p\}$ *on $\mathbb{R}P^n$;*

c) $\mathrm{Conj}(p)=\mathrm{Cut}(p)$ *on $\mathbb{K}P^n$ for $\mathbb{K}\neq\mathbb{R}$;*

d) *the multiplicity of the first point conjugate to p along any geodesic γ through p is $n-1$ for $\mathbb{R}P^n$ and $a-1$ for $\mathbb{K}P^n$ if $\mathbb{K}\neq\mathbb{R}$.*

Proof. The first equality in a) follows from 3.33 and 3.34. Now let q be a point such that $d(p,q)=\frac{\pi}{2}$ and let S be the projective line defined by p and q. Writing $T_q\mathbb{K}P^n=T_qS\oplus H$ it is enough to verify (according to Theorem 3.25) that $\exp_q H=\{q|d(p,q)=\frac{\pi}{2}\}$. In fact, let γ and θ be geodesics such that $\dot\gamma(0)=X$ in H and $\dot\theta(0)=Y$ in T_qS. By 3.26 $\exp_q(\mathbb{R}X\oplus\mathbb{R}Y)$ is isometric to $\mathbb{R}P^2$ and is totally geodesic. Hence, for all s, $d(p,\gamma(s))=\frac{\pi}{2}$ and $\exp_q H\subset\mathrm{Cut}(p)$.

3.36. Conversely, let r be a point such that $d(p,r)=\frac{\pi}{2}$ and let γ be a geodesic from q to r. Then $\dot\gamma(0)=X+Y$ with Y in T_qS and X in H such that $X\neq 0$ if $q\neq r$. If $Y\neq 0$, then r belongs to $\gamma(\mathbb{R})\subset \exp_q(\mathbb{R}X\oplus\mathbb{R}Y)$. By 3.26 the geodesic θ with $\dot\theta(0)=X$ must go through r. Then by 3.33, $\dot\gamma(0)$ belongs to $X\cdot\mathbb{K}\subset H$ and thus $Y=0$.

Assertions b), c), d) follow directly using 3.34 from the characterization of conjugate points and their multiplicities (see for example [MR 1] p. 77). □

F. The Topology of Projective Spaces

3.37 Proposition. *For each point m, $\mathbb{K}P^n$ is obtained from* $\mathrm{Cut}(m)\simeq\mathbb{K}P^{n-1}$ *by attaching an na-cell via the application* $\exp_m$.

Proof. Let $E_{na}=\{X \mid X\in T_m\mathbb{K}P^n, \|X\|\leqslant\frac{\pi}{2}\}$ and let S^{naM1} be the corresponding sphere. Then $\exp_m$ restricted to $(E_{na}-S^{na-1})$ is a homeomorphism and $\exp_m(S^{na-1}) = \mathrm{Cut}(m)$ by 3.32 and 3.35. □

3.38. Denoting by C^i an i-cell, one then has by iteration

$$\mathbb{K}P^n=C^0\cup C^a\cup\ldots\cup C^{(n-1)a}\cup C^{na}.$$

3.39 Proposition. a) *$\mathbb{C}P^n$ and $\mathbb{H}P^n$ are simply connected;*
b) $\pi_1(\mathbb{R}P^n)\simeq\mathbb{Z}_2 (n>1)$.

Proof. The statement in a) is a simple consequence of 3.38 and of the following fact: set $X=A\cup C^i$, $i\geqslant 2$, then the inclusion map $(A,x_0)\subset(X,x_0)$ for x_0 in A induces an epimorphism $\pi_1(A,x_0)\to\pi_1(X,x_0)$ (see also [SR] p. 146).
For the case of $\mathbb{R}P^n$, we refer to [SR] p. 74. □

3.40. We now turn to the homology and cohomology of projective spaces.
The computation of the homology and cohomology groups follows classically from the cellular decomposition 3.38 (see [GG] p. 90, [A-B] p. 179, and [GO 2] p. 217). *A* complete treatment of the cohomology algebra is given in [SR] p. 264, using Thom-Gysin sequences of sphere bundles. The special cases of $\mathbb{C}P^n$ or $\mathbb{H}P^n$ are treated in [MI-ST] p. 160 or p. 243.

3.41 Theorem. *The algebra $H^*(\mathbb{R}P^n,\mathbb{Z}_2)$ is a polynomial algebra over $\mathbb{Z}_2$ with one generator b of degree* 1, *truncated by the relation $b^{n+1}=0$.*

3.42 Theorem. *The algebra $H^*(\mathbb{K}P^n,\mathbb{Z})$ (with $\mathbb{K}=\mathbb{C}$ or $\mathbb{H}$) is a polynomial algebra over $\mathbb{Z}$ with one generator b of degree a, truncated by the relation $b^{n+1}=0$.*

In this particular situation, we illustrate the multiplicative structure of the cohomology using differential forms and the fact that each nonzero cohomology group is isomorphic to $\mathbb{Z}$.

Proof. We first consider the *Kähler form* ω on $\mathbb{C}P^n$. It is defined for $X=\pi(x,u)$, $Y=\pi(x,v)$ by

3.43 $$\omega(X,Y)=-\frac{i}{2}(\langle u,v\rangle-\langle v,u\rangle)=Im\langle u,v\rangle.$$

The form ω is well defined because $\langle u\lambda, v\lambda\rangle=\langle u,v\rangle$ if $|\lambda|=1$ and is a skew form which is (by 3.17) invariant under $U(n+1,\mathbb{C})$. The symmetric structure now implies that $D\omega=0$ and then ω is harmonic ([BE-L], p. 26).

Let $\omega^m=\omega\wedge\ldots\wedge\omega$ (m times), $1\leqslant m\leqslant n$, and $X_l=\pi(x,u_l)$, $Y_l=\pi(x,u_l\cdot i)$, $1\leqslant l\leqslant n$, be vectors of an adapted basis. Then it is clear that

$$\omega^m(X_1,Y_1,\ldots,X_m,Y_m)=m!.$$

If one sets $\omega_0=\frac{1}{\pi}\omega$, one gets

3.44 $$\int_{\mathbb{C}P^m}\omega_0^m=\frac{m!}{\pi^m}\cdot\mathrm{Vol}(\mathbb{C}P^m)=1 \quad \text{for} \quad 1\leqslant m\leqslant n,$$

where $\mathbb{C}P^m$ is some complex projective space imbedded in $\mathbb{C}P^n$. The volume of $\mathbb{K}P^n$ is computed for example in [B-G-M] p. 18–19–112.

3.45. We now consider *the fundamental* 4-*form* on $\mathbb{H}P^n$. In this case the Formula (3.43) is not convenient. Let $\mathscr{A}$ denote the antisymmetrization operator. For $X_l=\pi(x,u_l)$ with l in $\{1,2,3,4\}$ one sets

3.46 $$\begin{aligned}\varpi(X_1,X_2,X_3,X_4)&=\tfrac{1}{4}\mathscr{A}(\langle u_1,u_2\rangle\cdot\langle u_3,u_4\rangle)\\&=\tfrac{1}{4}\sum_{\sigma\in\mathfrak{S}_4}\varepsilon(\sigma)\langle u_{\sigma(1)},u_{\sigma(2)}\rangle\langle u_{\sigma(3)},u_{\sigma(4)}\rangle.\end{aligned}$$

3.47. The number $\varpi(X_1,X_2,X_3,X_4)$ is real, since

$$\begin{aligned}\overline{\langle u_{\sigma(1)},u_{\sigma(2)}\rangle\langle u_{\sigma(3)},u_{\sigma(4)}\rangle}&=\overline{\langle u_{\sigma(3)},u_{\sigma(4)}\rangle}\cdot\overline{\langle u_{\sigma(1)},u_{\sigma(2)}\rangle}\\&=\langle u_{\sigma(4)},u_{\sigma(3)}\rangle\cdot\langle u_{\sigma(2)},u_{\sigma(1)}\rangle\\&=\langle u_{\sigma\tau(1)},u_{\sigma\tau(2)}\rangle\cdot\langle u_{\sigma\tau(3)},u_{\sigma\tau(4)}\rangle,\end{aligned}$$

where $\tau(1)=4$, $\tau(2)=3$, $\tau(3)=2$, $\tau(4)=1$. Now by (3.46) and since $\varepsilon(\tau)=1$, it follows that

$$\bar{\varpi}(X_1,X_2,X_3,X_4)=\varpi(X_1,X_2,X_3,X_4).$$

3.48. We now check that ϖ is well defined. If λ belongs to Sp 1, then

$$\begin{aligned}\langle u_1\lambda,u_2\lambda\rangle\cdot\langle u_3\lambda,u_4\lambda\rangle&=\bar{\lambda}\cdot\langle u_1,u_2\rangle\cdot\lambda\cdot\bar{\lambda}\cdot\langle u_3,u_4\rangle\cdot\lambda\\&=\bar{\lambda}\cdot\langle u_1,u_2\rangle\cdot\langle u_3,u_4\rangle\cdot\lambda.\end{aligned}$$

Then $\mathscr{A}(\langle u_1\lambda, u_2\lambda\rangle\cdot\langle u_3\lambda, u_4\lambda\rangle)=\bar{\lambda}\cdot\mathscr{A}(\langle u_1, u_2\rangle\cdot\langle u_3, u_4\rangle)\cdot\lambda$
$=\mathscr{A}(\langle u_1, u_2\rangle\langle u_3, u_4\rangle)$ by 3.47.

3.49. Therefore ϖ is an exterior differential 4-form on $\mathbb{H}P^n$ which is invariant under $Sp(n+1)$. It again follows that $D\varpi=0$. We set

$$\varpi_0=\frac{1}{\pi^2}\varpi.$$

For some imbedding of $\mathbb{H}P^m$ in $\mathbb{H}P^n$ $(1\leqslant m\leqslant n)$ one has

$$\int_{\mathbb{H}P^m}\varpi_0^m=1,$$

which is a consequence of the following two facts:

$$\mathrm{Vol}(\mathbb{H}P^m)=\frac{\pi^{2m}}{(2m+1)!}$$

and

3.50 $$\varpi^m(X_{1,1},X_{1,2},X_{1,3},X_{1,4},\dots,X_{m,1},X_{m,2},X_{m,3},X_{m,4})=(2m+1)!,$$

where $(X_{l,s})$ $(1\leqslant s\leqslant 4, 1\leqslant l\leqslant m)$ are vectors of an adapted basis. A proof by induction of (3.50) may be found in [BO] p. 441 with a detailed study of $\mathbb{H}P^n$ (compare also [TN]).

3.51. We now come back to the $\mathbb{Z}$-cohomology rings of $\mathbb{C}P^n$ and $\mathbb{H}P^n$. Call $R^*(M)$ the de Rham cohomology ring of a compact manifold M, r the isomorphism

$$r: R^*(M)\to H^*(M,\mathbb{R})$$

and

$$j: H^*(M,\mathbb{Z})\to H^*(M,\mathbb{R})$$

the homomorphism induced by the natural coefficient homomorphism from $\mathbb{Z}$ to $\mathbb{R}$. Let $j'=r^{-1}\circ j$ and $[\alpha]$ be the class of a differential form α.

From (3.44) and (3.49), it follows that there exist c_1 in $H^2(\mathbb{C}P^n,\mathbb{Z})$ and c_2 in $H^4(\mathbb{H}P^n,\mathbb{Z})$ such that $[\omega_0]=j'(c_1)$ and $[\varpi_0]=j'(c_2)$. Now $[\omega_0^m]=j'(c_1^m)\neq 0$ and $[\varpi_0^m]=j'(c_2^m)\neq 0$. Hence, c_1^m (resp. c_2^m) is a generator of

$$H^m(\mathbb{C}P^n,\mathbb{Z})\text{ (resp. } H^m(\mathbb{H}P^n,\mathbb{Z}))$$

which is isomorphic to $\mathbb{Z}$, by the cellular decomposition 3.38. Hence, c_1 (resp. c_2) generates $H^*(\mathbb{C}P^n,\mathbb{Z})$ [resp. $H^*(\mathbb{H}P^n,\mathbb{Z})$] with the relation $c_1^{n+1}=0$ (resp. $c_2^{n+1}=0$). □

3.52. A remark is in order. Among the C_π-manifolds and in view of Bott-Samelson theorem 7.23 the CROSS'es have the further property that their cohomological structure can be pictured geometrically.

At the same time some of their topological properties are implied by the positivity of curvature only (see for example [G-K-M] and [CL]).

G. The Cayley Projective Plane

3.53. In this section we wish to arrive at a better understanding of some exceptional aspects of the Cayley plane and we shall try to give a survey of some points of view which should help us to grasp it. We begin by describing the deep links between projective plane geometries and alternative algebras following [HL] p. 347–420.

3.54. The axioms for a projective plane.

Definition. A projective plane P is a set whose elements are called *points* and which has particular subsets called *lines*, satisfying the following conditions:

1) any two distinct points are contained in a unique line,
2) any two distinct lines intersect, in exactly one point,
3) there exist four points which determine six distinct lines.

Axioms 1) and 2) are called the incidence axioms; the addition of axiom 3) permits the introduction of coordinate systems.

3.55 Definition. Isomorphisms of a projective plane P are called *collineations*. They are bijective maps of P onto itself which transform each line into a line.

The collineations of P form a group.

3.56 Definition. A *perspective* is a collineation which fixes a point (called its *center*) and a line pointwise (called its *axis*); more precisely every line through the center is invariant.

Assume we are given a line L, a point O and two distinct points M_1 and M_2 different from O, which do not lie on L and are such that O, M_1, M_2 lie on a line.

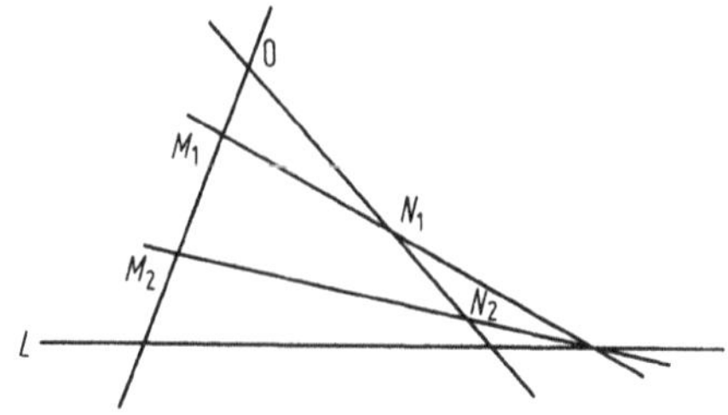

Fig. 3.56

One sees that there exists *at most one* perspective θ, with center O and axis L, such that $\theta(M_1)=M_2$. If such a perspective exists, then the following configuration holds (Desargues' theorem).

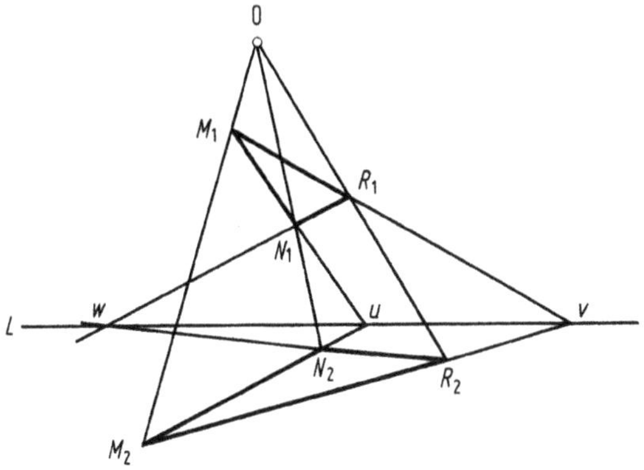

Fig. 3.57

The points $u = M_1N_1 \cap M_2N_2$, $v = M_1R_1 \cap M_2R_2$ and $w = N_1R_1 \cap N_2R_2$ lie on the axis L.

This property is not a consequence of Axioms 1), 2), 3). The collineations group must be big enough in view of the following elementary geometric fact:

3.57 Proposition. *The validity of Desargues theorem in P (one says that P is Arguesian) is equivalent to the property that each set $\{O, L, M_1, M_2\}$, as above, defines a perspective.*

In 1933 Ruth Moufang [MG] studied a definition weaker than Desargues' property:

3.58 Definition. A plane P is said to be a *translation plane* or a *Moufang plane*, if for every line L and for every point O belonging to L, all the possible perspectives with axis L and center O exist. (These perspectives are called *elations.*)

In fact, in order that P be a Moufang plane it suffices that all the elations defined with the four points and the six lines given by Axiom 3) exist.

3.59. We now give coordinates for a projective plane.

Consider the four points O, A, B, C and the following picture permitted by Axioms 1), 2), 3). The line AB is called the *line at infinity* and the points which are not on AB, the finite points. With each finite point on OC one associates a distinguished symbol, say α, and as coordinates (α, α). Denote by $(0, 0)$ and $(1, 1)$ the coordinates of O and C.

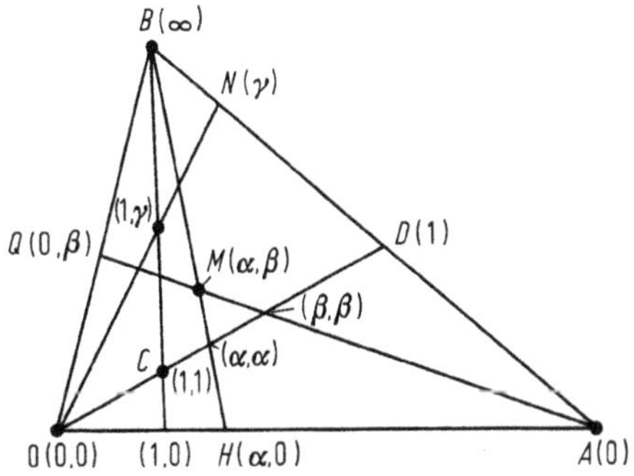

Fig. 3.59

Let M be a finite point such that $AM \cap OC$ (resp. $BM \cap OC$) has coordinates (β, β) [resp. (α, α)]. Then (α, β) are given as coordinates to M. Now let N be an infinite point and put $R = BC \cap ON$ one associates the unique coordinate (γ) to N if the coordinates of N are $(1, \gamma)$. The symbol (∞) is given to B.

3.60. In a geometric way one defines an *addition law* and a *multiplication law* over the set, say $\mathscr{A}$, of the symbols. Let α and β be in $\mathscr{A}$. Put $H = (\alpha, 0)$, $Q = (0, \beta)$ and $S = QD \cap HB = (\alpha, y)$.

One defines $y = \alpha + \beta$.

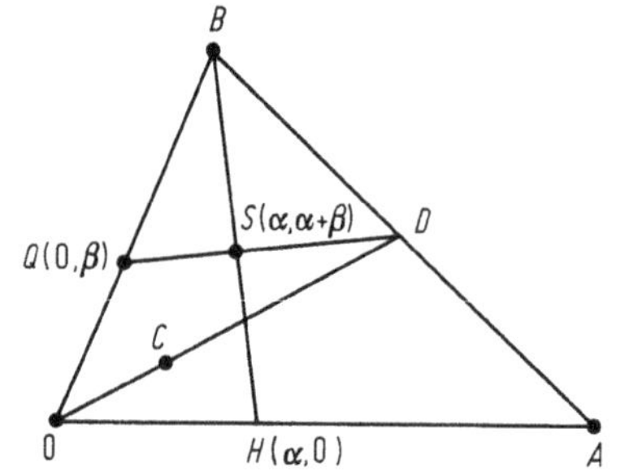

Fig. 3.60.1

Similarly, put $N(\gamma)$ for an infinite point and $U = ON \cap HB = (\alpha, z)$.

One defines $z = \gamma \cdot \alpha$.

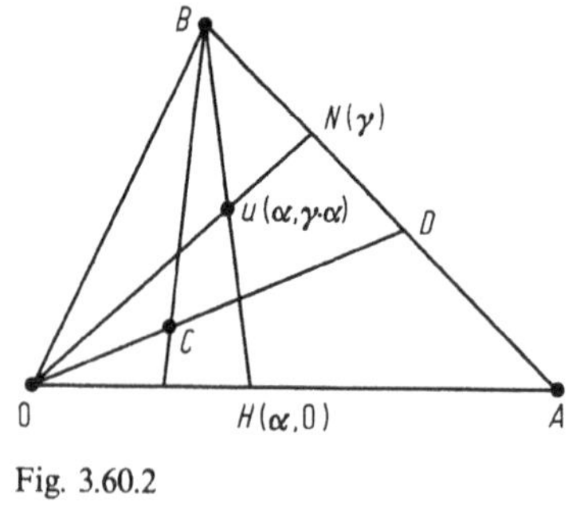

Fig. 3.60.2

Now, the main question is the following: When are the equations $y = \gamma \,.\, x + \beta$ precisely the equations of the lines in P_2 not containing B?

The relationship between plane projective geometry and algebra is quite remarkable as shown by the following

3.61 Theorem. *Let* $(\mathscr{A}, +, \cdot)$ *be the coordinate system of* P *which corresponds to* $(0, A, B, C)$. *Then* P *is a Moufang plane if and only if* $(\mathscr{A}, +, \cdot)$ *is an* alternative division ring *and the equations of the finite lines are* $y = \gamma, x + \beta$ *or* $x = \alpha$ *for* α, β, γ *in* $\mathscr{A}$. *Moreover, this algebra does not depend (at least up to isomorphism) on the coordinate system.*

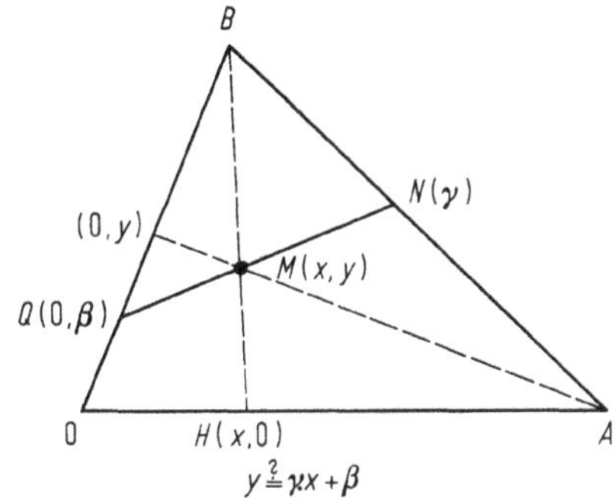

Fig. 3.61

See [HL] and [PT] for details concerning the numerous contributions to this result.

We recall that $(\mathscr{A}, +, \cdot)$ is an alternative division ring if $(\mathscr{A}, +)$ is an Abelian group, the left and right distributivity property holds, each element different from 0 has an inverse and the following alternative law holds: for each α, β in $\mathscr{A}$

$$(\alpha\cdot\beta)\cdot\beta=\alpha\cdot(\beta\cdot\beta) \quad \text{and} \quad (\beta\cdot\beta)\cdot\alpha=\beta\cdot(\beta\cdot\alpha).$$

In fact one can prove that each α, β span an associative algebra (see [SD] p. 108, [SC] p. 29). For example, consider the intersection of the lines: $y=\alpha\cdot x+\beta$ and $y=\alpha'\cdot x+\beta'$; then $(\alpha-\alpha')\cdot x=\beta'-\beta$, so that $(\alpha-\alpha')^{-1}\cdot[(\alpha-\alpha')\cdot x]$ $=(\alpha-\alpha')^{-1}\cdot(\beta'-\beta)$ and $[(\alpha-\alpha')^{-1}\cdot(\alpha-\alpha')]\cdot x=x=(\alpha-\alpha')^{-1}\cdot(\beta'-\beta)$.

Hence, the alternative law suffices to prove in 3.61 that the intersection of two lines is a point and similarly to find the unique equation of a line which contains two given points.

The Arguesian planes, among Moufang planes, have an illuminating algebraic characterization.

3.62 Theorem. *A projective plane is Arguesian if and only if its coordinatizing alternative division ring is associative (and hence a field).*

For the Arguesian geometry and the corresponding algebra one should read [AR].

Now the classification of projective planes endowed with a real manifold structure by their coordinate system follows from the celebrated classification of the real division algebras (see [MR 2]); these planes are then $\mathbb{R}P^2$, $\mathbb{C}P^2$, $\mathbb{H}P^2$ and $\mathbb{C}\mathrm{a}\,P^2$. The exceptional character of $\mathbb{C}\mathrm{a}\,P^2$ is also shown by the following striking structure theorem [B-K].

3.63 Theorem. *Each alternative division ring (of characteristic different from 2) is either associative or a Cayley algebra over its center.*

3.64. We recall that a *Cayley algebra* is an eight-dimensional vector space over a (commutative) field F with basis elements $e_0=1, e_1, \ldots, e_7$ for which a multiplication table involving three non zero scalars ([AT] p. 136) is given. It need not generally be a division algebra. In our case, one has $F=\mathbb{R}$ and there is up to isomorphism only one Cayley algebra, the classical one $\mathbb{C}\mathrm{a}$, which is such that $e_i \cdot e_0 = e_0 \cdot e_i = e_i$, with $e_i \cdot e_j$ given by the table

e_0	e_1	e_2	e_3	e_4	e_5	e_6	e_7
e_1	$-e_0$	e_3	$-e_2$	e_5	$-e_4$	$-e_7$	e_6
e_2	$-e_3$	$-e_0$	e_1	e_6	e_7	$-e_4$	$-e_5$
e_3	e_2	$-e_1$	$-e_0$	e_7	$-e_6$	e_5	$-e_4$
e_4	$-e_5$	$-e_6$	$-e_7$	$-e_0$	e_1	e_2	e_3
e_5	e_4	$-e_7$	e_6	$-e_1$	$-e_0$	$-e_3$	e_2
e_6	e_7	e_4	$-e_5$	$-e_2$	e_3	$-e_0$	$-e_1$
e_7	$-e_6$	e_5	e_4	$-e_3$	$-e_2$	e_1	$-e_0$

Moreover, $\mathbb{C}\mathrm{a}$ is a *normed* alternative division algebra over $\mathbb{R}$. To $x = a + \sum_{i=1}^{i=7} b_i \cdot e_i$ one associates the conjugate Cayley number $\bar{x} = a - \sum_{i=1}^{i=7} b_i e_i$ and the norm $|x|$ such that

$$|x|^2 = \bar{x} \cdot x = x \cdot \bar{x} = a^2 + \sum_{i=1}^{i=7} b_i^2 .$$

Likc in the quaternionic case one has $|x \cdot y| = |x| \cdot |y|$, $(\overline{x \cdot y}) = \bar{y} \cdot \bar{x}$, and $x^{-1} = \frac{1}{|x|^2} \cdot \bar{x}$.

Concerning hereditary properties we notice that any pair of numbers which do not commute generates an algebra which is isomorphic to $\mathbb{H}$.

To conclude this fast train excursion through geometric algebra we note that in an elementary way a Cayley geometry (that is a non-associative, non Arguesian one) is essentially the geometry of a plane ([HL], p. 352).

3.65 Proposition. *It is possible to imbed a projective plane P in a three-dimensional projective geometry if and only if P is Arguesian.*

The reason why $\mathbb{R}P^2$, $\mathbb{C}P^2$, $\mathbb{H}P^2$ are Arguesian is now clear.

3.66. The Cayley plane as a C^∞-manifold.

Theorem. *The Moufang plane P coordinatized by* $\mathbb{C}\mathrm{a}$ *is naturally a* C^∞*-manifold.*

Proof. First we give a basis for the topology of P; according to 3.59, one has $P = \mathbb{C}\mathrm{a}^2 \cup \mathbb{C}\mathrm{a} \cup \{(\infty)\}$. Clearly, one requires that each open set in $\mathbb{C}\mathrm{a}^2$ belongs to this topology. We denote by $V(m_0, A, \varepsilon)$ the set

$$\{(x, mx) \text{ and } (m) \mid |x| > A, |m - m_0| < \varepsilon\}$$

and we take it as an open neighborhood of m_0. We also denote by $W(\infty, A, B)$ the set $\{(x, mx), (m) \mid |m| > A, |xm| > B\}$ and take it as an open neighborhood of ∞. We then consider the topology which is generated by the topology of $\mathbb{C}\mathrm{a}^2$ and all of the sets $V(m_0, A, \varepsilon)$ and $W(\infty, A, B)$.

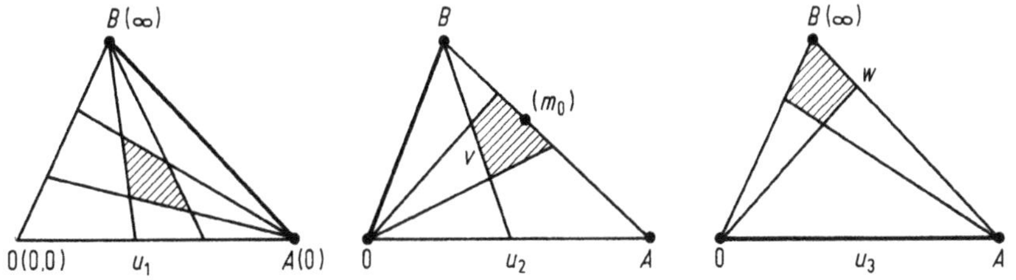

Fig. 3.66

Now we define charts (U_i, φ_i), $i = 1, 2, 3$ by setting $U_1 = \mathbb{C}\mathrm{a}^2$ and $\varphi_1 = Id$;

$$U_2 = \{(x, y) \mid x \neq 0\} \cup \{(m) \mid m \in \mathbb{C}\mathrm{a}\},$$

$$\varphi_2(x, y) = (x^{-1}, y \cdot x^{-1}) \quad \text{and} \quad \varphi_2(m) = (0, m);$$

$$U_3 = \{(x, y) \mid y \neq 0\} \cup \{(m) \mid m \in \mathbb{C}\mathrm{a}^*\} \cup \{(\infty)\},$$

$$\varphi_3(x, y) = (xy^{-1}, y^{-1}), \quad \varphi_3(m) = (m^{-1}, 0) \quad \text{and} \quad \varphi_3(\infty) = (0, 0).$$

It is easy to verify that $\varphi_i : U_i \to \mathbb{C}\mathrm{a}^2$ $(i = 1, 2, 3)$ is a homeomorphism. One needs only the alternative law and the norm properties. It is also clear that $\varphi_1 \circ \varphi_2^{-1}$, $\varphi_1 \circ \varphi_3^{-1}$ and $(\varphi_2 \circ \varphi_3^{-1})(u, v) = (vu^{-1}, u^{-1})$ define a diffeomorphism between open sets in $\mathbb{C}\mathrm{a}^2$ considered as $\mathbb{R}^{16}$. □

This atlas may also be used for $\mathbb{R}P^2$, $\mathbb{C}P^2$, $\mathbb{H}P^2$ when we work with the corresponding division algebras. Hence, for some imbedding $\mathbb{R} \subset \mathbb{C} \subset \mathbb{H} \subset \mathbb{C}\mathrm{a}$, one has inclusions $\mathbb{R}P^2 \subset \mathbb{C}P^2 \subset \mathbb{H}P^2 \subset \mathbb{C}\mathrm{a}P^2$ of submanifolds endowed with the induced topologies.

These definitions and manifold structures clearly coincide with the ones given in 3.2 and 3.3. It is also possible to fall back on the symmetric space structure starting from $\mathbb{K}P^2$ viewed as a Moufang plane, but it is a less intrinsic and more delicate study. In this direction notice that a very geometrical characterization of the

Cayley-Moufang plane as the exceptional CROSS $F_4/Spin\,9$ has been investigated by Tits. Some arguments in [TS] have, however, to be more detailed.

3.68. Basic ideas of Tits' study. In the Cayley plane P one associates to each point $a=(x,y)$ a line $p(a)$, called the *polar line* of a whose equation is

$$Y=-(\bar{y}^{-1}\cdot\bar{x})\,X-\bar{y}^{-1} \quad \text{if} \quad y\neq 0. \tag{3.69}$$

To complete this polarity one sets that $X=-\bar{x}^{-1}$ is the polar line of $(x,0)$, $x\neq 0$; $Y=-\bar{m}^{-1}X$ is the polar line of the point at infinity (m), $m\neq 0$; $Y=0$ is the polar line of (∞); $X=0$ is the polar line of (0); the infinite line D_∞ is the polar line of $(0,0)$.

The relation (3.69) is equivalent to:

$$\bar{y}Y+\bar{y}[(\bar{y}^{-1}\bar{x})X]+1=0,$$

or to

$$y=-(\bar{Y}^{-1}\cdot\bar{X})x-\bar{Y}^{-1}.$$

Hence, the relation $b\in p(a)$ is symmetric. The points a and b are called *conjugate* points, with respect to the Hermitian conic $|x|^2+|y|^2+1=0$.

Let G be the group of collineations which leave this conjugation invariant. This group acts transitively on P. The proof uses the fact that some algebraic, involutive correspondance without fixed point between x and X in $\mathbb{C}\mathrm{a}$ such that $x\to 0$ corresponds to $X\to\infty$, must have the following form: $\bar{X}x+k^2=0$ for some real number k.

The isotropy group H at $(0,0)$ is isomorphic to $Spin\,(9)$. In fact, each element u of H acts on D_∞ as an element of $SO(9)$ and this induced action determines u up to the symmetry with respect to $(0,0)$, which belongs to H.

Now G is a simple Lie group. This fact is obtained as a direct consequence of the fact that each non-trivial connected invariant subgroup of G must act transitively on P. Moreover, $\dim G=\dim P+\dim H=16+36=52$. A glance at E. Cartan's decomposition of simple Lie groups gives as the unique solution the compact real form of the exceptional group F_4.

It follows that *P is isomorphic to the exceptional* CROSS *$F_4/Spin\,9$ of E. Cartan* [CN 2].

3.70. In a natural way it is possible to carry out this construction for $P'\simeq \mathbb{H}P^2$. Then using the concept of a symmetric subspace as in 3.25 one gets an hereditary property similar to those given by Theorem 3.25 and Corollary 3.26. It follows that $\mathbb{C}\mathrm{a}\,P^2$ is a C_π-manifold, and that theorems corresponding to 3.30, 3.34, and 3.35 hold for $\mathbb{C}\mathrm{a}\,P^2$ (see [WF 2], [BN]).

3.71. Moreover, the cellular decomposition of $\mathbb{C}\mathrm{a}P^2$ (see 3.38) is given by

$$\mathbb{C}\mathrm{a}P^2=C^0\cup C^8\cup C^{16}.$$

This is clear from the very definition of $\mathbb{C}aP^2$ as a Moufang plane. It is interesting to ask whether the fundamental form ω defined in [B-G] furnishes the $\mathbb{Z}$-cohomology ring of $\mathbb{C}aP^2$ as in 3.51.

We mention that the identification between $\mathbb{C}aP^2$ and the CROSS $F_4/Spin\ 9$ seems to have been given first in 1950 by A. Borel [BL 3] following E. Cartan and G. Hirsch [HH]. Here each line in $F_4/Spin\,9$ is the conjugate locus of some point, and hence is an eight sphere. Then one can prove the incidence axioms of a plane geometry. Though we know of the existence of such non classical geometries, the topological restriction (3.71) forces a homogeneous space to be homeomorphic to $\mathbb{C}aP^2$.

3.72. Lastly we mention the very important and independent Freudenthal-Jordan approach to $\mathbb{C}aP^2$. This study is more algebraic and more symmetric than 3.68 because one does not distinguish particular points (the line at infinity). It uses the following algebraic characterization of the compact exceptional group F_4. Let $\mathcal{M}_3^+$ be the set of (3.3) Hermitian matrices whose entries are Cayley numbers, endowed with the symmetric product

$$A \circ B = \tfrac{1}{2}(A \cdot B + B \cdot A).$$

The space $\mathcal{M}_3^+$ has the structure of a Jordan algebra (i.e., a commutative algebra in which associativity is replaced by the weaker property $(A^2 \circ B) \circ A = A^2 \circ (B \circ A)$ for each A, B; see [AT] p. 144).

3.73 Theorem (Chevalley-Schafer) [C-S]. *The group of automorphisms of $\mathcal{M}_3^+$ is isomorphic to F_4.*

This theorem is also proved in the basic paper by Freudenthal [FL 1] using differential techniques. Freudenthal takes particular idempotent matrices in $\mathcal{M}_3^+$ as points and lines to define the Cayley plane $\mathbb{C}aP^2$. The latter is then proved to be isometric to $F_4/Spin\,9$. See also [SP] and [FL 2].

3.74. This algebraic approach has been widely generalized to other Cayley algebras. In fact, the corresponding octave planes are not C^∞-manifolds. See [FR] for references.

Concerning the non-compact dual space of $\mathbb{C}aP^2$, see [S-V], [TS], [MW] and [BN].

Chapter 4. Some Examples of C- and P-Manifolds: Zoll and Tannery Surfaces

A. Introduction

4.1. The main purpose of this chapter is to study Zoll's Surfaces, and to point out some facts and examples related to them.

Except in Section E, which indicates how to construct examples in higher dimensions, we restrict ourselves to 2-dimensional manifolds. However, in view of examples of embedded surfaces, which may be non regular, such as Tannery's Pear and Gambier's examples (see 4.27 and 4.29), we did not restrict ourselves to $C_{2\pi}$-manifolds (see 7.8). Unfortunately, the Riemannian metrics which we consider as P_l-metrics (see 7.8) are not necessarily C^1 Riemannian metrics. The reader who is only concerned with smooth metrics may skip differentiability conditions which we often detail in Section B. Otherwise he will notice (thinking in terms of Riemannian metrics g on S^2, and also in terms of the meridian z of an embedded surface of revolution) that:

4.2. to say that g is degenerate at m means that z is singular at m (non-differentiable on the left and on the right), see Figure 4.2(a); to say that g is nondegenerate at m means that z has left and right derivatives at m, see Figure 4.2(b); to say that g is continuous at m means that z is C^1 at m, see Figure 4.2(c).

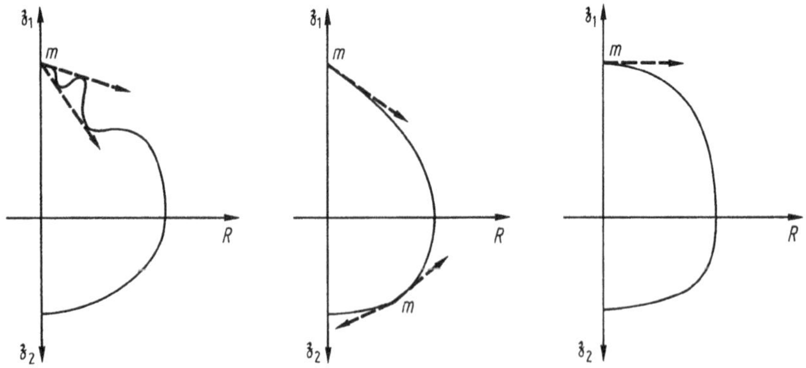

a b c

Fig. 4.2a–c

In Section B we give a characterization of P_l-surfaces of revolution. Notice that the geodesics are not as regular as we wright expect, because they cannot be C^1 geodesics at precisely those points of M where g is not continuous. However, we still call them P_l-manifolds, having in mind that the geodesics are as regular as they can be, in view of the regularity of the manifold.

In Section C we give a characterization of embedded Zoll surfaces (whose shadow line is always C^1) and of embedded Tannery surfaces. We give some examples of meridians.

Then in Section D we show that the non-trivial Zoll surfaces are not Blaschke manifolds, by taking an example for which we compute the middle, conjugate and cut loci of a point m of M, and draw their pictures in T_mM and in M (see 4.37 and 4.38).

In Section F we present A. Weinstein's result to the effect that, if g_0 and g_1 are two Zoll metrics on S^2, then their geodesic flows are conjugate under a symplectic diffeomorphism (see also Chapter VIII to learn how A. Weinstein uses this result to get information about the spectrum of Zoll surfaces).

In Section H we give a sketch of how V. Guillemin proves the existence of many Zoll metrics near the standard one, a fact which was established at a formal level by Funk some sixty years ago. V. Guillemin uses powerful tools in analysis: the Radon transform (see G), the Nash-Moser implicit function theorem and the theory of Fourier Integral Operators.

B. Characterization of P-Metrics of Revolution on S^2

4.3. Particularizing our study to a 2-dimensional manifold M, we first notice that M is diffeomorphic to S^2, or $\mathbb{R}P^2$, since $\pi_1(M)$ is finite (see 7.30).

We assume that M is diffeomorphic to S^2 in view of L. W. Green's Theorem (cf. [GN 2], [BR 3] p. 292 and 5.59).

We do not study all of the P_l and C_l metrics on S^2 but only those having S^1 as an effective isometry group (which can be called *metrics of revolution*). H. Hopf's theorem tells us that $\chi(M)$ can be written as the sum of the indices at the zeroes of any vector field with isolated singularities. An infinitesimal isometry on S^2 has necessarily isolated zeroes and the index at any of its singularities is equal to 1. So we have exactly two fixed points, say N and S (the so-called *North* and *South poles*). We denote by $\theta \mapsto R_\theta$ the S^1 action.

The fact that S^1 acts as an isometry group implies that the geodesic flow is integrable and so the equations of geodesics are explicitely solvable. Two first integrals suffice to compute everything explicitely. We always have the Energy as First Integral. Here there is a second one: Clairaut's First Integral (see 4.7). [There necessarily exists a second one as soon as there exists a continuous group of isometries, according to Souriau's theory of moments (see [SU] p. 105).]

4.4. We want to describe conditions for (M, g) to be a P_l-manifold. For this purpose, we shall choose nice parametrizations. Let γ_0 be a geodesic from N to S, with length $\varrho_{\gamma_0}(N, S) = L$. The isometries R_θ transform a geodesic into a geodesic. Set $U = M \setminus \{N, S\}$. For every m in U there exists a unique θ in S^1, such that m belongs to $R_\theta(\gamma_0) = \gamma_\theta$. As a matter of fact, we will usually choose θ from the interval $[0, 2\pi[$.

We define $u=\varrho_{\gamma_\theta}(m, N)$ and denote by (u, θ) the point $m_{(u,\theta)}$. We also need two other charts, namely the exponential charts centered at the fixed points:

$$U_N=\{N\}\cup\left\{(u,\theta)\in U \,\middle|\, u<\frac{L}{2}\right\}=B\left(N,\frac{L}{2}\right)$$

$$N\mapsto(0,0)$$

$$(u,\theta)\mapsto(x=u\cos\theta,\ y=u\sin\theta)$$

and

$$U_S=\{S\}\cup\left\{(u,\theta)\in U \,\middle|\, u>\frac{L}{2}\right\}=B\left(S,\frac{L}{2}\right)$$

$$S\mapsto(0,0)$$

$$(u,\theta)\mapsto(x=(L-u)\cos\theta,\ y=(L-u)\sin\theta).$$

4.5. Now $\{U, U_N, U_S\}$ together with the associated coordinate functions is a parametrization of M.

Thanks to the fact that the curves γ_θ are geodesics parametrized by arc length (together with Gauss lemma 1.96) g may be written on the cylinder U as

$$g=du^2+a^2(u)d\theta^2,$$

for some function a, $a:]0,L[\to\mathbb{R}^+$. The order of differentiability of g in U is exactly the order of differentiability of a in $]0,L[$.

4.6 Proposition. *Let M be diffeomorphic to S^2, let N and S be two points of M, and take the parametrization $\{U, U_N, U_S\}$ as defined above. Let g be a Riemannian metric on U, defined as $g=du^2+a^2(u)d\theta^2$ for some function $a:]0,L[\to\mathbb{R}^+$.*

i) *The metric g extends to a Riemannian metric on M if and only if a extends to a C^0 function $a:[0,L]\to\mathbb{R}^+$, which is differentiable at 0 and L and satisfies the conditions $a(0)=a(L)=0$ and $a'(0)\neq 0$, $a'(L)\neq 0$.*

ii) *The metric g extends to a C^k Riemannian metric on M if and only if a extends to a C^k function $a:[0,L]\to\mathbb{R}^+$, which is $(k+1)$-differentiable at 0 and L and satisfies*

$$a(0)=a(L)=0,\qquad a'(0)=1,\qquad a'(L)=-1$$

and $a^{(2p)}(0)=a^{(2p)}(L)=0$ for every integer p such that $2p\leqslant k+1$.

Proof. We just have to check the change of charts on $U\cap U_N$ and $U\cap U_S$.

The fact that g is nondegenerate at N is equivalent to the fact that $\dfrac{a^2(u)}{u^2}$ has a non-zero limit when u tends to 0. This proves i). The other conditions are obtained by defining $a^2:\mathbb{R}\to\mathbb{R}^+$ $(a^2(u+2L)=a^2(u)=a^2(-u))$ as a direct consequence of (for instance) Proposition 2.7 of [K-W 1]: "Let f be a C^∞ function of $\mathbb{R}^2$. Then a necessary and sufficient condition in order that there exists a C^∞ function $\tilde{f}$ on $\mathbb{R}^2$ such that

$$f(r,\theta)=\tilde{f}(r\cos\theta, r\sin\theta)$$

is

i) $f(-r,\theta)=f(r,\theta+\pi)$ for all r,θ

ii) $r^k\left(\frac{\partial^k f}{\partial r^k}\right)_{(0,\theta)}$ is a homogeneous polynomial of degree k in $x=r\cos\theta$ and $y=r\sin\theta$ for every integer $k\geqslant 0$." □

4.7 Proposition (Clairaut's First Integral). *Let S^2 be the two-sphere, equipped with the parametrization $\{U, U_N, U_S\}$ and the coordinate functions (u,θ) on U. Let g be a C^1 metric of revolution on U written as*

$$g=du^2+a^2(u)d\theta^2$$

for some function a from $]0,L[$ to $\mathbb{R}^+$. Then

i) the meridians are geodesics through N and S and any other geodesic is entirely contained in U;

ii) for each geodesic γ which is not a meridian, there exist real numbers u_1 and u_2 in the interval $]0,L[$ with $a(u_1)=a(u_2)$ such that γ is entirely contained between the parallels $(u=u_1)$ and $(u=u_2)$ and, along γ, the following relation is satisfied (ε_1 is ± 1 according to the orientation of γ)

$$\varepsilon_1\frac{d\theta}{ds}=\frac{a(u_1)}{a^2(u)}. \tag{4.7}$$

Proof. By 4.4 the γ_θ are geodesics through N and S, and by local uniqueness of geodesics, every geodesic through N or S is in fact a γ_θ. The curve γ_θ followed by $\gamma_{\theta+\pi}$ gives a simply closed geodesic of length $2L$ (possibly only piecewise C^1).

In order to prove ii) we first compute the equations of geodesics in U. The Christoffel coefficients (see 1.51) are

$$\Gamma^\theta_{u\theta}=\frac{a'(u)}{a(u)} \quad\text{and}\quad \Gamma^u_{\theta\theta}=-a'(u)a(u).$$

The other Christoffel coefficients are zero. A geodesic $\gamma: t\mapsto(u(t),\theta(t))$ in U satisfies [see (1.51)]

$$\begin{cases}\dfrac{d^2u}{dt^2}-a'(u)a(u)\left(\dfrac{d\theta}{dt}\right)^2=0\\[2ex] \dfrac{d^2\theta}{dt^2}+2\dfrac{a'(u)}{a(u)}\dfrac{du}{dt}\dfrac{d\theta}{dt}=0.\end{cases} \tag{4.8}$$

The second equation gives Clairaut's First Integral

$$a^2(u)\frac{d\theta}{dt}=c \quad (c \text{ a constant}) \quad \text{along } \gamma.$$

Let γ be parametrized by arc length. Then

$$1=\left(\frac{du}{ds}\right)^2+a^2(u)\left(\frac{d\theta}{ds}\right)^2,$$

so that

$$a^2(u)\left(\frac{d\theta}{ds}\right)^2\leqslant 1$$

and

$$\varepsilon_1 a^2(u)\frac{d\theta}{ds}=a^2(u)\left|\frac{d\theta}{ds}\right|=|c|\leqslant a(u).$$

If $|c|$ equals the maximum of a on $[0, L]$, then γ is a parallel curve ($u=Cst$). Otherwise, $\frac{du}{d\theta}=Cst\neq 0$, and γ would go through N or S.

Hence, $|c|<\sup\{a(u),\ u\in]0,L[\}$ and there exists a maximal interval $[u_1,u_2]\subset]0,L[$ such that $a(u_1)=a(u_2)=|c|$.

Then, along the geodesic γ, we have

$$u_1\leqslant u\leqslant u_2. \quad \square$$

Let us now deal with other coordinate functions on U, using the fact that M is a P_l-manifold. We suppose that a is piecewise C^2 on $]0,L[$.

4.9 Lemma. *The fact that all geodesics are closed, with a common period l, implies that there exists a unique u (say u_0) with $a'(u_0)=0$ and $a(u_0)=1$. Then $a''(u_0)<0$ (or left and right second derivatives are negative). The parallel $u=u_0$ is called the* equator *of M. Its length is $2\pi a(u_0)$. We normalize by taking $a(u_0)=1$.*

Proof. Using (4.8), we notice that the parallels ($u=C_{st}$) are not geodesics unless $a'(u)=0$. The length of a parallel is $2\pi a(u)$. Let u_0 be such that $a'(u_0)=0$. Let γ be the geodesic ($u=u_0$) and let J be the nonzero Jacobi field (see 1.94) along γ which is zero at the point $(u_0,0)$ of γ (notice that $J(l)=0$). The sectional curvature (see 1.86) at any point of U depends only on u, and we have

$$\sigma(u_0)=-\frac{a''(u_0)}{a(u_0)}.$$

The Jacobi field J is a solution of the differential equation

$$J''(s) - \frac{a''(u_0)}{a(u_0)} J(s) = 0.$$

This, together with the properties of J, implies that $a''(u_0) < 0$.

This shows that any point u at which $a'(u) = 0$ is a local maximum of a so that there can be only one such u.

Notice that it suffices that the left and right second order derivatives of a at u_0 be negative. □

4.10 Proposition. *Let g be a P_l-metric of revolution on S^2 equipped with the parametrization (U, U_N, U_S) and the coordinate functions (u, θ) on U, so that g may be written on U as $g = du^2 + a^2(u)d\theta^2$ for some function a from $]0, L[$ to $]0, 1]$. We suppose that a is C^1, so that g is C^1 except perhaps at N and S.*

It is possible to define a new system of coordinates (r, θ) on U by setting $a(u) = \sin r$. Then g may be written on U as

4.10 $$g = [f(\cos r)]^2 dr^2 + \sin^2 r d\theta^2,$$

where f is a function from $]-1, 1[$ to $\mathbb{R}^+$. The function f extends to a function from $[-1, 1]$ to $\mathbb{R}^+$; g is a C^k metric at N (resp. S) *if and only if f is C^k on $[-1, 1]$ and satisfies $f(1) = 1$* [resp. $f(-1) = 1$].

Proof. Thanks to 4.6 and since g is a metric on S^2, a extends to $[0, L]$ with $a(0) = a(L) = 0$. Using Lemma 4.9, we know that $\sin r$ and $a(u)$ have similar variations:

r	$0 \quad \frac{\pi}{2} \quad \pi$
$\cos r$	$1 \; + \; 0 \; - \; -1$
$\sin r$	$0 \nearrow 1 \searrow 0$

u	$0 \quad u_0 \quad L$
$a'(u)$	$>0 \; + \; 0 \; - \; <0$
$a(u)$	$0 \nearrow 1 \searrow 0$

We define b from $[0, L]$ to $[0, \pi]$ as

$$b(u) = \begin{cases} \operatorname{Arc\,sin} a(u) & u \in [0, u_0] \\ \pi - \operatorname{Arc\,sin} a(u) & u \in [u_0, L]. \end{cases}$$

We define c from $[-1, +1]$ to $[0, L]$ as

$$c(v) = \begin{cases} (a \upharpoonright [0, u_0])^{-1}(\sqrt{1 - v^2}) & v \in [0, 1] \\ (a \upharpoonright [u_0, L])^{-1}(\sqrt{1 - v^2}) & v \in [-1, 0]. \end{cases}$$

The functions b and c are piecewise differentiable and have the same order of differentiability as a. We have $b(u)=r$, $c(\cos r)=b^{-1}(r)=u$, and $a(u)=a[c(\cos r)]=\sin r$. Hence, we may write

$$a'[c(\cos r)]du=\cos r\,dr \quad \text{or} \quad b'(u)du=dr.$$

We define a function f from $]-1,+1[$ to $\mathbb{R}$ as

$$\begin{cases} f(v)=\dfrac{v}{a'[c(v)]} & \text{if} \quad v\neq 0 \\ f(0)=f_0. \end{cases}$$

Then g may be written on U as

$$g=[f(\cos r)]^2dr^2+\sin^2 r\,d\theta^2.$$

In fact, thanks to Part i) in Proposition 4.6, f is a function from $[-1,+1]$ to $\mathbb{R}^+$, with

$$f(1)=\frac{1}{a'(0)} \quad \text{and} \quad f(-1)=-\frac{1}{a'(L)}.$$

If $a''(u_0)$ exists, then $b'(u_0)$ exists and $b'(u_0)=\sqrt{-a''(u_0)}$. It follows that f is continuous at 0 if we choose $f_0=\dfrac{1}{\sqrt{-a''(u_0)}}$.

The conditions $a'(0)=1$ and $a'(L)=-1$ are replaced by $f(1)=1$ and $f(-1)=1$. The other conditions (namely $a^{(2p)}(0)=0$ and $a^{(2p)}(L)=0$) are also replaced by $f(1)=1$, and $f(-1)=1$, as one can see by applying the chain rule to the expression

$$a'(u)=\sqrt{1-a^2(u)}\,f(\sqrt{1-a^2(u)}). \quad \square$$

4.11 Theorem (Darboux—see [DX], p. 6). *Let (S^2,g) be a surface of revolution equipped with the parametrization (U,U_N,U_S) and the coordinate functions (r,θ) on U, such that g may be written as in* (4.10).

A necessary and sufficient condition in order that all geodesics of g be closed is that for every i one has

4.11 $$\int_i^{\pi-i}\frac{\sin i\cdot f(\cos r)}{\sin r(\sin^2 r-\sin^2 i)^{1/2}}dr=\frac{p}{q}\pi$$

where p and q are integers.

Then, apart from the equator, which is simply closed and of length 2π, every geodesic γ in U consists of $2q$ geodesic segments between two consecutive points of contact with the parallels $(r=i)$ and $(r=\pi-i)$. Its length is $2qL$, and it turns p times.

Compare with Figure 4.12.

Proof. The equator has the equations

$$r=\tfrac{\pi}{2} \quad \text{and} \quad s=(\theta-\theta_0).$$

For any other geodesic γ in U, we rewrite formula (4.7) using r and θ. Clairaut's First Integral is $\sin^2 r \dfrac{d\theta}{ds}=c$, and we may set $|c|=\sin i$ for an i in $]0, \tfrac{\pi}{2}[$.

The geodesic γ is contained between the parallels $(r=i)$ $(r=\pi-i)$, and is described by the following formulas

4.12
$$\begin{cases} \dfrac{d\theta}{ds}=\varepsilon_1 \dfrac{\sin i}{\sin^2 r} \\[2ex] \dfrac{d\theta}{dr}=\varepsilon_1 \varepsilon \dfrac{\sin i \cdot f(\cos r)}{\sin r(\sin^2 r-\sin^2 i)^{1/2}} \\[2ex] \dfrac{ds}{dr}=\varepsilon \dfrac{\sin r \cdot f(\cos r)}{(\sin^2 r-\sin^2 i)^{1/2}}. \end{cases}$$

The sign number ε_1 determines the orientation of the geodesic.

The sign of $\varepsilon=\pm 1$ is fixed on a segment of geodesic going from $r=i$ to $r=\pi-i$, and it changes whenever r equals i or $\pi-i$.

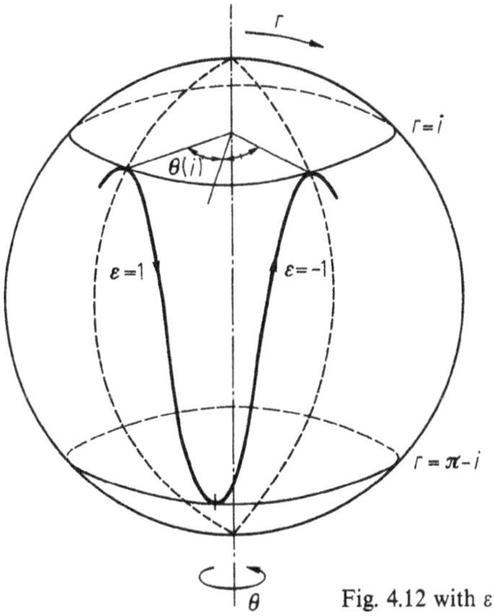

Fig. 4.12 with $\varepsilon_1=1$

The angle $\theta(i)$ between two consecutive points of contact with the extreme parallels is

$$\theta(i)=\varepsilon_1 \varepsilon \int_i^{\pi-i} \frac{\sin i \cdot f(\cos r)}{\sin r(\sin^2 r-\sin^2 i)^{1/2}}\, dr.$$

A geodesic γ_i is closed if and only if

$$\varepsilon_1\varepsilon\theta(i)=\int_i^{\pi-i}\frac{\sin i\cdot f(\cos r)}{\sin r(\sin^2 r-\sin^2 i)^{1/2}}\,dr=\frac{p(i)}{q(i)}\pi$$

for some rational number $\frac{p(i)}{q(i)}$.

The geodesic γ_i is then closed at the angle $2p(i)\pi$, and its length is $2q(i)\varepsilon s(i)$, where

$$\begin{aligned}\varepsilon s(i)&=\int_i^{\pi-i}\frac{\sin r\cdot f(\cos r)}{(\sin^2 r-\sin^2 i)^{1/2}}\,dr\\&=\varrho_{\gamma_i}\left[(i,\theta_0),\left(\pi-i,\theta_0+\frac{p(i)}{q(i)}\pi\right)\right].\end{aligned}$$

If all geodesics are closed, then $\varepsilon_1\varepsilon\theta$ is a continuous function from an interval of $\mathbb{R}$ ($]0,\frac{\pi}{2}[$) to $\mathbb{R}$. Its range is contained in $\mathbb{Q}\pi$, so it has to be a constant, say $\frac{p(i)}{q(i)}=\frac{p}{q}$ in $\mathbb{Q}$.

The function $\varepsilon s(i)=\varrho_{\gamma_i}\left[(i,\theta_0),\left(\pi-i,\theta_0+\frac{p}{q}\pi\right)\right]$ is also a constant, and, when i tends to 0 it admits $\varrho(N,S)=L$ as a limit. Hence, $\varepsilon s(i)=L$. □

4.13 Theorem (Characterization of P_l-surfaces of revolution, so-called Tannery surfaces).

i) *The metric g is a $P_{2p\pi}$-metric of revolution on S^2 if and only if g may be written in the parametrization $[U,(r,\theta)]$ as*

$$g=\left[\frac{p}{q}+h(\cos r)\right]^2 dr^2+\sin^2 r\,d\theta^2,$$

where $\frac{p}{q}$ is rational, and h is a function from $]-1,+1[$ to $\left]-\frac{p}{q},+\frac{p}{q}\right[$ which is odd on $]-1,+1[-\{0\}$.

Then the length of the equator is 2π, the length of any meridian is $2\frac{p}{q}\pi$ and the length of any other geodesic is $2p\pi$.

ii) *The metric g is nondegenerate at N* (resp. S) *if and only if h has a left limit at* 1 (resp. *a right limit at* -1).

Moreover, g is C^k at N (resp. S) *if and only if h extends to a C^{k-1} function in a neighbourhood of* 1 (resp. -1) *and* $h(1)=1-\frac{p}{q}$ $\left(\text{resp. } h(-1)=1-\frac{p}{q}\right)$.

Finally, g is C^k at the equator if and only if h is C^k in a neighbourhood of 0 *(and then $h(0)=0$).*

Proof. We know by Proposition 4.10 that if g is a P_l-metric of revolution on S^2, then g may be written as $g=[f(\cos r)]^2dr^2+\sin^2 r\,d\theta^2$.

The condition (4.11) in Darboux Theorem is satisfied for (S^2, can) with

$$\int_i^{\pi-i} \frac{\sin i}{\sin r(\sin^2 r - \sin^2 i)^{1/2}}\, dr = \pi$$

for any i in $]0, \frac{\pi}{2}]$.

Let us define a function h from $]-1, +1[$ to $\mathbb{R}$ by $f(v) = \frac{p}{q} + h(v)$.

Condition (4.11) is satisfied for (S^2, g) if and only if

$$\int_i^{\pi-i} \frac{\sin i \cdot h(\cos r)}{\sin r(\sin^2 r - \sin^2 i)^{1/2}}\, dr = 0$$

for any i in $]0, \frac{\pi}{2}]$.

We prove now that this is equivalent to saying that h is an odd function on $]-1, +1[-\{0\}$.

Let $h^e(v) = \dfrac{h(v) + h(-v)}{2}$ be the even part of h.

The function h is odd if and only if h^e equals 0. Write

$$\int_i^{\pi-i} \frac{\sin i \cdot h(\cos r)}{\sin r(\sin^2 r - \sin^2 i)^{1/2}}\, dr = 2 \sin i \cdot H(i),$$

$$H(i) = \int_i^{\frac{\pi}{2}} \frac{h^e(\cos r)}{\sin r(\sin^2 r - \sin^2 i)^{1/2}}\, dr.$$

Then clearly h^e equals 0 implies that $H(i) = 0$.

4.14. For the converse, we follow an idea of D. Singer and H. Gluck (see [G-S 1]). For a in $]0, \frac{\pi}{2}]$, let us define

$$I(a) = \int_a^{\frac{\pi}{2}} \frac{\sin i \cos t \cdot H(i)}{(\sin^2 i - \sin^2 a)^{1/2}}\, di.$$

Set

$$T_a = \{(r, i) \mid r \in [a, \tfrac{\pi}{2}], i \in [a, \tfrac{\pi}{2}], i \leqslant r\}.$$

The function

$$\frac{1}{(\sin^2 r - \sin^2 i)^{1/2}(\sin^2 i - \sin^2 a)^{1/2}}$$

is Lebesgue-integrable on T_a, and by Fubini's Theorem,

$$I(a)=\int_a^{\frac{\pi}{2}}\frac{h^e(\cos r)}{\sin r}\left(\int_a^r\frac{\sin i\cos i}{(\sin^2 r-\sin^2 i)^{1/2}(\sin^2 i-\sin^2 a)^{1/2}}\,di\right)dr$$
$$=\int_a^{\frac{\pi}{2}}\frac{h^e(\cos r)}{\sin r}\left(\int_0^{\infty}\frac{dx}{1+x^2}\right)dr,$$

where we define x as $x=(\sin^2 i-\sin^2 a)^{1/2}\,(\sin^2 r-\sin^2 i)^{-1/2}$ for a in $]0,\frac{\pi}{2}]$.

Therefore, $I(a)=\frac{\pi}{2}\int_a^{\frac{\pi}{2}}\frac{h^e(\cos r)}{\sin r}\,dr.$

Then the fact that $H(i)$ vanishes for all i in $]0,\frac{\pi}{2}]$ implies that $I(a)=0$ for all a in $]0,\frac{\pi}{2}]$, so that $h^e=0$.

Following Formula (4.10), we have

$$\frac{p}{q}+h(\cos r)>0.$$

This, together with the fact that h is an odd function, implies that the range of h is contained in the interval $\left]-\frac{p}{q},\frac{p}{q}\right[$.

We can compute the length of a meridian, which is

$$2\int_0^{\pi}\left[\frac{p}{q}+h(\cos r)\right]dr=2L.$$

The fact that h is odd implies that $L=\frac{p}{q}\pi$. Hence, $2qL$, which is the length of a geodesic (see 4.11) equals $2p\pi$.

Part ii) of the theorem is clear if we write the conditions of Proposition 4.6 and Proposition 4.10 in terms of h. □

4.15 Remarks. A continuous $P_{2p\pi}$-metric of revolution on S^2 satisfies $p=q=1$. Therefore, it is an $SC_{2\pi}$-metric.

For any function $h:[0,1]\mapsto\left]-\frac{p}{q},\frac{p}{q}\right[$, we may construct a $P_{2p\pi}$-metric on S^2 which is continuous at S. It looks like Tannery's-pear (see [TY], or [GR] and 4.27 for the classical example $p=2$, $q=1$ and $h(v)=v$).

We will call g a *Zoll metric* on S^2 if g is continuous and if (S^2,g) is a $C_{2\pi}$-surface.

4.16 Corollary. (Characterization of Zoll surfaces of revolution). *The metric g is a Zoll metric of revolution on S^2 if and only if g can be written in the parametrization $[U,(r,\theta)]$ as $g=[1+h(\cos r)]^2dr^2+\sin^2 r\,d\theta^2$, where h is an odd function from $[-1,+1]$ to $]-1,+1[$, whose value at 1 is 0. Furthermore, (S^2,g) is a $SC_{2\pi}$-manifold.*

Remark. Corollary 4.16 gives a proof of L. Green's Theorem 5.59 on the rigidity of $\mathbb{R}P^2$ as a C_π-manifold, in the *particular case* where $(\mathbb{R}P^2, g)$ admits S^1 as a group of isometries.

4.17. There is a wide choice of functions h. In particular, notice that there is no condition on the derivatives of h, i.e., on the sectional curvature of the P-surface constructed from h. Recall that the expression of the sectional curvature is

$$\sigma(r) = \frac{1}{\left[\frac{p}{q} + h(\cos r)\right]^3}\left[\frac{p}{q} + h(\cos r) - \cos r \cdot h'(\cos r)\right]. \tag{4.17}$$

For example, if we take $h(\cos r) = \cos r \sin(2k+1)r$, then the $P_{2\pi}$-surface associated to h is a C^∞ Zoll surface, and σ can be made less than any given negative number by a suitable choice of k.

C. Tannery Surfaces and Zoll Surfaces Isometrically Embedded in ($\mathbb{R}^3$, can)

Let us try to find a C^1-embedding of (S^2, g) in ($\mathbb{R}^3$, can) (where g is a P_l-metric of revolution)

$$\begin{cases} x = R(r)\cos\theta \\ y = R(r)\sin\theta \\ z = z(r), \end{cases}$$

so that $g = dR^2 + R^2 d\theta^2 + z'^2(r)dr^2$, $R = \sin r$ and $z'^2(r) = \left[\frac{p}{q} + h(\cos r)\right]^2 - \cos^2 r$.

4.18 Proposition. *A P_l-surface of revolution (Tannery Surface) may be isometrically C^1-embedded in* ($\mathbb{R}^3$, can) *if and only if*

$$\frac{p}{q} + h(\cos r) \geqslant |\cos r|. \tag{4.18}$$

Proof. This is just the condition that $z'^2(r) \geqslant 0$, see 4.30. □

Compare with Figures 4.18 to understand the embedding condition on h. We therefore have $\frac{p}{q} \geqslant 1$.

Remark. It follows from Figure 4.18 (b) that if a $C_{2\pi}$-manifold is embedded and non singular, at N (or S), then it is always continuous at N and S.

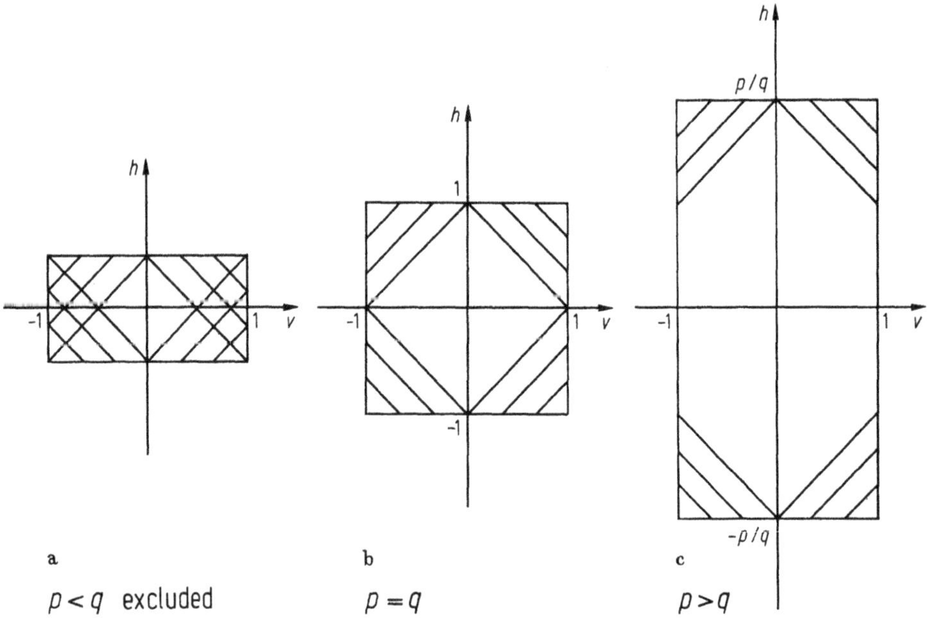

Fig. 4.18a–c

4.19 Remark. The following theorem is a deep one: "every Riemannian metric on S^2 with positive sectional curvature is induced by an embedding in $\mathbb{R}^3$" (the problem of H. Weyl, see [SK 2] p. 226 for details and references).

In our special case it is very simple to check that if the sectional curvature of a Tannery metric (or a Zoll metric) is positive, then necessarily Condition (4.18) is satisfied. In fact,

$$\frac{d}{dr}\left[\frac{\frac{p}{q}+h(\cos r)}{\cos r}\right]^2 = 2\,\frac{\left[\frac{p}{q}+h(\cos r)\right]^4}{\cos^3 r}\sin r\cdot\sigma(r)$$

[see (4.17)]. Hence, $\sigma(r)>0$ implies the following behaviour:

r	0		$\frac{\pi}{2}$		π
$\frac{d}{dr}\left[\frac{\frac{p}{q}+h(\cos r)}{\cos r}\right]^2$	$+$				$-$
$\left[\frac{\frac{p}{q}+h(\cos r)}{\cos r}\right]^2$	$\frac{p}{q}+h(o)$	$\nearrow$	$+\infty \mid +\infty$	$\searrow$	$\frac{p}{q}-h(0)$

with $h(0)$ in $\left]-\frac{p}{q},+\frac{p}{q}\right[$.

However, the converse is not true and there exist Zoll surfaces isometrically embedded in ($\mathbb{R}^3$, can) whose sectional curvature is not everywhere positive (see 4.25).

We now construct the embedding, and describe the embedded Tannery and Zoll Surfaces of revolution. Under the Condition (4.18), we have

$$z(r)=z_0-\int_0^r \sqrt{\left[\frac{p}{q}+h(\cos\varrho)\right]^2-\cos^2\varrho}\, d\varrho$$

It is more convenient to write z as a function of $R\,(=\sin r)$ to obtain a meridian of the surface. We define z_1 and z_2 from $[0,1]$ to $\mathbb{R}^+$ by

$$z_1(R)=z(\operatorname{Arc}\sin R)$$
$$z_2(R)=-z(\pi-\operatorname{Arc}\sin R).$$

Hence, z_1 and z_2 are two decreasing functions with $z_i(1)=0$.

4.20 Proposition. *Let (S^2,g) be a $P_{2p\pi}$-surface of revolution, so that g may be written on U as*

$$g=\left[\frac{p}{q}+h(\cos r)\right]^2 dr^2+\sin^2 r\, d\theta^2,$$

where h satisfies the Condition (4.18). *Then:*

i) *(S^2,g) is embedded in ($\mathbb{R}^3$, can) as a surface of revolution whose meridian curve is described by*

$$z_1(R)=\int_R^1 \sqrt{\frac{\left[\frac{p}{q}+h(\sqrt{1-u^2})\right]^2}{1-u^2}-1}\, du$$

and

$$z_2(R)=\int_R^1 \sqrt{\frac{\left[\frac{p}{q}-h(\sqrt{1-u^2})\right]^2}{1-u^2}-1}\, du\,;$$

ii) *if g is a nondegenerate metric at N (*resp. *S), then z_1 (*resp. *z_2) is differentiable at $z_1(0)$ [*resp. *$z_2(0)$] with*

$$z_1'(0)=-\sqrt{\left[\frac{p}{q}+h(1)\right]^2-1}$$

$$\left(\text{resp.}\quad z_2'(0)=-\sqrt{\left[\frac{p}{q}-h(1)\right]^2-1}\right).$$

One can compare this result with Figure 4.2(*b*).

*If g is a C^k-metric at N (*resp. *S), then z_1 (*resp. *z_2) is a C^k-function at* 0 [resp. 0] *and $z_1'(0)=0$* [resp. $z_2'(0)=0$]. *One can compare this with Figure* 4.2(c).

If g is a C^k-metric at the equator, then z_1 and z_2 are C^{k+1}-functions in a neighbourhood of 1, *and the two pieces of meridian curve have the same curvature at* 1 $\left(\sigma\left(\frac{\pi}{2}\right)=\frac{q^2}{p^2}\right)$.

Proof. This result is a direct consequence of the definition of z_1 and z_2, together with the conditions on h in Theorem 4.13. The standard computation for curves gives the curvature at 1 of z_1 as $\dfrac{1}{\left[\dfrac{p}{q}+h(0)\right]^2}$ and the curvature at 1 of z_2 as $\dfrac{1}{\left[\dfrac{p}{q}-h(0)\right]^2}$. □

4.21 Remark 1. The Condition (4.18) implies that

$$z_1(R)\leqslant 2\sqrt{\frac{p}{q}}\int_R^1\sqrt{\frac{\frac{p}{q}-\sqrt{1-u^2}}{1-u^2}}\,du=2\sqrt{\frac{p}{q}}\int_{\operatorname{Arc\,sin}R}^{\frac{\pi}{2}}\sqrt{\frac{p}{q}-\cos\varrho}\,d\varrho$$

and the same condition holds for z_2.

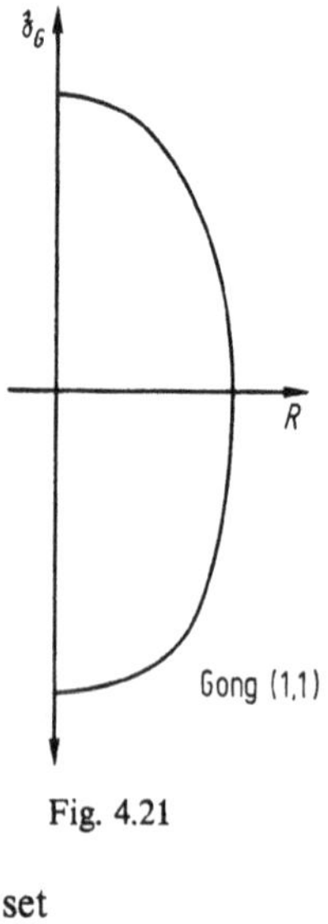

Fig. 4.21

We will set

$$z_G(R)=2\sqrt{\frac{p}{q}}\int_{\operatorname{Arc\,sin}R}^{\frac{\pi}{2}}\sqrt{\frac{p}{q}-\cos\varrho}\,d\varrho\,.$$

The surface defined by the meridian curve z_G and its symmetric with respect to the R axis is called the *Gong* (p,q) (see [GR] p. 96).

In the case of the Zoll surfaces, $p=q=1$ and $z_G(R)=4\left(\sqrt{1+\sqrt{1-R^2}}-1\right)$.

The Gong (1.1) is an algebraic surface of degree 4 (compare with Fig. 4.21).

4.22 Remark 2. In [BR 3] p. 150, in [GR] and in [QC], the construction of a meridian is given via the relation

$$\sqrt{1+z_1'^2(R)} + \sqrt{1+z_2'^2(R)} = \frac{2}{\sqrt{1-R^2}}.$$

These authors follow [ZL].

A function λ from $[0,1]$ to $\mathbb{R}^+$ is introduced. It is a solution of the equation

$$\sqrt{1+z_1'^2(R)} = \frac{1}{\sqrt{1-R^2}} + \lambda(R).$$

Notice that λ satisfies $h(\cos r) = \cos r \cdot \lambda(\sin r)$ and may be defined by

$$\lambda(v) = \frac{h(\sqrt{1-v^2})}{\sqrt{1-v^2}}.$$

Now let us try to find all surfaces in ($\mathbb{R}^3$, can) which are P_l-surfaces.

4.23 Proposition. *Let us consider a piecewise C^1 curve in $\mathbb{R}^2$, which is described by two half curves z_1 and z_2 from $[0,1]$ to $\mathbb{R}^+$, with $z_i(1)=0$.*

i) *The surface of revolution generated by this curve is a Tannery surface if and only if the following conditions are satisfied:*

4.23

a) *There exist integers p, q such that*

$$\sqrt{1+z_1'^2(R)} + \sqrt{1+z_2'^2(R)} = \frac{2p}{q\sqrt{1-R^2}}$$

for each R in $[0,1[$;

b) *the curve is contained in Gong (p,q);*

c) $$-2\sqrt{\frac{p}{q}}\sqrt{\frac{\frac{p}{q}-\sqrt{1-R^2}}{1-R^2}} \leqslant z_1'(R) \leqslant 0.$$

The function h associated with this Tannery surface is defined as

$$h:]0,1[\to \left]-\frac{p}{q}, +\frac{p}{q}\right[$$

with $$h(v) = v\sqrt{1+z_1'^2(\sqrt{1-v^2})} - \frac{p}{q}.$$

ii) *The associated metric g on S^2 is nondegenerate at N* (resp. *S*) *if and only if $z_1'(0)$* [resp. $z_2'(0)$] *exists.*

The metric g is C^k at N (resp. *S*) *if and only if z_1 is C^{k+1} at* 0 (resp. *z_2 is C^{k+1} at* 0) *and* $z_1'(0)=0$ (resp. $z_2'(0)=0$).

The metric g is C^1 at the equator (that is to say the embedded surface has a well defined curvature at the equator) if and only if

$$z_1'(R) \sim -\frac{p}{q\sqrt{1-R^2}} \quad \textit{when} \quad R \to 1.$$

Proof. We note that if z_1 is the equation of a half meridian, then z_1 and h are related by the equation

$$z_1'(R) = -\sqrt{\frac{\frac{p}{q} + h(\sqrt{1-R^2})}{1-R^2} - 1}\,.$$

By 4.13 we know the conditions on h in order that h give a P_l-metric on S^2. We write these conditions in terms of z_1 and z_2 to obtain the conditions (4.23). If conditions (4.23) hold we then prove that there exists a unique h turning S^2 into a P_l-surface of revolution. □

4.24 Remark. One of the hemispheres may be chosen almost arbitrarily, the other then being determined by compensation. In particular, we have no condition on z_1'' (i.e., upon the curvature). For instance, we may have the situations of Figures 4.24 (the figures are drawn in the case $p=q=1$) for z_1 and h.

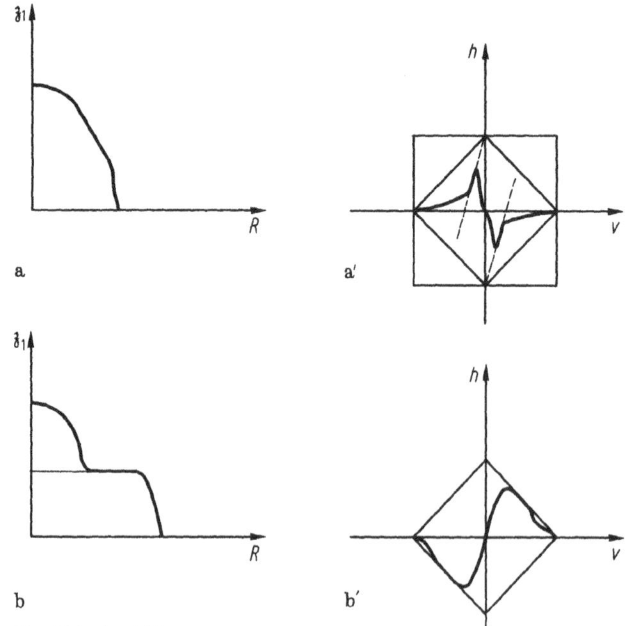

Fig. 4.24a, b, a', b'

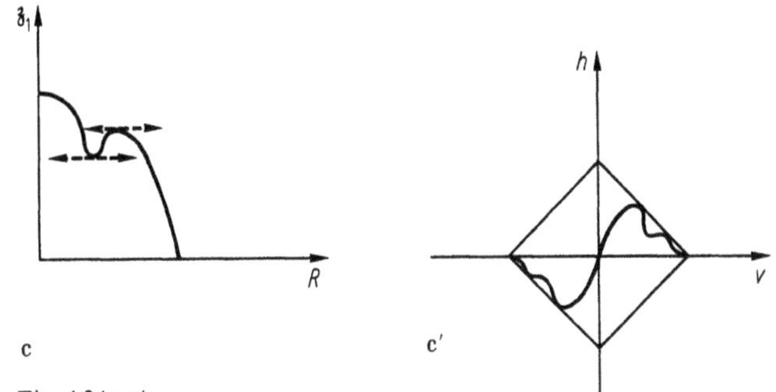

Fig. 4.24c, c′

4.25 Examples of C^1 Zoll surfaces*. See [QC] who drew the following meridians, using

$\lambda(R)=\frac{1}{2}R^2$, see Figure 4.25(a) where $\sigma>0$

$\lambda(R)=\frac{3}{4}R^4$, see Figure 4.25(b) where $\sigma>0$

$\lambda(R)=\frac{3}{2}R^4$, see Figure 4.25(c) where σ takes negative values.

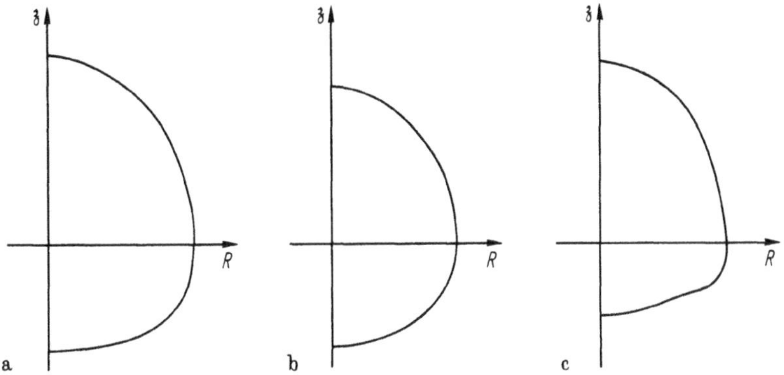

Fig. 4.25a–c

See also the following meridians, as an illustration of 4.24(c), using

$$\lambda(R)=\left(\frac{1}{\sqrt{1-R^2}}-1\right)\left(\frac{1}{2}+\frac{1}{2}\sin\left(4\pi R-\frac{\pi}{2}\right)\right), \quad \text{see Figure 4.25(d)}$$

$$\lambda(R)=\left(\frac{1}{\sqrt{1-R^2}}-1\right)\left(-1.875\frac{5^2}{2^2}\left(R-\frac{1}{2}\right)^2\left(\frac{5^2}{2^2}\left(R-\frac{1}{2}\right)^2-2\right)-0.875\right),$$

see Figure 4.25(e).

---

* When we talk about a C^k Zoll surface or a C^k Tannery surface, we take k as the order of differentiability of the metric g on S^2. Without other specification, g is non-singular at N or S in all the Examples (4.25 to 4.29).

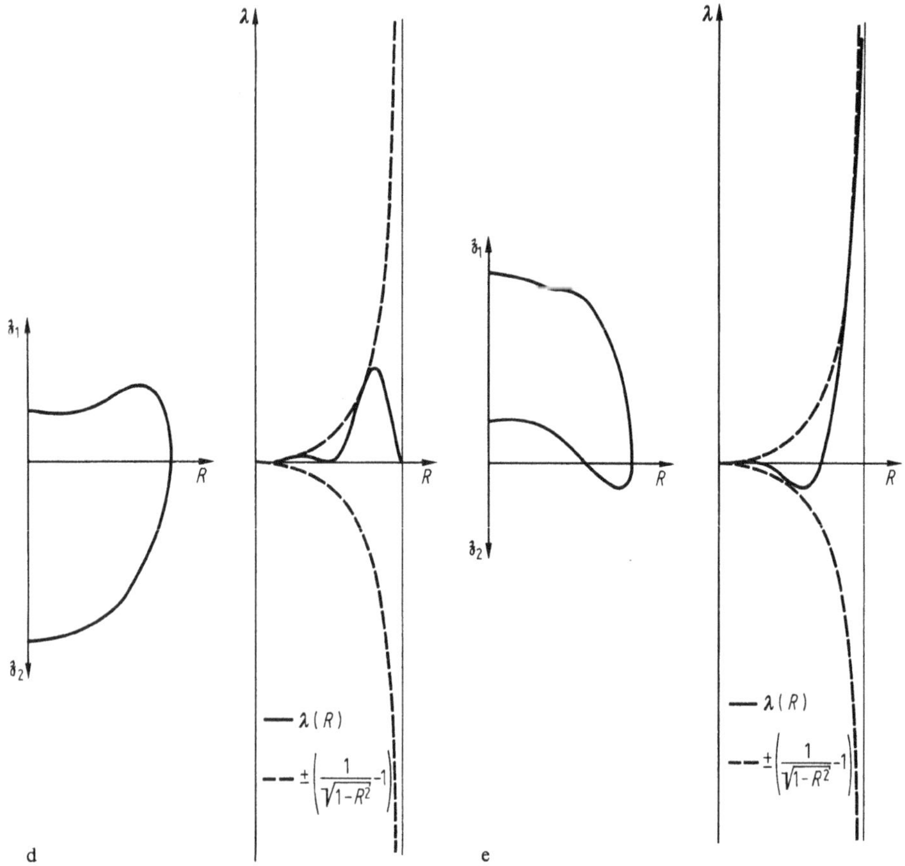

Fig. 4.25d–e

4.26. An example of a Zoll surface of revolution which is not C^1 at the equator is given by taking $\lambda(R)=\frac{1}{2}\left(\frac{1}{\sqrt{1-R^2}}-1\right)$ (see Fig. 4.26).

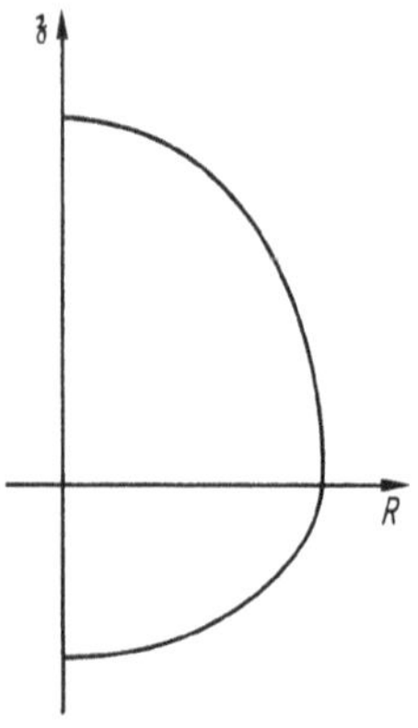

Fig. 4.26

4.27. *An Example of a Tannery surface: Tannery's Pear* (see [GR] p. 105 and [TY]). Let us choose $p=2$ and $q=1$. Then there is a very simple h satisfying Condition (4.18), namely $h(v)=v$ [see Fig. 4.27(a)]. Here N is a conic point and S is a regular point i.e., g is nonsingular at N and continuous at S [see Fig. 4.27(b)].

We apply Proposition 4.20 and get:

$$z_1(R)=4\left(1-\sqrt{1-\sqrt{1-R^2}}\right);$$

$$z_2(R)=4\left(\sqrt{1+\sqrt{1-R^2}}-1\right).$$

Then the meridian curve admits a global parametrization by r in $[0,\pi]$ as follows:

$$\begin{cases} R(r)=\sin r \\ z(r)=4\left(1-\sqrt{2}\sin\frac{r}{2}\right) \end{cases}$$

[compare with Fig. 4.27(c), which pictures Tannery's Pear with a geodesic.].

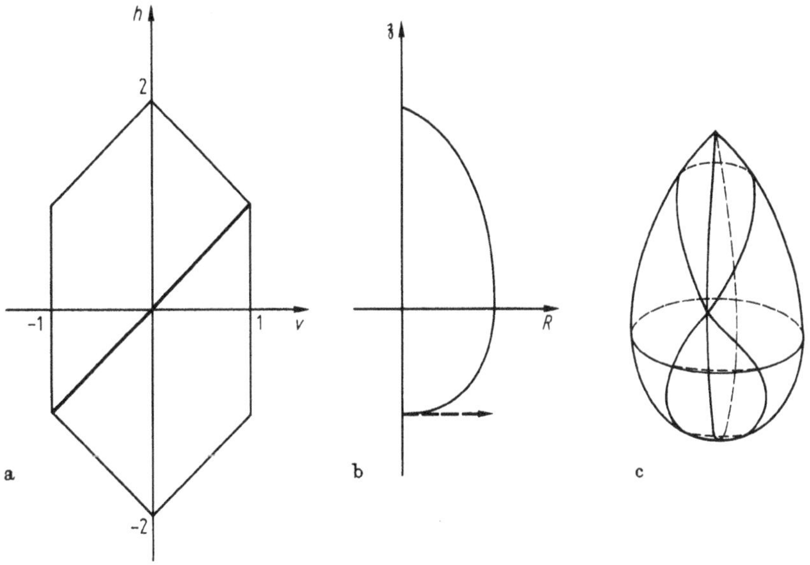

Fig. 4.27a–c. Tannery Pear

4.28. *Examples of P_l-surfaces with two conic points.*

This is the generic case for Tannery surfaces. For each p and q there is a meridian curve symmetric with respect to the R axis: the meridian obtained by taking $h\equiv 0$. Notice that if the P_l-surface is in fact a $C_{2\pi}$ surface, then N and S are both regular, and the symmetric model simply (S^2, can).

4.29. We now give an example of a P_l-surface which admits N as a conic point, S as a regular point, and whose metric is nonsingular but is not continuous at the equator. There are two possible least periods for the geodesics (see [GR] p. 110 and Fig. 4.29).

Remark. Using examples such as 4.29, it is possible to construct some P_l-surfaces which are not surfaces of revolution. We will see in 4.H that there are many Zoll surfaces which are not surfaces of revolution.

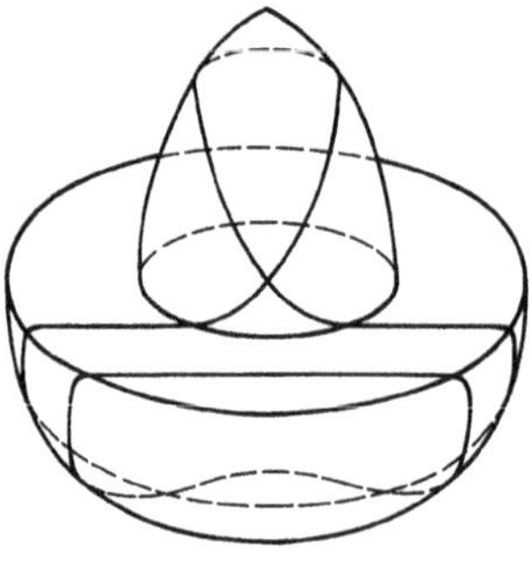

Fig. 4.29

4.30. In the construction of an embedding in 4.18, for $z'(r)$ one takes the positive square root of a positive C^k function $h(r)$. But $\sqrt{h(r)}$ is only C^0 if h vanishes up to order 2 at some point. It follows that in general $z(r)$ is only C^1 and one obtains a C^1-embedding. If we want to have a C^∞-embedding, we have to be more careful in the construction of a square root. For example if $h(r)=r^2$, we take $z'(r)=r$ and not $z'(r)=|r|$. More precisely by using Taylor's formula with integral remainder, it is not difficult to show that if h is a positive C^∞-function having only zeroes of finite order, then there exists a (nonnecessarily positive) C^∞-function g such that $g^2=h$ [this applies for example in the construction of 4.24.c) and c')]. But, as proved in [GL], it is not always true that a positive C^∞-function admits a C^∞ square root. In general it only admits a C^1-square root, which in our problem provides a C^2-embedding.

D. Geodesics on Zoll Surfaces of Revolution

We now return to the non-embedded case and on (S^2, g) consider the parametrization introduced in Theorem 4.13. We use the formulas (4.12) for geodesics in U. Without other specification, we always choose $\varepsilon_1=1$ for a geodesic, and, with an initial point $(r_0, 0)$, we choose $\varepsilon=1$ for $\left(\dfrac{d\theta}{dr}\right)_{(r_0,0)}$.

We denote by S_0 the piece of geodesic (between i and $\pi-i$) containing the initial point $(r_0, 0)$ and by S_1 the other side (see Fig. 4.31). As a matter of fact, we choose r_0 in $[0, \frac{\pi}{2}]$ and obtain the other situations by changing h into $-h$.

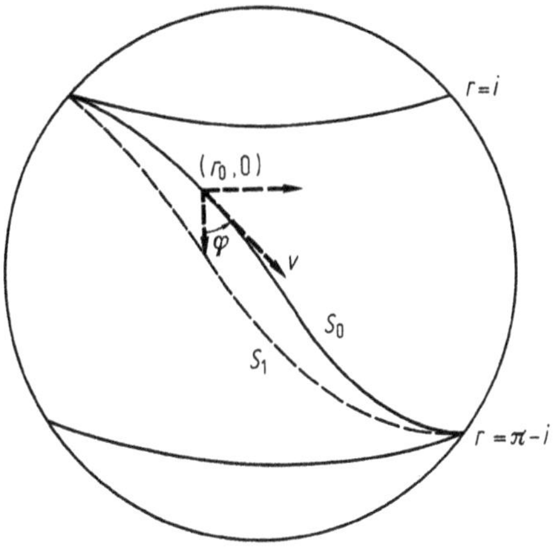

Fig. 4.31

4.31. Let v be a unit vector in $T_{(r_0,0)}S^2$, and φ its angle with the meridian $\theta=0$. The unique geodesic γ tangent to v at r_0 and parametrized by arc length is the one given by

$$\textbf{4.31}\qquad \begin{cases}\sin i=\sin r_0\sin\varphi\\ \cos\varphi=\varepsilon|\cos\varphi|\end{cases}.$$

Without any other specification, we restrict ourselves to φ in $[0,\frac{\pi}{2}]$ in our computations.

4.32 Proposition (Arc length s along a geodesic.) *Let γ be the geodesic defined by r_0 and φ. Let m denote a point of γ. We have the following cases:*

(1) $\qquad \varphi=0$

if m is in S_0,

$$s(m)=r(m)-r_0+\int_{r_0}^{r(m)} h(\cos\varrho)d\varrho$$

if m is in S_1,

$$s(m)=2\pi-r_0-r(m)+\int_{r_0}^{\pi} h(\cos\varrho)d\varrho+\int_{r(m)}^{\pi} h(\cos\varrho)d\varrho\,;$$

(2) $\qquad \varphi=\frac{\pi}{2}$ *and* $r_0=\frac{\pi}{2}$, $\quad s(m)=\theta(m)\,;$

(3) $\qquad$ *in the other cases,* $\quad \sin i=\sin r_0\sin\varphi$

$\qquad m=(r_m,\theta_m)\,;$

if m is in S_0

$$s(m)=\operatorname{Arc}\cos\left(\frac{\cos r_m}{\cos i}\right)-\operatorname{Arc}\cos\left(\frac{\cos r_0}{\cos i}\right)+\int_{r_0}^{r_m}\frac{h(\cos\varrho)\sin\varrho}{(\sin^2\varrho-\sin^2 i)^{1/2}}\,d\varrho\,.$$

and if m is in S_1

$$s(m)=2\pi-\operatorname{Arc}\cos\left(\frac{\cos r_0}{\cos i}\right)-\operatorname{Arc}\cos\left(\frac{\cos r_m}{\cos i}\right)$$

$$+\int_{r_0}^{\pi-i}\frac{h(\cos\varrho)\sin\varrho}{(\sin^2\varrho-\sin^2 i)^{1/2}}\,d\varrho+\int_{r_m}^{\pi-i}\frac{h(\cos\varrho)\sin\varrho}{(\sin^2\varrho-\sin^2 i)^{1/2}}\,d\varrho\,.$$

Proof. We leave this to the reader as an exercise (use Formulas (4.12)). □

Using the formulas in Proposition 4.32, it is easy to find the mid-point m' of γ. We have $m'=(r_{m'},\theta_{m'})$ with $s(m')=\pi$. It suffices to notice that m' is in S_1, because $1+h(\cos r)>0$.

Now let us compute the Jacobi fields on a Zoll surface. Let γ be the geodesic defined by the point $(r_0,0)$ and the angle φ, and let Y be the normal Jacobi field along γ (see 1.94) such that $Y(0)=0$ and $\dot{Y}(0)=v$, considered as a function of the arc length s. Let $Z(r)=Y(s(r))$. In fact, we will denote by Z_0 the Jacobi field Y on S_0, considered as a function of r, and by Z_1 the Jacobi field on S_1.

4.33 Proposition. (The Jacobi field along γ determined by v.)

(1) *If* $r_0=\frac{\pi}{2}$ *and* $\varphi=\frac{\pi}{2}$, *then*

$$Y(s)=C\sin s\,.$$

(2) *If* $r_0\neq\frac{\pi}{2}$ *and* $\varphi=\frac{\pi}{2}$, *then*

$$Z(r)=C(\sin^2 r-\sin^2 i)^{1/2}\,.$$

(3) *If* $\varphi\neq\frac{\pi}{2}$, *then*

$$Z_0(r)=-\frac{\cos r}{\cos^2 i}+\left[\int_{r_0}^{r}\frac{h(\cos\varrho)\sin\varrho}{(\sin^2\varrho-\sin^2 i)^{3/2}}\,d\varrho\right](\sin^2 r-\sin^2 i)^{1/2}$$

$$=-\frac{\cos r(1+h(\cos r))}{\cos^2 i}+\left[\frac{(\cos r_0)(1+h(\cos r_0))}{\cos^2 i(\sin^2 r_0-\sin^2 i)^{1/2}}\right](\sin^2 r-\sin^2 i)^{1/2}$$

$$-\left[\int_{r_0}^{r}\frac{\cos\varrho\sin\varrho\, h'(\cos\varrho)}{\cos^2 i(\sin^2\varrho-\sin^2 i)^{1/2}}\,d\varrho\right](\sin^2 r-\sin^2 i)^{1/2},$$

$$Z(i)=-\frac{1+h(\cos i)}{\cos i}$$

and

$$Z_1(r)=l_1(r_0)(\sin^2 r-\sin^2 i)^{1/2}+Z_0(r),$$

with

$$l_1(r_0) = -2\frac{\cos r_0(1+h(\cos r_0))}{\cos^2 i(\sin^2 r_0 - \sin^2 i)^{1/2}} + 2\int_{r_0}^{r}\frac{\cos\varrho\,\sin\varrho\cdot h'(\cos\varrho)}{\cos^2 i(\sin^2\varrho - \sin^2 i)^{1/2}}\,d\varrho\,.$$

Proof. The formulas in Proposition 4.33 follow directly from the equations of Jacobi fields [see (1.89)], using the expression (4.17) for the sectional curvature, together with the remark that Y and Y' at $s(i)$ are obtained from Z_0 as well as from Z_1. We then set $Z(r) = l(r)\,(\sin^2 r - \sin^2 i)^{1/2}$, and we show that l satisfies

$$l'(r) = \frac{C\sin r(1+h(\cos r))}{(\sin^2 r - \sin^2 i)^{3/2}}\,. \qquad \square$$

4.34. In higher dimensions (see E), the computations are not so simple. It is then useful to make a general remark about Jacobi fields, which allows us to obtain their equation. However, the initial point $(r_0, 0)$ plays no role, it is always taken to be $(i, 0)$ or $(\pi - i, 0)$.

4.35 Proposition (Determination of conjugate points). *Let $m_c = (r_c, \theta_c)$ denote the first conjugate point of $m_0 = (r_0, 0)$ along the geodesic γ defined by m_0 and φ.*

Then m_c is the unique conjugate point of m_0 distinct from m_0. In fact, m_c is in S_1, and if $s_c = s(m_c)$, then we have

(1) *if $r_0 = \frac{\pi}{2}$ and $\varphi = \frac{\pi}{2}$, then $m_c = (\frac{\pi}{2}, \pi)$ and $s_c = \pi$;*
(2) *if $r_0 \neq \frac{\pi}{2}$ and $\varphi = \frac{\pi}{2}$, then $m_c = (\pi - r_0, \pi)$ and $s_c = \pi$;*
(3) *if $\varphi \neq \frac{\pi}{2}$, then r_c is the unique solution of $l_1(r_c) = 0$, so that*

$$\theta_c = 2\pi - \operatorname{Arc\,cos}\frac{\operatorname{tg} i}{\operatorname{tg} r_0} - \operatorname{Arc\,cos}\frac{\operatorname{tg} i}{\operatorname{tg} r_c}$$
$$+ \int_{r_0}^{\pi-i}\frac{\sin i\cdot h(\cos\varrho)}{\sin\varrho(\sin^2\varrho - \sin^2 i)^{1/2}}\,d\varrho + \int_{r_c}^{\pi-i}\frac{\sin i\cdot h(\cos\varrho)}{\sin\varrho(\sin^2\varrho - \sin^2 i)^{1/2}}\,d\varrho$$

and

$$s_c = 2\pi - \operatorname{Arc\,cos}\frac{\cos r_0}{\cos i} - \operatorname{Arc\,cos}\frac{\cos r_c}{\cos i}$$
$$+ \int_{r_0}^{\pi-i}\frac{h(\cos\varrho)\sin\varrho}{(\sin^2\varrho - \sin^2 i)^{1/2}}\,d\varrho + \int_{r_c}^{\pi-i}\frac{h(\cos\varrho)\sin\varrho}{(\sin^2\varrho - \sin^2 i)^{1/2}}\,d\varrho\,.$$

Proof. This is a direct consequence of the fact that $1 + h(\cos r) > 0$. Therefore, $l_0(r)$ and $l_1(r)$ are monotonic functions on $]i, \pi - i[$. $\square$

4.36. The computations for the cut point of a point $m_0 = (r_0, 0)$ of (S^2, g) along a geodesic γ are more complicated because it is necessary to study the intersection of two different geodesics for every choice of ε_1 and ε [see (4.12)] (we have to determine the first time that such an intersection occurs). Hence, we have to solve one of the systems of equations [(a) or (b)] and [(c) or (d)], where

(a) is $T(r,i)-T(r,j)=0$,

(b) is $T(r,i)+T(r,j)-2\pi=0$,

(c) is $S(r,i)-S(r,j)=0$,

(d) is $S(r,i)+S(r,j)-2\pi=0$.

Here $T(r,i)$ denotes the angle θ of the point with parameter r along the geodesic γ defined by r_0 and φ with $\sin i=\sin r_0 \sin\varphi$ and $S(r,i)$ denotes the arc length s at the same point. Notice that the intersection points at same length always occur in S_1.

4.37. Let (S^2,g) be the Zoll surface of revolution characterized by $h(\cos r)=\cos r\dfrac{\sin^2 r}{2}$.

On $T_{(\frac{\pi}{2},0)}S^2$ we give the configurations of mid-points, conjugate locus and cut locus of $(\frac{\pi}{2},0)$ (see Fig. 4.37).

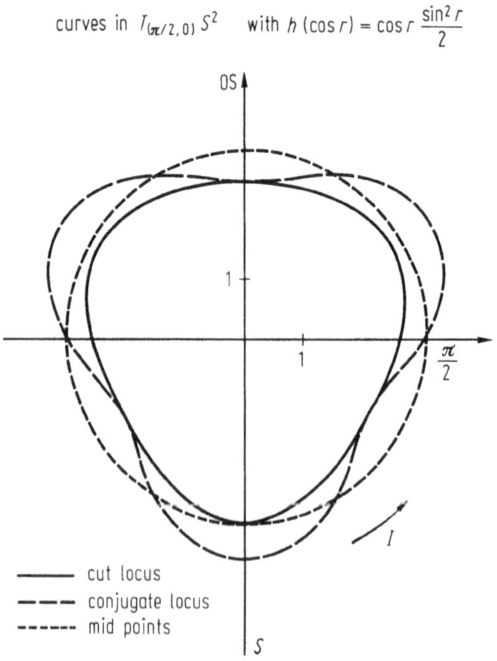

Fig. 4.37

4.38. We now give the same curves on S^2, in the parametrization $[U,(r,\theta)]$ centered at the point $(\pi-r_0,\pi)$, that is to say, at $(\frac{\pi}{2},\pi)$.

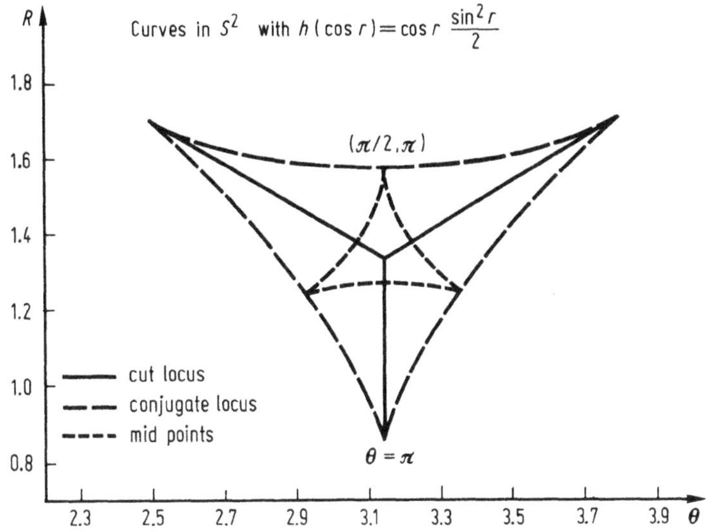

Fig. 4.38

E. Higher Dimensional Analogues of Zoll Metrics on S^2

4.39. The d-dimensional sphere S^d embedded in $\mathbb{R}^{d+1}$ can be parametrized in the following way:

$$\begin{cases} x^1 = \cos\theta^1 \\ x^2 = \sin\theta^1 \cos\theta^2 \\ \vdots \\ x^i = \left(\prod_{j=1}^{i-1} \sin\theta^j\right) \cos\theta^i \\ \vdots \\ x^{d+1} = \prod_{i=1}^{d} \sin\theta^i \end{cases}$$

for θ^i in $[0, \pi]$ $(i = 1, \ldots, d-1)$ and θ^d in $[-\pi, \pi]$.

This parametrization, which is a generalization of the standard one, of S^2 in $\mathbb{R}^3$ is nonsingular on an open dense subset of S^d (this process is nothing but the trigonometric expression of the construction of S^d by suspension of S^{d-1}).

4.40. In these coordinates it is convenient to think of the x^1-axis of coordinates as being vertical (so that, in particular, the north pole N has coordinates $(1, 0, \ldots, 0)$ and the south pole S coordinates $(-1, 0, \ldots, 0)$). One easily checks that the Riemannian metric can_d induced on S^d by the Euclidean metric on $\mathbb{R}^{d+1}$ has the following inductive expression:

$$\mathrm{can}_d = (d\theta^1)^2 + (\sin\theta^1)^2 \, \mathrm{can}_{d-1} \, .$$

One can say that can_d is obtained from can_{d-1} by a certain "Riemannian suspension".

Also notice that the parameter θ^1 is nothing but the Riemannian distance to the north pole.

We now want to give examples (due to A. Weinstein, unpublished) of higher dimensional analogues of Zoll metrics.

The idea is to imitate the two-dimensional construction by suspending can_{d-1} in the way that Funk did.

4.41 Proposition. *Let h be a C^∞-odd-function defined on $[-1,+1]$ with values on $]-1,1[$ and such that $h(-1)=h(1)=0$. Then the metric*

$$g=(1+h(\cos\theta^1))^2(d\theta^1)^2+(\sin\theta^1)^2\mathrm{can}_{d-1}$$

extends to a $C_{2\pi}$-metric on S^d.

Proof. This is fairly technical. In order to check that a geodesic is closed with length 2π, we almost have to solve explicitly its differential equation and verify that the point at length 2π is precisely the point we started from.

The meridians (curves with θ^i constant for $i=2,\ldots,d$) are automatically closed with length 2π since the chosen system of coordinates is clearly a normal one centered at the north pole (see 1.55). The same is true for geodesics in the equator defined by $\theta^1=\frac{\pi}{2}8$ since there the metric is the standard one on S^{d-1}. In order to develop the equations of a geodesic γ, we need the Christoffel symbols: except for the coefficients Γ^1_{ii}, $i=1,\ldots,d$, which involve the function f, all the others are the standard ones. Hence, the equations of a geodesic are the same as those for the standard metric except for the equation in θ^1, which can be advantageously replaced by the energy first integral.

Therefore, one can solve inductively as in the standard case, starting with $\frac{d\theta^d}{ds}$ and then finding integrals for $\frac{d\theta^i}{ds}$ $(i=2,\ldots,d)$. Then by using the energy integral the arc length s can be computed as a function of θ^1. If θ^1_0 denotes an extremal value of θ^1 along γ, then $\pi-\theta^1_0$ is another one (θ^1 comes in the formula via its sine).

The quantity $s(\pi-\theta^1_0)-s(\theta^1_0)$ appears as an integral, on an interval symmetric with respect to $\frac{\pi}{2}$, of the sum of an even function of $\cos\theta^1$ (which contributes as in the standard case) and of an odd function of $\cos\theta^1$ which does not contribute anything. The same phenomenon will occur for the variations of the parameters θ^i $(i=2,\ldots,d)$ along γ. Inductively, the values $\theta^i(\pi-\theta^1_0)-\theta^i(\theta^1_0)$ can be computed as integrals in θ^1 of the same type as above and so are equal to the standard value. We have proved at the same time that the geodesic γ is closed and has length 2π. □

4.42. We claim that the Riemannian metric we have just described admits $SO(d)$ (acting as the isotropy group of the poles of S^d for can_d) as isometry group. Indeed, the metric that we constructed is obtained from can_d by a modification involving only a symmetric two-form defined in terms of the distance for the standard metric to the fixed point set of $SO(d)$.

The existence of this large group of isometries explains why the explicit computation of the geodesics is possible. According to Souriau's moment theory (cf. [SU] p. 105), we have a first integral with values in $\mathfrak{so}(d)$ $\left(\text{recall that }\dim \mathfrak{so}(d) = \frac{d(d-1)}{2}\right)$, allowing a step-by-step integration.

From a Riemannian point of view, this first integral can be viewed as follows: if X is an infinitesimal isometry and γ a geodesic, then the function $g(\dot{\gamma}, X)$ is a constant along γ.

4.43. An infinitesimal version of this vector-valued integral can be explained as follows: if X is a globally-defined infinitesimal isometry, then X as a vector field along a geodesic γ is necessarily a Jacobi field. Indeed, X is the transverse field to the variation of γ by the curves $\varphi^t(\gamma)$ (where (φ^t) is the flow of isometries generated by X) which are geodesics. So if we have sufficiently many infinitesimal isometries we can solve the Jacobi equation (which is, as we noticed in 1.87, the linearized equation of the geodesic flow) by quadrature.

In our example one can easily check that $\frac{\partial}{\partial \theta^d}$ is an infinitesimal isometry. Hence, if along a geodesic γ we take the part of $\frac{\partial}{\partial \theta^d}$ orthogonal to $\dot{\gamma}$, then we have a normal Jacobi field. The Jacobi equation has its degree automatically lowered by one and hence can be solved explicitly.

Then we proceed inductively.

4.44. Another fact related to the preceding one is the value of the curvature tensor for our Zoll metric. One can directly check that R is diagonal in the sense that

$$R\left(\frac{\partial}{\partial \theta^i}, \frac{\partial}{\partial \theta^j}\right)\frac{\partial}{\partial \theta^k} = 0$$

as soon as $i \neq j \neq k$.

The solutions of the Jacobi equation along a geodesic γ can be computed explicitly using parallel fields along γ by taking adequate functional combinations of the coordinate fields $\frac{\partial}{\partial \theta^i}$.

F. On Conformal Deformations of P-Manifolds: A. Weinstein's Result

4.45. Using Moser's method of time dependent vector fields ([MO]), A. Weinstein in ([WN 4]) proves some results on the deformation of Zoll metrics. The following proposition is a particular case of A. Weinstein's result.

4.46 Proposition. *Let $g_t = \exp(\varrho_t)\, g_0$ be a smooth family of $P_{2\pi}$-metrics on a compact manifold M (i.e., if (ζ_t^s) denotes the geodesic flow for the metric g_t, then $\zeta_t^{2\pi} = \mathrm{Id}$) such that $\varrho_0 = 0$. We set $\dot\varrho = \left.\frac{d\varrho_t}{dt}\right|_{t=0}$. Then*

i) *for any closed geodesic $\gamma:[0,2\pi]\to M$ of the metric g_0, $\int_0^{2\pi} \dot\varrho(\gamma(s))ds = 0$;*

ii) *there exists a smooth family of homogeneous canonical diffeomorphisms $\chi_t : \mathring{T}^*M \to \mathring{T}^*M$ such that $q_t \circ \chi_t = q_0$ (q_t denotes the norm on $\mathring{T}^*M$ associated with the metric g_t). In particular, taking the symplectic gradients ξ_{q_t} of the q_t's, we have $\chi_t^*(\xi_{q_t}) = \xi_{q_0}$ (one can say that the χ_t's intertwine the geodesic flows on M).*

4.47. Part i) of 4.46 is proved for example in [ML 2] and in 5.86 for a more general deformation (not only a conformal one). In [GU], V. Guillemin indicates how to deduce i) from ii). Here we shall give a more direct proof of Part ii). In the case of a conformal deformation we have $q_t = \exp(-\varrho_t/2) q_0$. If $\tilde\gamma(s)$ denotes an integral curve of ξ_{q_0} in $\mathring{T}^*M$ which lies in $\{q_0 = c\}$ and whose projection in M is $\gamma(s)$, then we have

$$\int_0^{2\pi} \dot q_0(\tilde\gamma(s))ds = -(c/2)\int_0^{2\pi} \dot\varrho_0(\gamma(s))ds = 0.$$

Before proving ii) by Moser's method, we need a

4.48 Lemma. *There exists a smooth family of operators $A_t : \mathscr{T} \mapsto \mathscr{T}$ (where $\mathscr{T}$ denotes the space of C^∞ functions f on $\mathring{T}^*M$ which are $\mathbb{R}^*$-homogeneous of degree 0) such that if f satisfies $\int_0^{2\pi} f(\tilde\gamma(s))ds = 0$ for all closed integral curves $\tilde\gamma$ of ξ_{q_t}, then $f = \xi_{q_t}\cdot A_t(f)$.*

4.49 *Proof.* We use the fact that there exists a unique linear operator $B: C^\infty(S^1)\to C^\infty(S^1)$ such that $B(1)=1$; if $\int_{S^1} f = 0$, $(Bf)' = f$ and $\int_{S^1} B(f) = 0$. The operator B is given by $B\left(\sum_{n\in\mathbb{Z}} a_n e^{inx}\right) = \sum_{n\in\mathbb{Z}\setminus 0} \frac{a_n}{in} e^{inx}$. B can also be given by the following formula: If s_0 is a point in S^1, $B(f)(s) = \int_{s_0}^{s} \tilde f(\sigma)d\sigma - \frac{1}{2\pi}\int_0^{2\pi}\left(\int_{s_0}^{u} \tilde f(\sigma)d\sigma\right)du$, where $\tilde f(s) = f(s) - \frac{1}{2\pi}\int_0^{2\pi} f(\sigma)d\sigma$.

Now for each closed integral curve γ of ξ_{q_t}, we take $A_t(f)\restriction\gamma = B(f\restriction\gamma)$ where we identify S^1 and γ by $s\mapsto\gamma(s)$ [the argument is correct even if γ is not simply periodic, because if $f: S^1\to\mathbb{C}$ has $\frac{2\pi}{N}$ as period, then $B(f)$ admits the same period].

We only have to verify that A_t is smooth. Using a partition of unity on $\mathring{T}^*M$, it is sufficient to prove that, for each f with sufficiently small compact support, $t\mapsto A_t(f)$ is smooth from $\mathbb{R}$ into $\mathscr{T}$. Let Σ be an hypersurface transversal to $\xi_{q_{t_0}}$. Then we have

$$A_t(f|(\zeta_t^s(\lambda)) = \int_0^s \tilde f_t(\zeta_t^\sigma(\lambda))d\sigma - \frac{1}{2\pi}\int_0^{2\pi} d\sigma \int_0^\sigma \tilde f_t(\zeta_t^u(\lambda))du$$

for λ in Σ and $\tilde{f}_t = f - \frac{1}{2\pi}\int_0^{2\pi} f(\zeta_t^s(\cdot))ds$. This formula proves the smoothness of $t \mapsto A_t(f)$ and then the lemma. $\square$

4.50. *Proof of Part* ii) *of* 4.46. We want to construct a smooth family χ_t of homogeneous canonical transformations such that $\chi_t^*(q_t) = q_0$. We define the Hamiltonian vector field ξ_t by $\frac{d}{d\tau}\chi_\tau\Big|_{\tau=t} = \chi_t^*(\xi_t)$. Hence, when we differentiate the preceding relation, we get

$$\xi_t \cdot q_t + \dot{q}_t = 0 .$$

Every homogeneous Hamiltonian vector field (not of degree 0) is globally Hamiltonian so that, we have $\xi_t = \xi_{K_t}$, where K_t is in $\mathcal{T}$. Using the commutation relation $\xi_{K_t} \cdot q_t = -\xi_{q_t} \cdot K_t$, we obtain $\xi_{q_t} \cdot K_t = \dot{q}_t$. Now, we take $K_t = A_t(\dot{q}_t)$ and use Part i) of Theorem 4.46. The vector field ξ_t which we have just constructed is integrable because of homogeneity. Integration of this time-dependent vector field ξ_t gives the χ_t's.

4.51. As a corollary of 4.46. i) we find the rigidity under conformal deformation of any $\mathbb{K}P^n$ by using standard results on the Radon transform on projective spaces (see [HN 2], pp. 169 ff.).

4.52. As another corollary, we find A. Weinstein's relation (2.17) between the volume and the period L of a P_L-manifold in the case of conformal deformations.

G. The Radon Transform on (S^2, can)

The main object of G is to prove the following theorem due to Funk (see [FK 1] or [SM]).

4.53 Theorem. *For the canonical metric on* S^2, *the following two properties are equivalent for continuous functions:*

i) *the function* f *is odd (i.e.,* $f(\tau m) = -f(m)$ *for every* m, *where* τ *denotes the antipodal map);*

ii) *for any closed geodesic* γ *(i.e.,* γ *is a great circle on* S^2*),* $\int_\gamma f = 0$.

As a corollary of 4.46 and 4.53, we get

4.54 Theorem (Funk). *If* $g_t = \exp(\varrho_t) \cdot g_0$ *is a smooth family of* $P_{2\pi}$*-metrics on* S^2, *with* $\varrho_0 = 0$ *and* $\dot{\varrho} = \frac{d\varrho_t}{dt}\Big|_{t=0}$, *then* $\dot{\varrho}$ *is an odd function on* S^2.

4.55. Following Funk's paper ([FK 1]), we want to give two alternative proofs of 4.53 based on the study of the Radon transform on S^2. We can identify the set of oriented great circles on S^2 with S^2 using the following picture (see also 2.8).

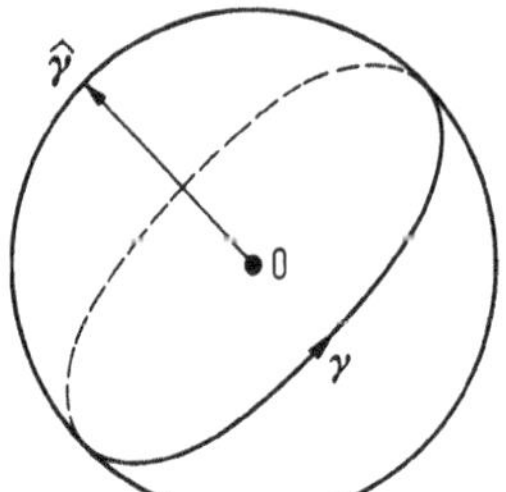

Fig. 4.55

We define an operator R on $C^0(S^2)$ (called the *Radon transform*) by $R(f)(\hat{\gamma}) = \int_\gamma f$. If $\mathscr{E}$ is a space of functions on S^2, we *denote* by $\mathscr{E}_+$ (resp. $\mathscr{E}_-$) the space of even (resp. odd) functions f in $\mathscr{E}$. The following properties of R are clear:

4.56. $R(C^0_-) = \{0\}$, $R(C) \subset C^0_+$ (because reversing the orientation of γ gives the same integral);

4.57. the operator R commutes with the $O(3)$ action on S^2.

4.58. The *first proof* given by Funk works only for nonpathological f, say those which are real analytic.

4.59. As a first remark, using 4.56 and the fact that $C^0(S^2) = C^0_+ \oplus C^0_-$, we restrict ourselves to proving Theorem 4.53 in the even case.

Let Γ be a subgroup of $SO(3)$ isomorphic to S^1. Using 4.57 one can show that for every function f one has $R\left(\int_\Gamma \varphi^*(f)d\varphi\right) = \int_\Gamma \varphi^*(Rf)d\varphi$. As a corollary, we get the fact that if $Rf = 0$, then $R\left(\int_\Gamma \varphi^*(f)d\varphi\right) = 0$. Now let us suppose that we can prove the following

4.60 Proposition. *For any even function f invariant under some subgroup Γ of $SO(3)$ isomorphic to S^1, $R(f) = 0$ implies $f = 0$.*

Then 4.53 follows. In this case, we obtain that the integral of f on every circle (not only on great circles) is zero. The function f being continuous, this proves that $f = 0$.

4.61 *Proof of* 4.60. Let $f: S^2 \to \mathbb{R}$ be analytic and invariant under rotation around the Oz axis (and of course even). Set $g(t) = f(\sqrt{1-t^2}, 0, t)$; g is an even analytic function on $]-1, +1[$ and on the great circle making an angle i with the equator we can compute the integral of f, $J(i)=4\int_0^{\pi/2} g(\sin\lambda\cdot\sin i)d\lambda$. If we write $\sin i=\alpha$, we obtain $J(i)=4I(\alpha)$, where

$$I(\alpha)=\int_0^{\alpha}(\alpha^2-t^2)^{-1/2}g(t)dt.$$

If $g\neq 0$, we have $g(t)=at^{2n}+O(t^{2n-1})$ with $a\neq 0$ and

$$I(\alpha)=a\alpha^{2n}\int u^{2n}(1-u^2)^{-1/2}du+O(\alpha^{2n-1}).$$

Hence, if $J(i)=0$, $I(\alpha)=0$ and $a=0$, then $g=0$. □

4.62 Remark. In fact, one can prove directly that for a continuous function f, $I(\alpha)=0$ implies $f=0$ by using an argument similar to 4.14. Computing $J(\beta)=\int_0^{\beta}\frac{\alpha I(\alpha)d\alpha}{\sqrt{\beta^2-\alpha^2}}$, we obtain by the Lebesgue-Fubini theorem, $J(\beta)=c\int_0^{\beta}g(t)dt$. This computation might give an explicit expression to invert the Radon transform.

4.63. The *second proof* given by Funk makes use of spherical harmonics. In fact, the same idea leads to a better understanding of the nature of R.

4.64 Proposition. *There exists an elliptic invertible Fourier Integral Operator B of order $-\frac{1}{2}$ on S^2, such that $R=\frac{Id+J}{2}\circ B=B\circ\frac{Id+J}{2}$, where $Jf(m)=f(\tau m)$. As a consequence,* $\operatorname{Ker}R=\operatorname{Ker}\frac{Id+J}{2}=C^0_-$.

4.65. We will only prove that B can be chosen to be an invertible operator of order $-\frac{1}{2}$, i.e., that B is a homeomorphism of $H^s(S^2)$ onto $H^{s+\frac{1}{2}}(S^2)$ for every real s. V. Guillemin proves in the appendix of [GU] that B is a Fourier Integral Operator by using the spectral decomposition of B.

4.66. Let us recall that $L^2(S^2)=\bigoplus_{k\geq 0}\mathscr{H}_k$, where $\mathscr{H}_k$ is the space of homogeneous harmonic polynomials of degree k in $\mathbb{R}^3$. Furthermore, the action of $SO(3)$ induces on each $\mathscr{H}_k$ an irreducible representation (for these classical results on spherical harmonics, see for example [B-G-M] pp. 159 ff.). The operator R commuting with the $SO(3)$ action (4.57) acts on each $\mathscr{H}_k$ by multiplication by a *constant* c_k. One easily checks that $c_{2k+1}=0$.

4.67 Lemma. $c_{2k}=(-1)^k 2\pi\cdot\frac{1\cdot 3\cdot 5\ldots(2k-1)}{2\cdot 4\cdot 6\ldots(2k)}$.

Proof. We only have to compute $R(f)(\hat{\gamma})$ for a homogeneous polynomial of degree $2k$, for example, $f(x,y,z)=\operatorname{Re}(x+iy)^{2k}$, $\gamma=\{x=0\}$, $\hat{\gamma}=(\pm 1,0,0)$: $f(\hat{\gamma})=1$ and $R(f)(\hat{\gamma})=(-1)^k\int_0^{2\pi}(\cos\theta)^{2k}d\theta$. A classical computation of this integral by integration by parts gives 4.67. □

4.68. Now let $d_s = i^s \int_0^{2\pi} |\cos\theta|^s d\theta$ and $B \upharpoonright \mathscr{H}_k = d_k \cdot Id_{\mathscr{H}_k}$. Clearly, B satisfies the relations of 4.64 and is bijective (because $d_s \neq 0$ for every s in $\mathbb{N}$). The main ingredient is now Wallis formula which asserts that $\frac{2\cdot 4\cdot 6 \ldots 2k}{1\cdot 3 \ldots 2k-1} \sim (\pi k)^{1/2}$ when k goes to infinity. As a consequence, we get that $d_s \sim c\cdot s^{-1/2}$ when s goes to infinity. We recall (cf. [B-G-M] p. 159) that $\Delta \upharpoonright \mathscr{H}_k = \lambda_k Id_{\mathscr{H}_k}$, where $\lambda_k = k(k+1)$. We only have to prove that $\Delta^{1/4} \circ R = R \circ \Delta^{1/4}$ is bounded on $L^2(S^2)$. But $\Delta^{1/4} \circ R \upharpoonright \mathscr{H}_k = [k(k+1)]^{1/4} \times d_k Id_{\mathscr{H}_k}$ and $|d_k| \sim ck^{-1/2}$. □

H. V. Guillemin's Proof of Funk's Claim

4.69. In 1913, Funk ([FK 1]) gave a method for obtaining all Zoll metrics on S^2, in the analytic form $g_t = (1 + t\Phi_1 + \ldots + t^n\Phi_n + \ldots)g_0$. But his proof was only *formal.* He did not succeed in proving the convergence of the series for t small. This was probably the birth of the Radon transform and sufficient motivation for Blaschke conjecture (5.F). Using powerful tools of analysis, V. Guillemin proves in [GU] the following beautiful theorem which is a modern complete proof of Funk's method.

4.70 Theorem. *For every odd function $\dot{\varrho}$ on S^2, there exists a smooth one-parameter family of C^∞-functions ϱ_t such that $\varrho_0 = 0$, $\left.\frac{d\varrho_t}{dt}\right|_0 = \dot{\varrho}$ and $\exp(\varrho_t)\cdot g_0$ is a Zoll metric for small t.*

4.71 Corollary. *There are plenty of Zoll metrics on S^2 with a trivial group of isometries. More precisely, there exists an open dense subset Ω of $C^\infty_-(S^2)$ such that for any $\dot{\varrho}$ in Ω, $\exp(\varrho_t)\cdot g_0$ admits no non-trivial isometry for t small.*

Proof. Let $\dot{\varrho}$ be in $C^\infty_-(S^2)$ and let $\exp(\varrho_t)\cdot g_0$ be the family of Zoll metrics constructed in 4.70. Suppose that there exists a sequence t_n going to 0, such that $g_n = \exp(\varrho_n)\cdot g_0$ admits a non-trivial isometry group I_n. Recall that there necessarily exists an isometry φ_n in I_n such that $\delta(\varphi_n) = \sup_{x\in S^2} d(x, \varphi_n(x)) \geq \mathrm{Inj}(g_n)/2$. Using the semi-continuity properties of Inj and of I_n [EN 1] we can find a sequence (φ_n) with φ_n in I_n, such that (φ_n) converges to φ_0 in $O(3)$, $\varphi_0 \neq Id$. We have $\varphi_n^*(g_0) = \exp(\varrho_n - \varphi_n^*(\varrho_n))g_0$. Then $\frac{1}{t_n}\varrho_n - \frac{1}{t_n}\varphi_n^*(\varrho_n) = \frac{1}{t_n}a(\varphi_n)$. If we let n go to infinity, we obtain that $\dot{\varrho} - \varphi_0^*(\dot{\varrho})$ is in E (where E is the 3-dimensional space of linear functions on $\mathbb{R}^3$). Here it occurs as $F'(T_eG)$, where G is the conformal group of S^2 and $F: \varphi \mapsto a(\varphi)$ is defined for φ in G by $\varphi^*(g_0) = \exp(a(\varphi))\cdot g_0$. Now $\dot{\varrho} = \varrho_1 + \varrho_2$ where ϱ_1 belongs to E and ϱ_2 to $E^\perp$. Furthermore, $\varphi_0^*(\dot{\varrho}_1 = \varrho_1' + \varrho_2'$, where ϱ_1' is in E, ϱ_2' is in $E^\perp$. We conclude that $\varphi_0^*(\varrho_2) = \varrho_2$. Corollary 4.71 will be proved if the following lemma is true. □

4.72 Lemma. *The set $\Omega_1 = \{f \in C^\infty_- \cap E^\perp \mid \nexists \varphi \in O(3) - \{Id\}, \varphi^*(f) = f\}$ is open and dense in $C^\infty_-(S^2) \cap E^\perp$.*

In fact, we take $\Omega = \Omega_1 + E$ which is also open and dense.

It is clear that Ω_1 is open by D. Ebin's theorem (see [EN 1], p. 34) on the semi-continuity of isometry groups [because $\varphi^*(f) = f$ is equivalent to the fact that φ is an isometry of $\exp(f) \cdot g_0$].

One can show that Ω_1 is dense by approximating every f in $C^\infty_-(S^2) \cap E^\perp$ by a sequence (f_n) in the same space such that (f_n) has the following properties: there exists a point m_0 in S^2 such that $f_n(m) < f_n(m_0)$ for each m in S^2 and f_n admits, in the normal coordinates at m_0, a Taylor expansion of the form $f(x^1, x^2) = f_n(m_0) + p_2(x^1, x^2) + p_3(x^1, x^2) + \ldots$, where $p_2 + p_3$ cannot be invariant under some nontrivial element in $O(2)$. Then it is easy to show that f_n has a trivial isometry group in $O(3)$.

4.73. *Sketch of the proof of* 4.70. Denote by $\mathscr{S}$ the group of homogeneous symplectic diffeomorphisms on $\mathring{T}^*S^2$ and suppose that we can prove the following

4.74 Proposition. *Let $F: \mathscr{S} \times C^\infty(S^2) \to \mathscr{T}$ be the map defined by*

$$F(\varphi, \mu) = \log(\varphi_*(q_0)) - \log q_0 - \mu \,.$$

Then there exist a neighbourhood $\mathcal{O}$ of 0 in $\mathscr{T}$ and a smooth map $G = (G_1, G_2)$ from $\mathcal{O}$ to $\mathscr{S} \times C^\infty(S^2)$ such that:

i) $G(0) = (\mathrm{Id}, 0)$;

ii) $G_2'(0)(\mu) = 0$ *for every μ in $C^\infty_-(S^2)$*;

iii) $F \circ G = \mathrm{Id}_{\mathcal{O}}$.

(We consider $C^\infty(S^2)$ as a subspace of $\mathscr{T}$ via the pull back by $p^*_{S^2}$, where $p^*_{S^2}: \mathring{T}^*S^2 \to S^2$ is the projection).

4.75. We want to show why 4.70 is a corollary of 4.74. Let (φ_t, μ_t) be $G(-t\dot{\varrho})$. Then we have $\varphi_0 = \mathrm{Id}$ and $\mu_0 = 0$ by i), $\dot{\mu}_0 = 0$ by ii), and $\varphi_t^*(q_0) = \exp(-t\dot{\varrho} + \mu_t) q_0$ by iii). Set $\varrho_t = t\dot{\varrho} - \mu_t$; the Hamiltonian function $\exp(-\varrho_t) q_0$ is associated with the metric $\exp(\varrho_t) \cdot g_0$. The Hamiltonian is conjugate under φ_t to q_0, which has a 2π-periodic flow, so it also has this property.

4.76. *Sketch of the proof of* 4.74. The main idea of the proof of 4.74 is to apply an implicit function theorem. But implicit function theorems for mappings between Frechet manifolds are very difficult and require some strong conditions not only on the derivatives of F at the point $\mathrm{Id} \times O$ but also in a neighbourhood of $\mathrm{Id} \times O$. V. Guillemin uses a special form of the Nash-Moser implicit function theorem due to F. Sergeraert ([ST]).

4.77. We want to show how to use the results of Section F on the Radon transform on (S^2, can) in order to get some information on $F'_{Id \times O}$ (which would give the result 4.74 if we were in the Banach manifold context).

Let us recall that the tangent space to $\mathscr{S}$ at Id is the space of homogeneous Hamiltonian vector fields ξ on $\mathring{T}^*S^2$. The derivative L of F at $\mathrm{Id} \times O$ is given by

$$L(\xi, v) = \frac{\xi \cdot q_0}{q_0} - v \,.$$

In order to prove that L is surjective and to find a right inverse to L [which will be $G'(0)$], we want to solve the equation $\xi \cdot q_0 - v = \mu$ on $q_0 = 1$, where μ is given in $\mathscr{T}$. Then we will extend the solution by homogeneity. Integration over a periodic geodesic γ gives $R(v)(\gamma) = -\int_{\tilde{\gamma}} \mu$. If we set $\psi(\gamma) = \int_{\tilde{\gamma}} \mu$, we notice that ψ is even (because $q_0(-\lambda) = q_0(\lambda)$). Using the notation of 4.64 we choose $G_2'(\mu) = -B^{-1}(\psi)$, which satisfy 4.74 ii) because for μ odd, $\psi = 0$. Now we want to solve $\xi \cdot q_0 = \mu + G_2'(\mu)$, or equivalently taking $\xi = \xi_K$, $-\xi_{q_0} \cdot K = \mu + G_2'(\mu)$. We take $K = -A_0(\mu + G_2'(\mu))$ (with the notation of 4.48) and $G_1'(\mu) = -\xi_{A_0(\mu + G_2'(\mu))}$. In fact, the preceding argument leads only to a formal proof of Funk's result.

4.78. The main difficulty in applying F. Sergeraert's implicit function theorem is obtaining uniform estimates of some right inverse of $F'_{\varphi,\mu}$, where (φ, μ) is in a neighbourhood of $\mathrm{Id} \times 0$. In order to overcome this difficulty, V. Guillemin studies the Radon transform associated with some $C_{2\pi}$-metric $\varphi^*(q_0)$ close to q_0, and he proves an analogue of Proposition 4.64 for those metrics, where the operator B_φ depends smoothly on φ. Then L. Hörmander's theory of Fourier Integral Operators allows him to obtain the desired non-trivial uniform estimates. □

Chapter 5. Blaschke Manifolds and Blaschke's Conjecture

A. Summary

5.1. Section B recalls very briefly some basic results on the metric structure of a Riemannian manifold, in particular, of the cut-locus. We define the distance function and the notion of a segment; recall that segments are necessarily geodesics and locally unique. We define the cut-value and the cut-point of a geodesic. We recall the strict triangle inequality and the acute angle property. Finally we define what a manifold with spherical cut-locus is.

5.2. In Section C we define the Allamigeon-Warner manifolds: a manifold is Allamigeon-Warner if first conjugate points always appear as they appear in a CROSS, i.e., for every initial unit vector the first conjugate point appears at the same distance and with the same multiplicity. Then we prove the Allamigeon-Warner theorem to the effect that such a manifold, if simply connected, has a conjugate locus which coincides with its cut-locus and is, moreover, a nice submanifold.

5.3. In Section D we define pointed Blaschke manifolds and Blaschke manifolds; for motivation the reader can see 0.10 and 0.30. Pointed Blaschke manifolds can be characterized by four equivalent properties; this equivalence is not so easy to prove and is described in detail so that the reader will not have to refer to the literature.

5.4. In Section E we use the results of Chapter 7 to get some properties of Blaschke manifolds; in particular, we detail the cases of extreme orders of conjugacy, which turn out to be the case of spheres and that of real projective spaces, linked together by a Riemannian covering. The spherical case in dimension 2 was precisely the one Blaschke was interested in.

Then Section F explains what we understand by Blaschke's conjecture and contains the proof of the conjecture when the dimension is equal to 2 (Green's theorem). We then add some remarks to explain why some ways to attack the conjecture in higher dimensions fail and propose some other ways to attack it.

Section G gives a partial result for Kähler-Blaschke structures on the complex projective space.

Section H proves Michel's theorem, i.e., the validity of an infinitesimal version of Blaschke's conjecture for the spherical case; this proof is new.

B. Metric Properties of a Riemannian Manifold

5.5. For these classical facts, good general references are [K-N 2], p. 96—102 and [KI]. For a Riemannian manifold (M, g) and points m, n in M we define

5.6. $\mathscr{C}(m, n)$ as the set of rectifiable curves from m to n

and

5.7. $\varrho(m, n) = \inf\{L(c) \mid c \in \mathscr{C}(m, n)\}$.

The function $\varrho: M \times M \to \mathbb{R}_+$ is in fact a metric on M and the one we will always use. The topology induced by this metric coincides with the manifold topology of M.

For balls in (M, g) with respect to our metric, we shall use the following notation:

5.8 $B(m, r) = \{n \in M \mid \varrho(m, n) < r\}$, $B'(m, r) = \{n \mid \varrho(m, n) \leqq r\}$.

We also set

5.9 $\operatorname{Diam}(g) = \sup\{\varrho(m, n) \mid m, n \in M\}$

and call it the *diameter* of (M, g).

Our metric is related to geodesics of Chapter 1, Section F as follows: "geodesics locally minimize distance and do this uniquely". More precisely, we saw in 1.53 that for any point m in M there exists an $\varepsilon > 0$ such that the restriction $\exp_m \restriction B(0_m, \varepsilon)$ is a diffeomorphism onto its image; moreover, $\exp_m(B(0_m, \varepsilon)) = B(m, \varepsilon)$ and for all n in $B(m, \varepsilon)$ there is one and only one curve from m to n with length equal to $\varrho(m, n)$, namely the geodesic $\gamma: s \mapsto \exp_m(su)$, where $u = \exp_m^{-1}(n) \cap B(0_m, \varepsilon)$ (this basic fact is quite lengthy to establish, see [K-N 1], p. 165 for example; it uses basically Gauss lemma, see 1.96).

Then it is natural to introduce the following

5.10 Definition. $\operatorname{Inj}(g, m)$ is the supremum of the positive real numbers ε such that $\exp_m \restriction B(0_m, \varepsilon)$ is a diffeomorphism onto its image. We call it the *injectivity radius* of (M, g) at m.

5.11 Definition. The *injectivity radius* of (M, g) is $\operatorname{Inj}(g) = \inf\{\operatorname{Inj}(g, m) \mid m \in M\}$.

It might happen that $\operatorname{Inj}(g) = 0$ or $\operatorname{Inj}(g) = \infty$; but when M is compact, $\operatorname{Inj}(g)$ is always finite and (strictly) positive.

This last assertion will result, for example, from the existence of convex balls (see [K-N 1], p. 166): the ball $B(m, \varepsilon)$ is said to be *convex* if for any n, p in $B(m, \varepsilon)$ there is one and only one curve from n to p with length equal to $\varrho(n, p)$; this curve is required to lie in $B(m, \varepsilon)$. Such convex balls always exist at any point.

5.12. We will also use the notion of *segments* in (M, g), i.e., curves which are parametrized by arc length and whose length is equal to distance; segments are necessarily geodesics and we *set*

5.13. $\begin{cases} \text{Seg}(m,n) \text{ equal to the set of geodesics from } m \text{ to } n \text{ which are param-} \\ \text{etrized by arc length and whose length is equal to } \varrho(m,n). \end{cases}$

5.14. We also recall the Hopf-Rinow theorem (see [K-N 1], p. 172): a Riemannian manifold (M,g) is said to be *complete* if one of the following equivalent properties holds:

(i) (M,g) is complete for the metric topology of ϱ;
(ii) for some point m in M, $\exp_m$ is defined on all T_mM;
(iii) for every point m in M, $\exp_m$ is defined on all T_mM.

The contents of the Hopf-Rinow theorem are precisely the equivalence of (i), (ii) and (iii). In particular, a compact manifold M admits only complete Riemannian metrics, compare with 1.52 and 1.58. Note that on any complete Riemannian manifold (M,g) one always has

$$\text{Seg}(m,n) \neq \emptyset \quad \text{for every} \quad m,n \quad \text{in} \quad M,$$

but that the converse is false (think for example of the unit disc in $\mathbb{R}^2$).

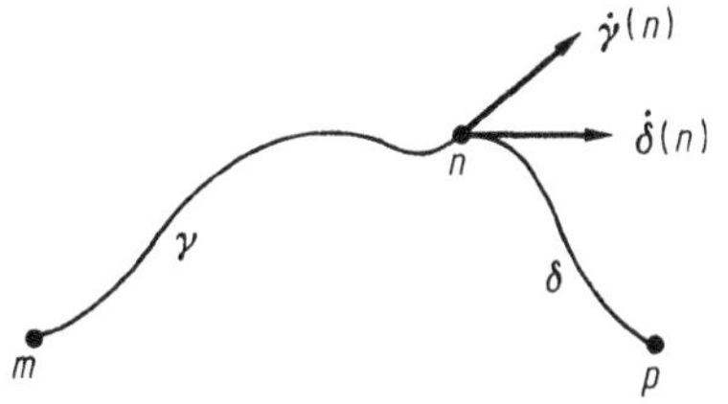

Fig. 5.15

5.15. A Riemannian manifold satisfies the *strict triangle inequality*, i.e., given points m, n in M, γ in $\text{Seg}(m,n)$ and δ in $\text{Seg}(n,p)$, if $\dot{\gamma}(n) \neq \dot{\delta}(n)$, then

$$\varrho(m,p) < \varrho(m,n) + \varrho(n,p).$$

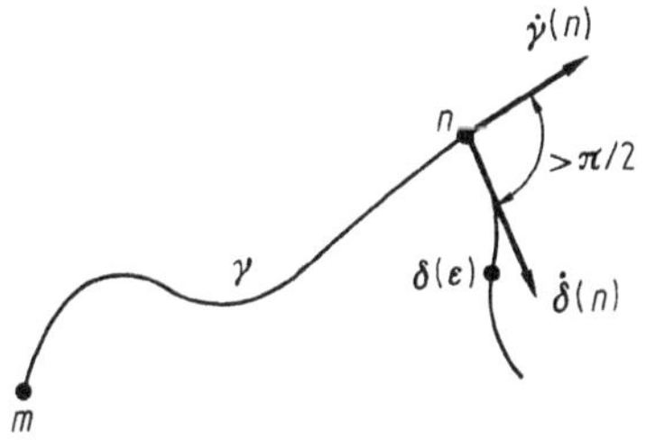

Fig. 5.16

5.16. We also have in a Riemannian manifold (M,g) the *acute angle property*: given points m, n in M, γ in $\text{Seg}(m,n)$ and a geodesic δ starting from n, if $(\dot{\gamma}(n), \dot{\delta}(n)) < 0$, then for $\varepsilon > 0$ small enough

$$\varrho(m, \delta(\varepsilon)) < \varrho(m,n), \qquad \text{(see Figure 5.16 above)}.$$

5.17. To investigate the uniqueness of segments in more detail, we use the following trick (all of the Riemannian manifolds we consider are complete): let u be an element of U_mM and $\gamma: s \mapsto \exp_m(su)$; we saw in 5.5 that, for s small enough, $\mathrm{Seg}(m=\gamma(0), \gamma(s)) = \{\gamma|_0^s\}$, i.e., it consists of exactly one element. The set

$$A = \{s \in \mathbb{R}_+ \mid \gamma|_0^s \in \mathrm{Seg}(\gamma(0), \gamma(s))\}$$

is necessarily $\mathbb{R}_+$ or a compact interval $[0, r]$. If $A = \mathbb{R}_+$, we say that *there is no cut-point on* γ; if $A = [0, r]$, we say that $\gamma(r)$ is *the cut-point* and r *the cut-value of* γ. The *cut-map* is

5.18 $\qquad \mu: UM \to \mathbb{R}_+^* \cup \{\infty\}$,

where $\mu(u) = r$ if $A = [0, r]$ and $\mu(u) = \infty$ if $A = \mathbb{R}_+$. The cut-map μ is continuous (see [K-N 2], p. 98, [KI], p. 100); and the link with 5.10 is

5.19 $\qquad \mathrm{Inj}(g, m) = \inf\{\mu(u) \mid u \in U_mM\}$.

5.20. Notice that for $s = \mu(u)$ we can still have $\mathrm{Seg}(\gamma(0), \gamma(s)) = \{\gamma|_0^s\}$, but then necessarily $\gamma(0)$ and $\gamma(s)$ are conjugate along γ (see 1.96, [K-N 2], p. 97, [KI], p. 116). Notice that compactness of M is equivalent to the finiteness of $\mu(u)$ for all u in UM. Finally, if $\gamma(r)$ is the cut-point of $\gamma(0) = m$ on γ, then $\gamma(0)$ is the cut-point of $\gamma(r)$ on the reversed geodesic $\gamma': s \mapsto \exp_{\gamma(r)}(s(-\dot\gamma(r)))$.

5.21. The *cut-locus of* m in M is

$$\mathrm{Cut}(m) = \{\exp_m(\mu(u)u) \mid u \in U_mM\}.$$

When $\dim M = 2$, its structure is well understood (see [MS 1], [MS 2]); otherwise not much is known about cut-loci: see [KI], p. 103, [WN 2], [BU], [G-S 1], [SA], [BR 2].

5.22 Definition. A Riemannian manifold (M, g) is said to have *spherical cut-locus at* m if for every u in U_mM the cut-value $\mu(u)$ is finite and does not depend on u (or, equivalently, $\varrho(m, n)$ does not depend on n when n runs through $\mathrm{Cut}(m)$).

We will characterize manifolds with spherical cut-locus in 5.43.

From now on in this chapter every Riemannian manifold is assumed to be complete.

C. The Allamigeon-Warner Theorem

5.23 Definition. Given a Riemannian manifold (M, g) (of dimension d) and a point m in M, we say that (M, g) is an *Allamigeon-Warner manifold at* m if there exists a length $l > 0$ and an integer k between 1 and d such that, for every geodesic γ starting from m, the first conjugate point of m along γ appears at length l and has index $k-1$ (see 1.96).

Examples of manifolds which have the Allamigeon-Warner property at every m are the CROSSes. The converse statement is more or less Blaschke's conjecture: see 0.33 and Section F.

5.24. *From now on in the present section we suppose that (M, g) is an Allamigeon-Warner manifold at m, with length l and index $k-1$;* we now deduce some consequences of that fact, the final aim being Theorem 5.29.

Set $S = S(0_m, l)$ and let f be $\exp_m \restriction S$; by assumption, f is a C^∞-map with constant rank equal to $d-k$. Then the family of $(k-1)$-planes $\ker(T_u f) = (T_u f)^{-1}(0)$ for u in S, gives rise to $(k-1)$-dimensional foliation of S; we *denote* by $\varphi(u)$ the maximal leaf containing u in S and by Φ the set of these maximal leaves.

5.25. We first study a given maximal leaf φ in Φ; the image $f(\varphi)$ is a single point n in M. The counter-image $f^{-1}(n)$ is compact—being closed in S—and consists of maximal leaves; hence φ, which is a connected component of $f^{-1}(n)$, is also compact. Notice that n does not necessarily belong to Cut(m), think of ($\mathbb{R}P^d$, can), for example.

5.26. Thanks to the "rank theorem" (see for example [DE 1], p. 42 where f is called a *subimmersion*) for a fixed u in φ, there exists a neighborhood V of u in S such that $f(V)$ is a submanifold of dimension $d-k$ of M.

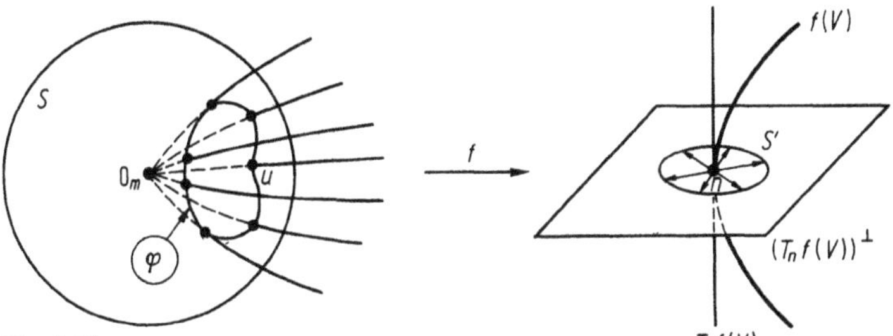

Fig. 5.26

5.27 Lemma. *For u in φ* denote *by γ_u the geodesic* $s \mapsto \exp_m\left(\frac{s}{|u|}\,u\right)$. *Then the map $\psi: u \mapsto \dot{\gamma}_u(l)$ from φ to $U_n M$ is a diffeomorphism between φ and the great sphere $(T_n f(V))^\perp \cap U_n M$ of $U_n M$* (cf. 5.36).

Proof. The map ψ is injective because a geodesic with given initial velocity vector is unique. It has maximal rank because its tangent map $T_u\psi$ can be described as follows: if Y is the Jacobi field along γ_u with $Y(0)=0$ and $Y'(0)=v$ for v in $\mathring{T}_u\varphi$, then $T_u\psi(v) = Y'(l) \neq 0$ (because $Y(l)=0$ and a Jacobi field with $Y(l)=0$ and $Y'(l)=0$ would be identically zero). Now notice that the image $\psi(\varphi)$ is contained in the great sphere $S' = (T_n f(V))^\perp \cap U_n M$, by the first variation formula 1.109. Then $\psi: S \to S'$ is a C^∞ map of maximal rank between two manifolds with the same dimension $k-1$. So the image $\psi(\varphi)$ is open and closed in S'; since S' is connected, $\psi(\varphi) = S'$. □

5.28. Notice that the counter-image $\psi^{-1}(\psi(V))$ is a neighborhood of u in S which is a union of maximal leaves.

5.29 Theorem (Allamigeon: [AN], Sections 3 and 4 and Warner: [WR 2], p. 208—209). *For an Allamigeon-Warner manifold (M, g) at m, the set Φ of maximal leaves (with the above notation) has a natural structure of a C^∞ $(d-k)$-dimensional manifold; the canonical projection $p: S \to \Phi$ makes S into a $(k-1)$-sphere bundle over Φ and f factors through an immersion $q: \Phi \to M$ such that $f = q \circ p$. If on $B'(0_m, l)$ we define the equivalence relation $\mathcal{R}$ by letting $v\mathcal{R}w$ for $v \neq w$ if and only if v and w are in S and in the same maximal leaf, then $\hat{M} = B'(0_m, l)/\mathcal{R}$ has a natural structure of a C^∞ d-dimensional manifold. Moreover, the topological structure of $\hat{M}$ is as follows: $\hat{M}$ can be written as $\hat{M} = D \cup_a E$, where D is the d-dimensional closed ball, E is a C^∞-closed k-disc bundle over a C^∞ $(d-k)$-dimensional compact manifold, whose boundary ∂E is diffeomorphic to S^{d-1} and $a: \partial D \to \partial E$ is an attaching diffeomorphism. Finally the restriction $\exp_m \restriction B'(0_m, l)$ factors via the canonical projection $\hat{p}: B'(0_m, l) \to \hat{M}$ through a map $\hat{q}: \hat{M} \to M$ which is a covering map and such that $\exp_m \restriction B'(0_m, l) = \hat{q} \circ \hat{p}$.*

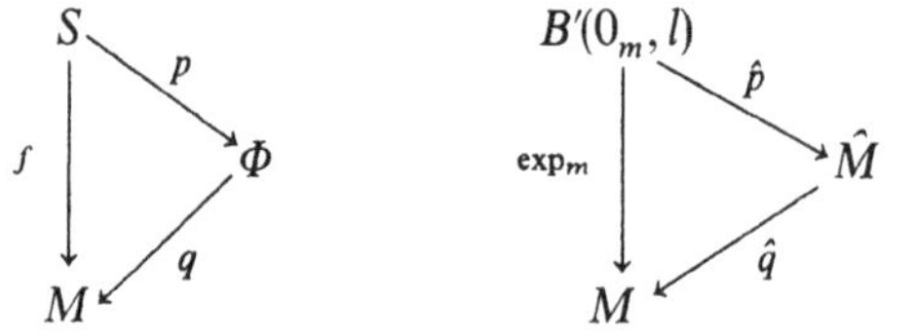

Proof. The manifold structure of Φ is precisely given by 5.28: locally at φ, the structure of Φ will be (by definition) that of the submanifold $f(V) \subset M$ (notice that in general $f^{-1}(f(\varphi))$ is larger than the maximal leaf φ; for example, this is the case in the quotient of $\mathbb{C}P^{2n-1}$ described in 5.32).

5.30. Now consider the normal geodesic neighborhood of radius ε of $f(V)$; for small enough ε, this is $\{n \in M \mid \varrho(n, f(V)) < \varepsilon\} = B(f(V), \varepsilon)$, and one knows that the exponential map $TM \to M$, when restricted to the set

$$\{v \in TM \mid p(v) \in f(V), |v| < \varepsilon, v \in (T_{p(v)} f(V))^\perp\}$$

is a diffeomorphism onto $B(f(V), \varepsilon)$.

Now by compactness one can find a positive ε and a finite number of neighborhoods V_i of the φ_i's in Φ such that $\bigcup_i f^{-1}(f(V_i)) = S$ and such that we have diffeomorphisms onto the $B(f(V_i), \varepsilon)$ for every i as above.

We now set $D = B'(0_m, l-\varepsilon)$ and $E = (B'(0_m, l) \setminus B(0_m, l-\varepsilon))/\mathcal{R}$; the conclusions of the theorem follow then from our very construction, using 5.27. The map $\hat{q}$ is a covering because M is compact and $\hat{q}$ has maximal rank everywhere. □

5.31 Corollary. *If (M, g) is a simply-connected Riemannian manifold which is at some point m an Allamigeon-Warner manifold, then $\hat{q}: \hat{M} \to M$ is a diffeomorphism. Moreover, $\mathrm{Cut}(m) = f(S)$ and, in particular, the cut-locus of m is spherical: for every n in $\mathrm{Cut}(m)$, one has $\varrho(m, n) = l$ (see 5.22).*

5.32. Some examples of non simply connected Allamigeon-Warner manifolds are given by quotients of CROSSes: firstly lens spaces (see 0.39) and, among them: $(\mathbb{R}P^d, \text{can})$, secondly the following quotient of $\mathbb{C}P^n$ for $n=2m-1$: define a map t from $\mathbb{C}^{2m}$ into itself by

$$t(z_1, z'_1, \ldots, z_m, z'_m) = (\bar{z}'_1, -\bar{z}_1, \ldots, \bar{z}'_m, -\bar{z}_m).$$

The map t induces an involutive isometry on $(\mathbb{C}P^{2m-1}, \text{can})$ without fixed point and the quotient is an Allamigeon-Warner manifold.

5.33. The converse to 5.29 and 5.31 will be considered in 5.43.

5.34 Remark. If (M, g) is an Allamigeon-Warner manifold at every point, then every maximal leaf φ is a great circle of S as one immediately sees by exchanging the roles of m and $n=f(\varphi)$. Notice that the classification of fibrations of spheres by great circles is an interesting but almost untouched subject, see very special cases in [BR 2], [ES].

D. Pointed Blaschke Manifolds and Blaschke Manifolds

For distinct points m and n in M we define the *link from m to n* to be

5.35 $$\Lambda(m, n) = \{\dot{\gamma}(n) \in U_n M \mid \gamma \in \text{Seg}(n, m)\}.$$

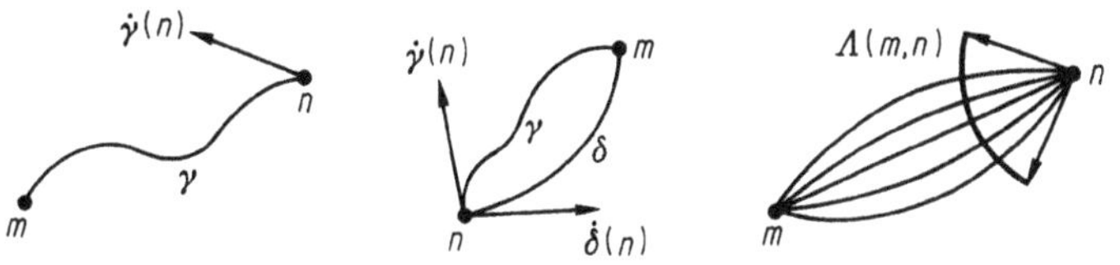

Fig. 5.35

5.36. A subset Θ of the unit sphere Σ of a Euclidean sphere space V is said to be a *great sphere* if there exists a vector subspace W of V such that $\Theta = W \cap \Sigma$. By definition, the *dimension of* Θ is $\dim \Theta = \dim W - 1$.

5.37 Definition. A compact Riemannian manifold (M, g) is said to be a *Blaschke manifold at the point m in M* if for every n in $\text{Cut}(m)$ the link $\Lambda(m, n)$ is a great sphere of $U_n M$. The manifold (M, g) is said to be a *Blaschke manifold* if it is a Blaschke manifold at every point in M.

5.38. In this definition we do not exclude two extreme cases: $\Lambda(m, n) = U_n M$ as in (S^d, can) and $\Lambda(m, n)$ consists only of two antipodal points as in $(\mathbb{R}P^d, \text{can})$ The CROSSes are examples of Blaschke manifolds. Other examples of Blaschke manifolds at a point m are manifolds which can be topologically written as $D\cup_a E$ (for instance the Eells-Kuiper exotic quaternionic projective planes, see [E-K], and exotic spheres see Appendix C) by 5.52. We will now see that Blaschke manifolds are at the same time Allamigeon-Warner manifolds and SC-manifolds (see 5.42) and that they have spherical cut-loci.

5.39 Proposition. *Let (M,g) be a Blaschke manifold at m. Then* $\dim\Lambda(m,n)$ *is a constant (say $k-1$) when n runs through* Cut(m). *This cut-locus* Cut(m) *is a $(d-k)$-dimensional submanifold of M and for every n in* Cut(m) *we have* $\Lambda(m,n)=U_nM\cap(T_n\,\mathrm{Cut}(m))^{\perp}$. *Moreover, (M,g) is a SL_l^m-manifold (see 7.7) and finally, the cut-locus* Cut(m) *is spherical at distance $l/2$.*

5.40 Lemma. *If (M,g) is a Blaschke manifold at m, then (M,g) is a SL_l^m-manifold, i.e. there exists an l such that every geodesic starting from m is a simple curve which goes back to m at time l. Moreover, all geodesic loops from m with length l have a common index, say $k-1$.*

Proof. The loop property is trivial from the definition. If γ is a geodesic starting from m, then it necessarily meets Cut(m) at some point n (since M is compact, cf. 5.20). But since $\Lambda(m,n)$ is a great sphere, both $\dot\gamma(n)$ and $-\dot\gamma(n)$ belong to $\Lambda(m,n)$. Hence, there exists a geodesic δ in Seg(n,m) with $\dot\delta(n)=\dot\gamma(n)$; in particular, $\gamma\cup\delta$ is a geodesic loop at m.

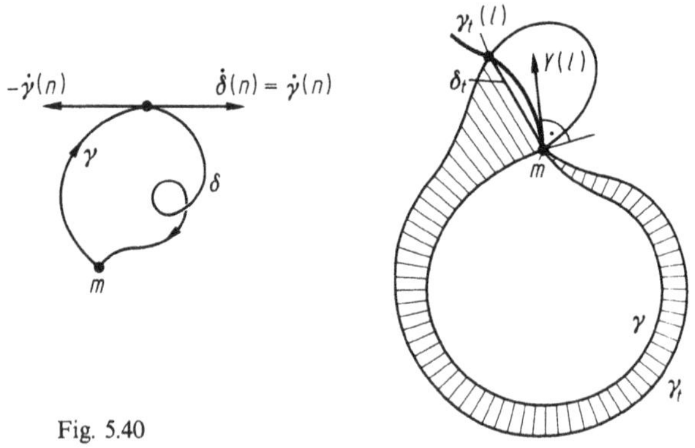

Fig. 5.40

We now have to prove that all these loops through m have the same length l. That this is not true in general for Y_l^m-manifolds i.e., manifolds all of whose geodesics through m come back to m at time l (see 7.7 and Example 0.39). But here we will use the fact that the length function of these loops is *continuous*, a fact which follows from the continuity of the cut-value (cf. 5.17) since $L(\gamma\cup\delta)=2\varrho(m,n)$. More precisely, let $\gamma:s\mapsto\gamma(s)$ be a geodesic loop through $m=\gamma(0)=\gamma(l)$ and consider a differentiable family of geodesics γ_t such that $\gamma_0=\gamma$. It yields a Jacobi field $Y:s\mapsto Y(s)$ along γ. We claim that $Y(l)=0$ for every such family γ_t so that if we can show the latter we are done since then $\gamma_t(l)=\gamma_0(l)=m$ for every t and for every such family.

We now prove the claim using reductio ad absurdum. If $Y(l)\neq 0$ the curve $t\mapsto\gamma_t(l)$ is non-singular at $m=\gamma_0(l)$ and for t small enough we get a unique geodesic δ_t in

$\mathrm{Seg}(m, \gamma_t(l))$ (cf. 5, Section B)). But by assumption, for some real number $s(t)$ one has $\gamma_t(s(t)) = m$, $s(t)$ is continuous in t and $s(0) = l$. Hence, $\gamma_t|_{s(t)}^l$ would be a second element in $\mathrm{Seg}(m, \gamma_t(l))$ since the angle between δ_t and $\gamma_t|_{s(t)}^l$ at $\gamma_t(l)$ is close to $\pi/2$ when t is close to 0.

The last assertion of the lemma follows from the continuity of conjugate points, see [WR 1], p. 599, [ME], p. 235 or 1.98, indeed a loop γ through m which is the union of two segments can only have conjugate points at its middle or at m itself (see [KI], p. 112). □

To finish the proof of 5.39 we consider two cases. Assume first that the common index $k-1$ is strictly positive. Then (M, g) is an Allamigeon-Warner manifold at m with length $l/2$ and index $k-1$. So we can apply 5.29. Here the covering map $\hat{q}: M \to M$ is a diffeomorphism since it is injective on all $B(m, l/2)$ by the very definition of the cut-locus. All the assertions of 5.39 now follow from those of 5.29.

Now suppose that k is equal to 1. First note that $\dim \Lambda(m, n) = 0$ for every n in $\mathrm{Cut}(m)$ because for a Blaschke manifold at m any point n in $\mathrm{Cut}(m)$ has an order of conjugacy along γ and with respect to m larger than or equal to $\dim \Lambda(m, n)$. Since n is not conjugate to m along any geodesic in $\mathrm{Seg}(m, n)$ and since $\Lambda(m, n)$ has only two elements, we can conclude that $\mathrm{Cut}(m)$ is a submanifold of M of codimension 1. Moreover, $\Lambda(m, n) = U_n M \cap (T_n \mathrm{Cut}(m))^{\perp}$ by the first variation formula (cf. 1.109). □

5.41 Note. It is not true a priori that the dimension of $\Lambda(m, n)$ is equal to the index of γ in $\mathrm{Seg}(m, n)$; this fact drops out as a consequence of the result.

5.42 Corollary. *If (M, g) is a Blaschke manifold, then (M, g) satisfies the assertions* 5.39 *at every point m in M. Moreover, (M, g) is an SC-manifold and* $\mathrm{Diam}(g) = \mathrm{Inj}(g)$.

Proof. We only have to apply 7.6(a). □

Now is the right time to characterize the pointed Blaschke manifold.

5.43 Theorem. *For a Riemannian manifold (M, g) and a point m in M consider the following assertions:*

(i) *(M, g) is a Blaschke manifold at m,*

(ii) *the cut-locus* $\mathrm{Cut}(m)$ *of m is spherical,*

(iii) *(M, g) is an Allamigeon-Warner manifold at m.*

For a differentiable manifold consider the following assertion:

(iv) *M can be written as $D \cup_a E$, where D is the d-dimensional closed ball, E a C^∞-closed k-disc bundle over a $(d-k)$-dimensional C^∞-compact manifold with boundary ∂E diffeomorphic to S^{d-1} and $a: \partial D \to \partial E$ an attaching diffeomorphism.*

Then:

1) (i) *and* (ii) *are equivalent and imply* (iii),
2) *if M is simply connected, then* (iii) *implies* (i) *and* (ii),
3) (i) *or* (ii) *implies* (iv); *conversely if M satisfies* (iv) *there exists a Riemannian metric g on M for which* (i) *and* (ii) *are true for the center m of D.*

Proof. 1) comes from 5.39 and 5.44 below.

2) is only a restatement of 5.31.

The direct part of 3) is 5.39 and the converse is 5.52. □

5.44 Proposition. *Let (M, g) be a Riemannian manifold and m a point in M. If the cut-locus* Cut(m) *is spherical, then (M, g) is a Blaschke manifold at m.*

This proposition was first proved by Omori in the real-analytic case (cf. [OM]) and later by Nakagawa and Shiohama (cf. [N-S 1] and [N-S 2]) in the differentiable case. The proof in the differentiable case is a little lengthy and technical. Notice that in [OM] and [N-S 1] and [N-S 2] a more general statement is proved; the point m is replaced by any submanifold N of M with spherical "cut-locus".

Proof. Our Riemannian manifold (M, g) now has spherical cut-locus at a given point m in M. We introduce the *notations* $\Lambda(n) = \Lambda(m, n)$ for n in Cut(m) and

5.45 for u in TM, $\quad v(u) = \dfrac{u}{\|u\|}$,

5.46 $\left\{\text{for } n \text{ in Cut}(m) \text{ we } \textit{denote} \text{ by } N(n) \text{ the set of } v \text{ in } U_nM \text{ such that there exists a sequence } (r(i))_{i\in\mathbb{N}} \text{ in Cut}(m)\setminus n \text{ with } \lim_{i\to\infty} r(i) = n \text{ and } v = \lim_{i\to\infty} v(\exp_n^{-1}(r(i)))\right.$.

The idea is to prove that $\Lambda(n)$ and $N(n)$ are great spheres in orthogonal complementary spaces of U_nM for all n in Cut(m) with in fact

$$N(n) = U_nM \cap T_n\text{Cut}(m).$$

That will be achieved by deriving successive conditions on the sets $\Lambda(n)$ and $N(n)$ from the sphericity of Cut(m). *We will not always repeat* "for every n in Cut(m)". Finally we *normalize* the common value $\varrho(m, n)$ for n in Cut(m) to be $\pi/2$.

We first deduce from the acute angle property 5.16 that

5.47 "for every u' in $\Lambda(n)$ and every u'' in $N(n)$, $(u', u'') \leq 0$."

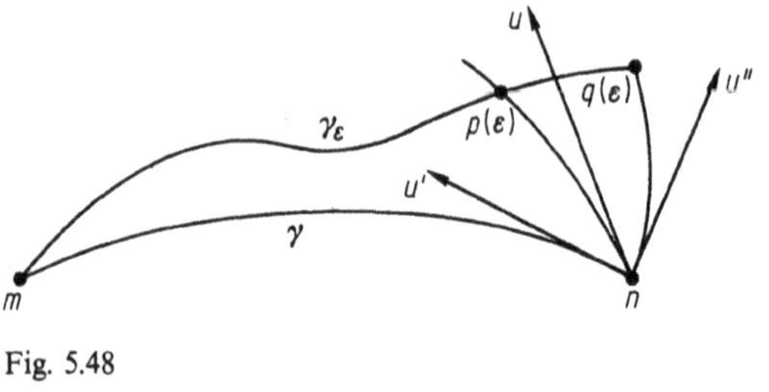

Fig. 5.48

5.48 Lemma. *For every u in U_nM there exist u' in $\Lambda(n)$ and u'' in $N(n)$ such that u belongs to the quadrant $\mathbb{R}_+u' + \mathbb{R}_+u''$ and $(u', u'') = 0$.*

Proof. We can assume that u is not in $N(n)$, otherwise the assertion of the lemma would be trivial. By assumption, $\varrho(m, p) < \pi/2$ for every p in M which is not in $\mathrm{Cut}(m)$. Hence, the very definition of $N(n)$ implies (if we set $p(\varepsilon) = \exp_n(\varepsilon u)$) that $\varrho(m, p(\varepsilon)) < \pi/2$ for every small enough positive ε. We will only work with such ε's. Call γ_ε the unique element of $\mathrm{Seg}(m, p(\varepsilon))$ and extend γ_ε beyond $p(\varepsilon)$ up to the point $q(\varepsilon) = \gamma_\varepsilon(\pi/2)$. We have $\lim_{\varepsilon\to 0} p(\varepsilon) = \lim_{\varepsilon\to 0} q(\varepsilon) = n$. By compactness we can extract from the set (γ_ε) a sequence $(\gamma_i)_{i\in\mathbb{N}}$ which converges to a geodesic and such that the sequence $\nu(\exp_n^{-1}(q(i)))$ has a limit, called u'', when i tends to infinity. Since γ is in $\mathrm{Seg}(m, n)$, $u' = -\dot\gamma(n)$ is in $\Lambda(n)$. If we can show that u is a positive linear combination of u' and u'', then 5.47 will imply that $(u', u'') = 0$.

To prove that u belongs to the quadrant $\mathbb{R}_+ u' + \mathbb{R}_+ u''$ we proceed as follows. By 1.53 the map $\mathrm{Exp}: TM \to M \times M$ is locally a diffeomorphism around 0_n and (n, n) so that for ε small enough we can lift the situation to TM. The inverse images of the pairs $(p(\varepsilon), q(\varepsilon))$ in $M \times M$ are tangent vectors in $T_{p(\varepsilon)}M$. We now take a vector bundle chart of TM, centered at 0_n and with values in the vector space V. The set-up is now the following. We have a sequence $(t_i)_{i\in\mathbb{N}}$ in $\mathbb{R}$ such that $\lim_{i\to\infty} t_i = 0$ and a fixed vector α and a sequence $(v_i)_{i\in\mathbb{N}}$ in V such that the two following conditions hold: the directions of the v_i's have a limit given by a vector β in V and the directions of the vectors $t_i\alpha + v_i$ have a limit given by some γ (when i goes to infinity, of course). What we have to prove is that γ is a linear combination of α and β, the matter of the sign being trivial. But this is elementary: the plane generated by α and v_i has a limit, when i goes to infinity, namely the plane P generated by α and β. Hence, the direction of $t_i\alpha + v_i$ has a limit which belongs to P. □

5.49 Lemma. *The subset $\Lambda(n)$ of U_nM is* convex *(i.e., for every u, v in $\Lambda(n)$ with $u \neq -v$ one has $\nu(\mathbb{R}_+ u + \mathbb{R}_+ v) \subset \Lambda(n)$). Moreover, there exists u in $\Lambda(n)$ such that $-u$ is in $\Lambda(n)$.*

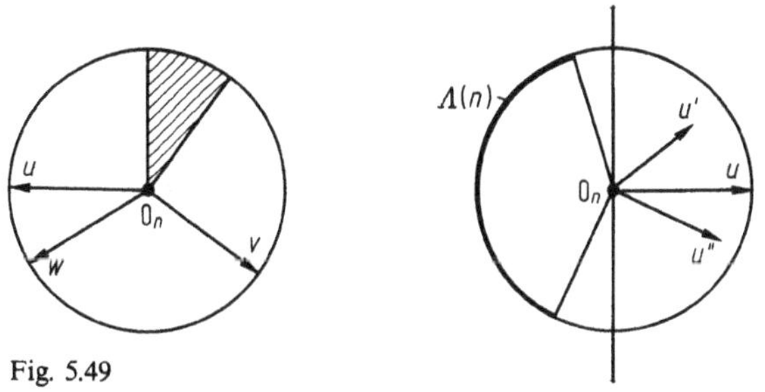

Fig. 5.49

Proof. Let w be in $\nu(\mathbb{R}_+ u + \mathbb{R}_+ v)$ with $v \neq \pm u$, i.e., w belongs to the arc of great circle defined by u and v. Using 5.48 we write $w = \alpha w' + \beta w''$ with $\alpha, \beta \geqslant 0$. Since $(w'', u) \leqslant 0$ and $(w'', v) \leqslant 0$ by 5.47 the only possibility is $w'' = 0$ and we are left with $w = w'$ in $\Lambda(n)$.

If now $-u$ were not in $\Lambda(n)$ for every u in $\Lambda(n)$, by convexity we would deduce that $\Lambda(n)$ is contained in an open hemisphere; if we denote by u in U_nM the south pole of that hemisphere, then 5.48 yields a contradiction. □

5.50 Lemma. *For every u in $\Lambda(n)$, there exists a v in $\Lambda(n)$ with $(u, v) < 0$.*

Proof. By 5.49 we can pick a point u_0 in $\Lambda(n)$ together with $-u_0$. If $(u, u_0) \neq 0$, we are trivially done. Suppose from now on that $(u, u_0) = 0$. We first prove that

5.51 $\begin{cases} \text{for every } v \text{ close enough to } -u \text{ and for every small} \\ \text{enough } \varepsilon \text{ one has } \varrho(m, \exp_n(\varepsilon v)) < \pi/2\,. \end{cases}$

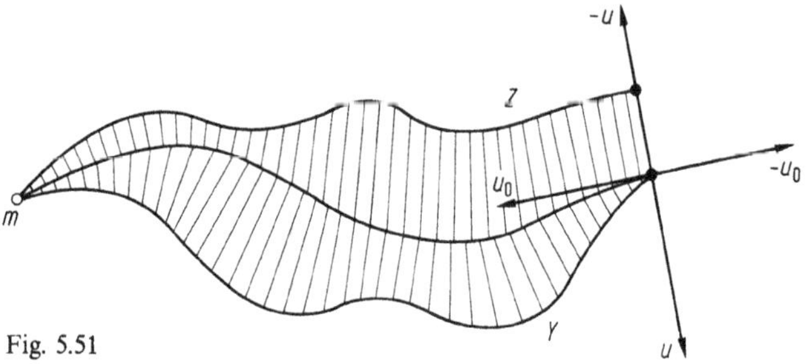

Fig. 5.51

By 5.49 there is a one-parameter family of geodesics in Seg(m, n) whose velocity vectors at n run through the great half-circle of U_nM defined by u, u_0, $-u_0$. In particular, along the geodesic γ in Seg(m, n) with $\dot{\gamma}(n) = -u_0$ there is a Jacobi field Y with $Y(0) = 0$, $Y(\pi/2) = 0$ and $Y'(\pi/2) = u$. Now pick any vector field Z along γ with $Z(0) = 0$ and $Z(\pi/2) = -u$. Then for some real number k any one-parameter family γ_α of curves defined by $\gamma_0 = \gamma$ and $\dfrac{\partial \gamma_\alpha}{\partial \alpha}(0) = Y + kZ$ will have lengths $L(\gamma_\alpha)$ with

$$\frac{d^2 L(\gamma_\alpha)}{d\alpha^2}(0) = I(Y + kZ, Y + kZ)\,,$$

where $I(.,.)$ is the index form along γ (see for example [K-N 2], p. 81 or [B-C], p. 220). Because Y is a Jacobi field vanishing at 0 and at $\pi/2$, we have $I(Y, Y) = 0$. Because Y is a Jacobi field we have also

$$I(Y, Z) = (Y'(\pi/2), Z(\pi/2)) - (Y'(0), Z(0)) = -1\,.$$

Hence,

$$\frac{d^2(\varrho(m, \exp_n(-\alpha u)))}{d\alpha^2}(0) \leqslant \frac{d^2 L(\gamma_\alpha)}{d\alpha^2}(0) < 0$$

for k small enough. Moreover, by continuity of the index form, we again have

$$\frac{d^2(\varrho(m, \exp_n(\alpha v)))}{d\alpha^2}(0) < 0$$

for every v close enough to $-u$.

Now 5.51 implies (via 5.46) that $-u$ is not in $N(n)$. If we apply 5.48 to $-u$ we get $-u = \alpha u' + \beta u''$ with $\alpha, \beta \geqslant 0$, u' in $\Lambda(n)$ and u'' in $N(n)$. Hence, $-u \neq u''$, $(u, u') = -(-u, u') = -\alpha < 0$. $\square$

Proof of 5.44. We have to prove that $\Lambda(n)$ is a great sphere. By 5.49 it is enough to show that for all u in $\Lambda(n)$ $-u$ is also in $\Lambda(n)$. We do this by induction with the following first step. By 5.49 there exists u_0 in $\Lambda(n)$ together with $-u_0$. *Set* $\Lambda^1(n) = \Lambda(n) \cap u_0^\perp$. We will show the existence of a vector u_1 which is in $\Lambda^1(n)$ together with $-u_1$. The other steps of the induction will be clear.

We argue as in the proof of 5.49 ad absurdum. Suppose that $-u$ is not in $\Lambda^1(n)$ for all u in $\Lambda^1(n)$. Then the convex set $\Lambda^1(n)$ (cf. 5.49) of the $(d-2)$-dimensional sphere $U_n M \cap u_0^\perp$ will be contained in an open hemisphere. Call u the south pole of that hemisphere and apply 5.48, $u = \alpha u' + \beta u''$ with $\alpha \geqslant 0$ and u' in $\Lambda(n)$. By 5.49 the great half circle defined by u_0, $-u_0$, u' lies entirely in $\Lambda(n)$ and its intersection $\bar{u}'$ with $u_0^\perp$ yields the desired contradiction. □

5.52 Proposition. *If M can be written as $D \cup_a E$ (with notation explained in 5.43), then there exists a Riemannian metric g on M such that (M, g) is a pointed Blaschke manifold at the center m of D.*

Proposition 5.52 is due to A. Weinstein, see [WR 2], p. 210. We will give a detailed proof of it in Appendix C, see C.4.

E. Some Properties of Blaschke Manifolds

5.53. From 5.10 and 5.39 it follows that every metric sphere

$$S(m, r) = \{n \in M \mid \varrho(m, n) = r\}$$

in a Blaschke manifold is a differentiable submanifold. In fact if we set $D = \mathrm{Diam}(g) = \mathrm{Inj}(g)$, the assertion comes from 5.10 for $r < D$ and from 5.39 for $r = D$.

5.54. Moreover, in a Blaschke manifold one can define the global geodesic symmetry σ_m around m in M by

$$\sigma_m : p \mapsto \exp_m(-\exp_m^{-1}(p)) .$$

The existence and differentiability of σ_m follows on the ball $B(m, D)$ from $D = \mathrm{Inj}(g)$ and on $M \setminus \{m\}$ from 5.39. All σ_m's are involutive diffeomorphisms.

5.55. As L. Green noticed in 1961, Blaschke's conjecture would be greatly helped if one could prove that for a Blaschke manifold (M, g) the subgroup of Diff(M), generated by the σ_m's when m runs through M, is *compact*. In fact it would solve it since then our Riemannian manifold would be homogeneous. Then we have only to apply 7.55.

5.56 Topology. Let (M, g) be a Blaschke manifold of dimension d with cut-loci of dimension $d-k$. We know by 7.23 that k can only take the values $1, 2, 4, 8, d$. We first discuss the cases $k=1$ and $k=d$. *Note that our k differs by* 1 *from that of Chapter* 7.

5.57 Proposition. *If $k=d$, for every point m in M, then the cut-locus* Cut(m) *consists of a single point (we* denote *it by m'). The map $\sigma: m \mapsto m'$ is involutive and is an isometry of (M, g). As a manifold, M is diffeomorphic to S^d. Moreover, the quotient Riemannian manifold $(M/\sigma, g/\sigma)$ is a Blaschke manifold with diameter half that of (M, g) with $k=1$ and the manifold M/σ is diffeomorphic to $\mathbb{R}P^d$.*

Conversely, if $k=1$, then M is diffeomorphic to $\mathbb{R}P^d$ and its universal Riemannian covering $(\tilde{M}, \tilde{g})$ is a Blaschke manifold with diameter twice that of (M, g), with $k=d$ and $\tilde{M}$ is diffeomorphic to S^d.

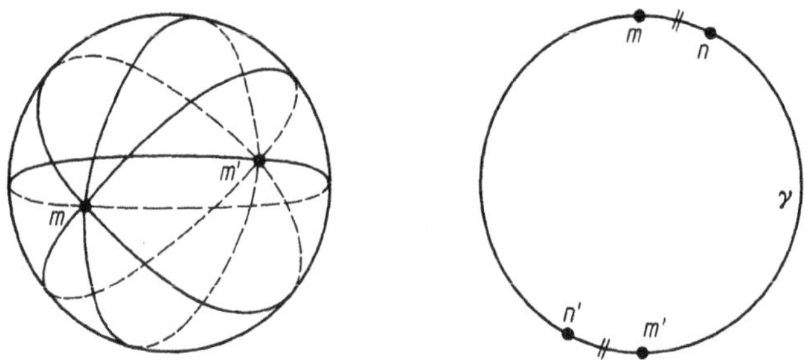

Fig. 5.57

Proof. We normalize (M, g) to have diameter equal to $\pi/2$. Pick points m and n in M and a geodesic γ in Seg(m, n). Extend γ to a closed geodesic—again called γ—with length π (thanks to 5.42). If $m=\gamma(0)$ and $n=\gamma(s)$, then their antipodal points are necessarily $m'=\gamma(\pi/2)$ and $n'=\gamma(s+\pi/2)$. Hence, $\varrho(m', n')=\varrho(m, n)=s$ since there exists a geodesic form m' to n' with length $s \leqslant \pi/2$.

Since σ is an isometry without fixed point we can consider the Riemannian quotient $(M/\sigma, g/\sigma)$. It is an $SC_{\pi/2}$-manifold and in particular, $\mathrm{Diam}(g/\sigma)=\pi/4$. To see that $\mathrm{Inj}(g/\sigma)=\pi/4$, we use 5.20 to the effect that if $\varrho(m, n)=\mathrm{Inj}(g/\sigma)$, then necessarily either m and n are conjugate along some geodesic γ in Seg(m, n) or there exist two distinct geodesics in Seg(m, n). From the assumption we made on (M, g), the conjugate point case is ruled out and we are left with two distinct geodesic segments which are necessarily the two parts of the closed geodesic of length $\pi/2$ through m and n. Then $\mathrm{Inj}(g/\sigma)=\pi/4$. We obtain that $(M/\sigma, g/\sigma)$ is a Blaschke manifold with a number k equal to 1. Then we apply the last assertion of 7.23 to the effect that M/σ is diffeomorphic to $\mathbb{R}P^d$. Hence its connected two-sheet covering M is necessarily diffeomorphic to S^d.

If conversely (M, g) is a Blaschke manifold with $k=1$, then M is diffeomorphic to $\mathbb{R}P^d$ (again by 5.42 and 7.23). Consider its universal two-sheet Riemannian covering $(\tilde{M}, \tilde{g})$ with $\tilde{M}$ diffeomorphic to S^d. If $\tilde{m}$ and $\tilde{m}'$ are two distinct points in $\tilde{M}$ with the same projection m in M, then a segment $\tilde{\gamma}$ in Seg($\tilde{m}$, $\tilde{m}'$) will project down onto a closed geodesic γ through m. But all closed geodesics through m are homotopic and will be lifted up into homotopic geodesics from $\tilde{m}$ to $\tilde{m}'$. This proves that $(\tilde{M}, \tilde{g})$ is a Blaschke manifold with $k=d$. □

5.58 Remark. The above result is false for pointed Blaschke manifolds, for example for any M which can be written as $D \cup_a E$ thanks to 3) of 5.42 and to the examples in

5.38. Even better there exists an exotic sphere which is a pointed Blaschke manifold at m with all of its geodesics through m closed, see C. 19.

To decide whether the property that M admits a Blaschke manifold structure implies more than to admit a SC^m-structure (see 7.7) is still an open question; the test manifolds could be the Eells-Kuiper exotic quaternionic projective planes, see [E-K].

F. Blaschke's Conjecture

We recall (cf. 0.33) Blaschke's conjecture: "every Blaschke manifold is isometric to a CROSS". The present section is devoted to the proof of L. Green's result and general considerations for the case $d>2$ and $k=d$. The case $k=2$ is considered in Section G).

5.59 Theorem (L. W. Green, [GN 2]). *If (M, g) is a two-dimensional Blaschke manifold, then (M, g) has constant curvature and in particular, is isometric (up to some constant) to* $(S^2$, can) *or* $(\mathbb{R}P^2$, can).

5.60 History. Before proving the theorem a little bit of history and some comments are in order. The above result was posed as a question in the first edition (1921) of [BE]. The second edition (1924) contained an appendix with a (fake) proof by Reidemeister and the third edition (1930) gave an explanantion of why Reidemeister's nice idea did not work. The conjecture (for $d=2$) was proved by L. Green in 1961.

In fact, Blaschke was not considering $\mathbb{R}P^2$ but only (S^2, g) with first conjugate points at constant distance, i.e., a Blaschke structure with $k=2$. He called these manifolds "Wiedersehensflächen". See also Appendix D.

5.61. We remark that Green's theorem really belongs to the realm of Riemannian geometry and not to that of Finsler or G-space geometry. In fact, Skorniakov proved in [SV] the following result: given any system Γ of simply-closed curves in $\mathbb{R}P^2$ which enjoys the set-theoretic properties of projective lines ("through two distinct points there goes exactly one line of Γ and two different lines of Γ meet at exactly one point"), one can put on $\mathbb{R}P^2$ a G-space metric whose geodesics are exactly the elements of Γ (for definitions and properties of G-spaces we refer the reader to the basic and wonderful book [BN]). It remains to exhibit a Skorniakov G-space which is not the underlying G-space of a Riemannian space. The possibility of doing so comes from the fact that a system Γ as above need not in general satisfy Desargues' theorem (which holds of course for the geodesics of $(\mathbb{R}P^2$, can) which are projective lines!). Contrarywise the idea on which Reidemeister's attempt to solve Blaschke's conjecture was based was to prove that geodesics satisfied Desargues' theorem (see [BR 10], 2.6.7 et 5.4.3) and then to apply Beltrami's theorem (see [DX], Chapter III). Needless to say, Blaschke, Reidemeister, etc... were working on surfaces in $\mathbb{R}^3$ and not on abstract two-dimensional Riemannian manifolds. If the reader looks at Reidemeister's attempt in [BE], p. 228—229 he will see how the introduction of the notions of abstract Riemannian manifold, and abstract unit tangent bundle was absolutely necessary to solve Blaschke's conjecture.

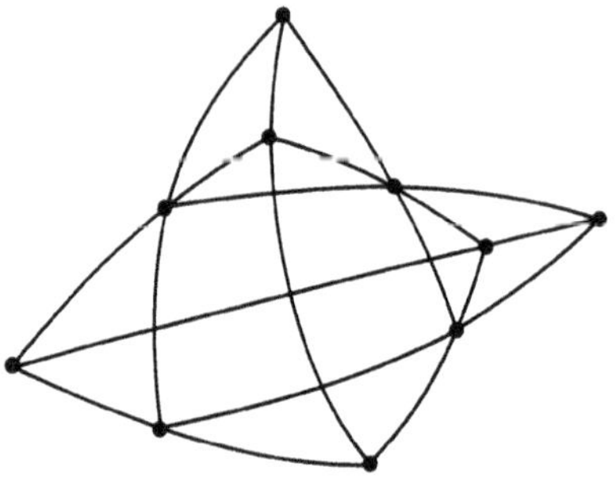

Fig. 5.61

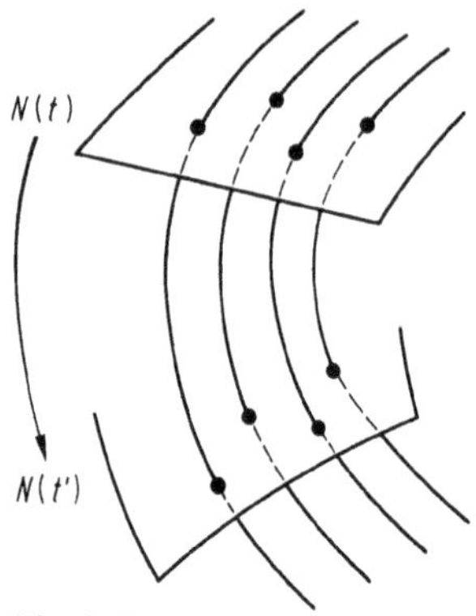

Fig. 5.62

5.62 *Focal points.* For a Blaschke manifold one can say that "on any geodesic the first conjugate point is at constant distance". One could ask the same question for the first focal point (i.e., suppose that a Riemannian manifold (M,g) is such that every Jacobi field Y with $Y(0)=0$ satisfies $(Y(\pi/2), Y'(\pi/2))=0$). This time it is not hard to see that this condition implies that (M,g) has constant sectional curvature equal to 1. The reason is simple: every $S(m,\pi/2)$ will be a totally geodesic submanifold because $(Y(\pi/2), Y'(\pi/2))$ is nothing but the second fundamental form of $S(m,\pi/2)$. Hence in our manifold we have a whole bunch of totally geodesic hypersurfaces. A classical result of Schur (see for example [BN], p. 309—332) for the more general case of G-spaces) then implies that the sectional curvature is constant according to the following observation: let $N(t)$ be a one-parameter family of totally geodesic submanifolds in (M,g). Then the bijections $N(t)\to N(t')$ given by the orthogonal trajectories to the $N(t)$'s are isometries between $(N(t), g \upharpoonright N(t))$ and $(N(t'), g \upharpoonright N(t'))$ by application of the first variation formula 1.109. For more on focal objects see [OT].

5.63 *How to Prove L. Green's Theorem?* We note the following results for a two-dimensional Riemannian manifold (S^2,g) (not necessarily a Blaschke manifold):

① if (S^2,g) is a Blaschke manifold with diameter equal to π, then $\mathrm{Vol}(g)=4\pi$;

② if (S^2,g) has no conjugate point before distance π on any geodesic, then $\mathrm{Vol}(g)\geqslant 4\pi$ and equality holds if and only if $g=\mathrm{can}$;

③ if $\lambda_1(g)$ is the first eigenvalue of Δ on (S^2,g) (see 8.45), then $\lambda_1(g)\,\mathrm{Vol}(g)\leqslant 8\pi$ and equality holds if and only if $g=\mathrm{can}$;

④ if (S^2,g) is a Blaschke manifold, then $\lambda_1(g)\geqslant 2$, and equality holds if and only if $g=\mathrm{can}$.

One sees that L. Green's theorem follows from the conjunction of ① and ②, or from the conjunction of ③ and ④ and ② (or ①).

As far as proofs are concerned, ① follows from A. Weinstein's theorem 2.24 or from 5.73. The result ② will be proved in 5.67 below. J. Hersch proved ③ in [HE]. Finally, ④ is proved in 8.47 and 8.49.

An inequality for the volume of manifolds with no conjugate point before distance a.

5.64 Proposition ([GN 2] p. 297). *Let (M,g) be a compact Riemannian manifold of dimension d and scalar curvature* Scal. *We suppose that along any geodesic of length a there are no conjugate points. Then*

$$\mathrm{Vol}(g) \geqslant \frac{a^2}{d(d-1)\pi^2} \int_M \mathrm{Scal}\, d\mu_g\,.$$

Moreover, equality holds if and only if (M,g) has constant sectional curvature π^2/a^2.

Proof. Let $\gamma:[0,a]\to M$ be a geodesic of (M,g). Since there are no conjugate points on γ between 0 and a, the index form $I(X,X)$ is positive or zero for any vector field X along γ with $X(0)=0$ and $X(a)=0$. Since we want to compare our situation with the case of constant sectional curvature π^2/a^2, let us take a vector field X which would be a Jacobi field in that case, i.e., X has the form $X(t)=\sin\frac{\pi t}{a}\,v(t)$ where $v(t)$ is parallel along γ and $(\dot\gamma, v)=0$. By 1.96 we have

$$I(X,X)=\int_0^\pi \left[\frac{\pi^2}{a^2}\cos^2\frac{\pi t}{a} - \sigma(\dot\gamma(t), v(t))\sin^2\frac{\pi t}{a}\right] dt \geqslant 0\,.$$

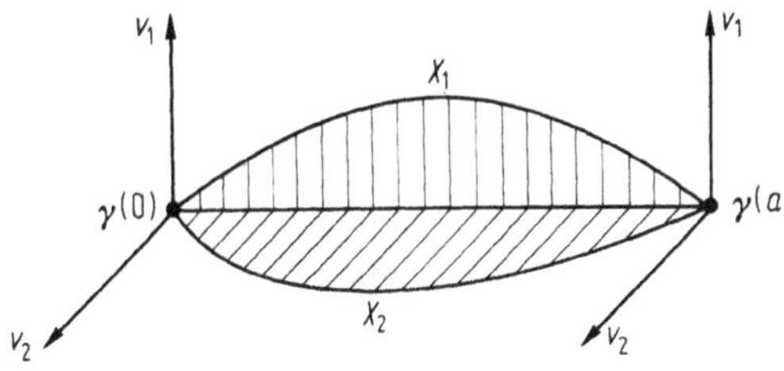

Fig. 5.64

Here $\sigma(\dot\gamma(t), v(t))$ is the sectional curvature of the plane spanned by $\dot\gamma(t)$ and $v(t)$ (cf. 1.86). If we take $d-1$ orthogonal fields of this type $X_1,\dots,X_{d-1}$, we get by definition of the Ricci curvature

5.65 $$(d-1)\frac{\pi^2}{a^2}\int_0^\pi \cos^2\frac{\pi t}{a}\,dt \geqslant \int_0^\pi \mathrm{Ric}(\dot\gamma(t),\dot\gamma(t))\sin^2\frac{t}{a}\,dt\,.$$

The idea now is to integrate (5.65) over all geodesics of length a, an integration which can be performed as follows: notice that $\dot\gamma(t)=\zeta^t(\dot\gamma(0))$, where ζ is the geodesic flow (Chapter 1.G). In fact we integrate on the product space $UM\times[0,a]$ for the canonical measure $d\mu_1\otimes dt$ and apply Fubini's theorem several times.

Using 1.125 we get (with the notation 1.M)

$$\frac{(d-1)\pi^2}{a^2} \int_{UM\times[0,a]} \cos^2\frac{\pi t}{a}\, d\mu_1 \otimes dt$$

$$\geqslant \int_{UM\times[0,a]} \sin^2\frac{\pi t}{a}\,\mathrm{Ric}(\zeta^t(u),\zeta^t(u))d\mu_1\otimes dt,$$

$$\int_{UM\times[0,a]} \cos^2\frac{\pi t}{a}\, d\mu_1\otimes dt$$

$$=\mathrm{Vol}(g_1)\int_0^a \cos^2\frac{\pi t}{a}\,dt=\beta(d-1)\,\mathrm{Vol}(g)\int_0^a\cos^2\frac{\pi t}{a}\,dt,$$

$$\int_{UM\times[0,a]} \sin^2\frac{\pi t}{a}\,\mathrm{Ric}(\zeta^t(u),\zeta^t(u))d\mu_1\otimes dt$$

$$=\int_{UM\times[0,a]} \sin^2\frac{\pi t}{a}\,\mathrm{Ric}(u,u)d\mu_1\otimes dt=\int_0^a\sin^2\frac{\pi t}{a}\,dt\int_{UM}\mathrm{Ric}(u,u)d\mu_1,$$

and

$$\int_{UM}\mathrm{Ric}(u,u)d\mu_1=\int_M\left(\int_{U_mM}\mathrm{Ric}(u,u)d\sigma\right)d\mu_g,$$

where σ is the canonical measure on U_mM (see (1.123)). A general formula for a quadratic form on a Euclidean space yields

$$\int_{U_mM}\mathrm{Ric}(u,u)d\sigma=\frac{\beta(d-1)}{d}\,\mathrm{Scal}(m).$$

Proposition 5.64 follows from the above computations if we notice that

$$\int_0^a\cos^2\frac{\pi t}{a}\,dt=\int_0^a\sin^2\frac{\pi t}{a}\,dt.$$

We are left with the proof of the if and only if part in the equality case. This is a consequence of the Jacobi equation (1.89) and of the fact that, for a geodesic $\gamma:[0,a]\to M$ without conjugate points between 0 and a, any vector field X with $X(0)=0$, $X(a)=0$ and $I(X,X)=0$ is necessarily a Jacobi field, see [B-C], p. 233, for example. □

5.66 Corollary. *Under the same assumption as in 5.64 and when $d=2$, one has*

$$\mathrm{Vol}(g)\geqslant\frac{2a^2}{\pi}\chi(M),$$

where $\chi(M)$ is the Euler characteristic of M.

This is a direct consequence of 5.64 via the Gauss-Bonnet formula (see [K-N 2], p. 358). □

5.67. For $M=S^2$ we have $\chi(M)=2$ and we get ② of 5.63.

Santalo's formula

5.68. A $(d-1)$-dimensional submanifold A of finite volume is given in (M,g): $\mathrm{Vol}(A, g \upharpoonright A) < \infty$. We set

$$UM \upharpoonright A = \bigcup_{n \in A} U_n M \quad \text{and} \quad UA = \bigcup_{n \in A} (T_n A \cap U_n M).$$

Then $W=(UM \upharpoonright A) \setminus UA$ is a $(2d-2)$-dimensional submanifold of UM (for details on Santalo's formula see [BR 3], p. 286—292.

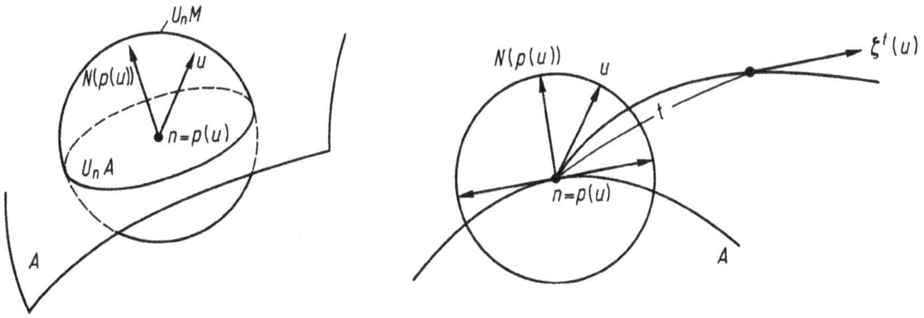

Fig. 5.68

The formula we are looking for concerns the subset $E_l = \bigcup_{t \in [0,l]} \zeta^t(W)$ of UM, where l is a given positive real number and (ζ^t) is the geodesic flow (see 1.47). We want to compute $\mathrm{Vol}(E_l)$ for the canonical measure $d\mu_1$ of UM (see (1.121)). To do so we introduce the map $f: \mathbb{R} \times W \to UM$ defined by $f(t,u)=\zeta^t(u)$, we endow $\mathbb{R} \times W$ with the natural product measure $dt \otimes d\bar{\mu}_A$ where dt is the Lebesgue measure on $\mathbb{R}$ and $d\bar{\mu}_A = d\mu'_1 \otimes d\mu_A$ where this measure is the infinitesimal product of the Riemannian measure $d\mu_A$ on $(A, g \upharpoonright A)$ and $d\mu'_1$ the collection of the Euclidean g-measures on the spheres $U_n M$ for n running through A.

5.69. To compute $\mathrm{Vol}(E_l)$ we have to evaluate the function defined by $f^*(d\mu_1) = \varphi(dt \otimes d\bar{\mu}_A)$. By Liouville's Theorem 1.56, we have for t in $[0,l]$ for u in W

5.70 $\qquad \varphi(t,u)=\varphi(0,u)$.

By a direct computation (using only the definitions of the various measures)

5.71 $\qquad \varphi(0,u)=|g(u, N(p(u)))|$,

where $N(p(u))$ is any unit vector orthogonal to $T_{p(u)}A$ (the hyperplane tangent to A at the foot $p(u)$ of u in W). From (5.70) and (5.71) we see that f has everywhere maximal rank. Hence, f is a local diffeomorphism and

$$\mathrm{Vol}(E_l) \leqslant \int_{[0,l]\times W} f^*(d\mu_1) = \int_{[0,l]\times W} \varphi dt \otimes \bar{\mu}_A$$

$$= \int_0^l dt \int_{a\in A} \left(\int_{u\in U_a M \setminus U_a A} |g(u, N(p(u)))| d\mu_1' \right) d\mu_A .$$

But $\int_{U_n M} |g(u, N(p(u)))| d\mu_1'$ is a universal constant, computable by elementary calculus and equal to $\frac{2}{d-1}\beta(d-1)$. We have thus proved the following

5.72 Proposition (Santalo, [SO], p. 488 or [BR 3], p. 290). *For every submanifold A and every $l > 0$,*

$$\mathrm{Vol}(E_l) \leqslant \frac{2}{d-1}\beta(d-1)\,\mathrm{Vol}(A, g \upharpoonright A).$$

Moreover, equality holds when f is injective.

5.73 Corollary. *If $(S^2, g) = (M, g)$ is a Blaschke manifold, then* $\mathrm{Vol}(g) = 4\pi$.

Let us take in 5.72 the submanifold A to be a closed geodesic γ length 2π and $l = \pi$. Then $\mathrm{Vol}(A, g \upharpoonright A) = L(\gamma) = 2\pi$ and f is injective by the definition of a Blaschke manifold (and Figure 5.73). Moreover, $E_\pi = UM \setminus UA$ and since UA has measure 0, we have $\mathrm{Vol}(E_\pi) = 8\pi^2 = \mathrm{Vol}(UM) = 2\pi \mathrm{Vol}(g)$ by (1.124). □

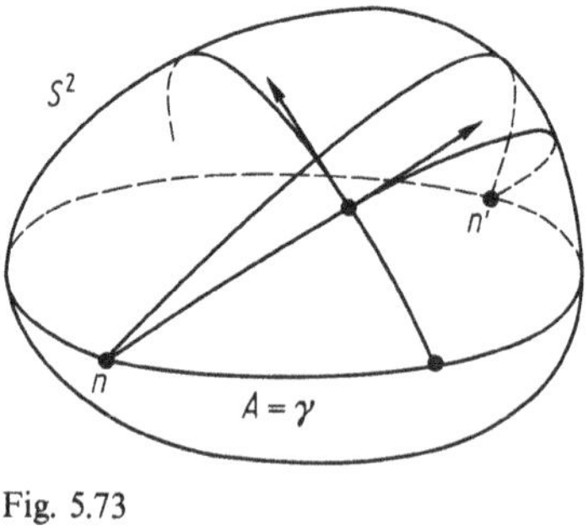

Fig. 5.73

Notes on Blaschke's Conjecture for the Sphere in Higher Dimensions

5.74. The inequality 5.64 is useless for $d \geqslant 3$ since for every $d \geqslant 3$ there exist Riemannian metrics on S^d with constant scalar curvature equal to any given number (see [K-W 2], p. 329). However, we will exploit the technique of 5.64 to get the partial result 5.81.

5.75. With Santalo's formula 5.72 we could hope to draw conclusions from a nice choice of the submanifold A. Take $A = S(m, \pi/2)$. We encounter two difficulties. The first one is that there is no reason for us to have equality in 5.72 (except if A were totally geodesic but there is no reason for that, compar with 5.62). The second one is that the inequality $\mathrm{Vol}(A, g \upharpoonright A) \geqslant \beta(d-1)$ is the one which would reasonably be expected (and not the inequality $\leqslant$!). Indeed, it is reasonable to expect that some day an extension of Pu's theorem to all dimensions will be obtained (see 5.79 below). So Santalo's generalized inequality would go in the wrong direction, at least in the present state of knowledge.

5.76. Notice also that the $S(m, \pi/2)$ are preserved by the antipody of 5.57 on a Blaschke manifold (S^d, g) but that they are not a priori totally geodesic and hence are not themselves Blaschke manifolds. If this were the case, an induction would prove the conjecture.

5.77. Looking at the generalization 8.49 of 5.63 we have $\lambda_1^-(g) \geqslant d$. If we could prove that $\lambda_1(g) = \lambda_1^-(g)$, and hence that $\lambda_1(g) \geqslant d$, we would not be through since the extension of Hersch's theorem ③ would be missing. Such an extension is unknown at the moment.

5.78. At least when d is even, an approach to Blaschke's conjecture on S^d would be to use A. Weinstein's theorem 2.24 which implies that $\mathrm{Vol}(g) = \beta(d)$ together with the following assertion

for every Riemannian manifold (M, g) of dimension d, one has

$$\mathrm{Vol}(g) \geqslant \frac{\beta(d)}{\pi^d} \mathrm{Inj}^d(g) \text{ with equality if and only if } (M, g) = (S^d, \mathrm{can}).$$

Such an inequality (except in dimension 2!) is completely open at the moment (see [BR 8], [BR 9] and [BR 11] for a discussion and see also Appendix D).

5.79. Finally, another approach would be to start again with A. Weinstein's result (d even) $\mathrm{Vol}(g) = \beta(d)$ and to have the following extension of Pu's theorem at our disposal:

$$\text{for every Riemannian metric on } \mathbb{R}P^d\text{: } \mathrm{Vol}^{d-1}(g) \geqslant 2 \frac{(\beta(d))^{d-1}}{(\beta(d-1))^d} \mathrm{Carc}_{d-1}^d(g),$$

(where $\mathrm{Carc}_{d-1}(g)$ denotes the infimum of the volumes $\mathrm{Vol}(N, g \upharpoonright N)$ of all submanifolds N in $\mathbb{R}P^d$ which are homologous to $\mathbb{R}P^{d-1} \subset \mathbb{R}P^d$) with equality if and only if $g = \mathrm{can}$.

Then we could work on $(S^d/\sigma, g/\sigma)$ as in 5.57. Such an extension looks completely open at the moment, see [BR 5].

G. The Kähler Case

5.80. All the above ideas concern Blaschke's conjecture on $M = S^d$. For the cases $k = 2, 4, 8$ (see 5.56) we know of no result except the following. The idea is to exploit

5.64 in a case where we know that $\int_M \mathrm{Scal}\, d\mu_g$ is an invariant (as it was by Gauss-Bonnet theorem when $d=2$). The Kähler situation yields exactly this invariance property.

5.81 Proposition ([BR 7]). *Let* $(\mathbb{C}P^n, g)$ *be a Kähler structure on* $\mathbb{C}P^n$, *whose underlying complex structure is the standard one of* $\mathbb{C}P^n$. *Let us moreover suppose that*

(i) $(\mathbb{C}P^n, g)$ *is a Blaschke manifold (say normalized with* $\mathrm{Diam}(g)=\pi/2$) *and that*

(ii) *its Weinstein's integer (f. 2.21) is equal to that of* $(\mathbb{C}P^n, \mathrm{can})$.

Then $g=\mathrm{can}$.

Remark. Some readers might prefer to replace (ii) by the stronger condition: (ii′) the metric g is close to can.

Proof. In fact, we do not exactly use 5.64 but the idea of the proof of 5.64 in a somewhat simpler way. By assumption we know that on every geodesic γ of length $\pi/2$ there are no conjugate points. Along γ we take the specific Jacobi field $X(t)=\sin 2t\, J(\dot\gamma(t))$, where J denotes complex multiplication by $\sqrt{-1}$ in $T\mathbb{C}P^n$ and we again integrate the inequality $I(X,X)\geqslant 0$ on the set of geodesics with length $\pi/2$ (here without any summation on the X's, as in 5.64). For every unit vector $u=\dot\gamma(0)$ we get

$$\text{5.82} \qquad 4\int_0^{\pi/2} \cos^2 2t\, dt \geqslant \int_0^{\pi/2} \sigma(\zeta^t(u), J(\zeta^t(u)))\sin^2 2t\, dt\,.$$

We remark that $\sigma(v, J(v))$ is the so-called *holomorphic curvature* of the complex line of $T\mathbb{C}P^n$ spanned by the vector v. We will be through if we can compute for a Kähler manifold the integral $\int_{U_mM} \sigma(v, J(v))d\sigma$ of the holomorphic curvature of all complex lines through a point m in M. An algebraic computation yield

$$\text{5.83} \qquad \int_{U_mM} \sigma(v, J(v))\, d\sigma = \frac{\beta(2n-1)}{n(n+1)}\mathrm{Scal}_g(m)\,.$$

From (5.82) and (5.83) we get by integration on UM (as in 5.64)

$$\text{5.84} \qquad \mathrm{Vol}(g)\geqslant \frac{1}{4n(n+1)}\int_{\mathbb{C}P^n} \mathrm{Scal}_g d\mu_g\,.$$

Note that equality can only occur if the holomorphic curvature is constant (i.e., if $g=\mathrm{can}$ by [K-N], p. 170 and 171).

Now by A. Weinstein's Theorem 2.21 and the assumption (ii) we have $\mathrm{Vol}(g)=\mathrm{Vol}(\mathrm{can})$. Let us denote the Kähler form of g (resp. can) by ω (resp. ω_0) on $\mathbb{C}P^n$. Since $\dim H^2(\mathbb{C}P^n, R)=1$ (cf. 3.42), one has $\omega=k\omega_0+d\alpha$ for some differential 1-form α. Since the volume form of g (resp. can) is $\frac{\omega^n}{n!}\left(\text{resp. } \frac{\omega_0^n}{n!}\right)$ (see for example

[B-G-M], p. 117) we get $\mathrm{Vol}(g)=k^n \mathrm{Vol}(\mathrm{can})$. Therefore, $k=1$. A classical result asserts that $\omega=\omega_0+d\alpha$ implies

$$\int_{\mathbb{C}P^n} \mathrm{Scal}_g d\mu_g = \int_{\mathbb{C}P^n} \mathrm{Scal}_{\mathrm{can}} d\mu_{\mathrm{can}}$$

(see for example [B-G-M], p. 118). If we apply (5.84) to $g=\mathrm{can}$ (in order to avoid an explicit computation) we get

$$\mathrm{Vol}(\mathrm{can})=\frac{1}{4n(n+1)} \int_{\mathbb{C}P^n} \mathrm{Scal}_{\mathrm{can}} d\mu_{\mathrm{can}}.$$

The comparison of the various formulas obtained implies that equality should hold in (5.84). Since g have constant holomorphic curvature, it is therefore the canonical metric. ☐

H. An Infinitesimal Blaschke Conjecture

5.85. We are interested in the study of infinitesimal variations of the C_l-property. We first derive the equations satisfied by the first jet of a family of C_l-metrics (this is a certain linearized C_l-condition).

5.86 Proposition. *If $t \mapsto g(t)$ is near 0 a curve of C_l-metrics, then $h=\frac{dg(t)}{dt}(0)$ satisfies the* zero energy condition, *i.e., "for any geodesic γ of $g(0)=g$ $\int_\gamma h(\dot\gamma,\dot\gamma)=0$."*

Proof. Let γ_t be a differentiable family of closed curves such that for each t near 0 the curve γ_t is a geodesic of $g(t)$. We will suppose that the parameter s of the curve γ_t belongs to $[0, l]$ and is normal for $g(t)$. The transverse vector field to the variation (γ_t) along γ_0 is denoted by X_0.

Since $t \mapsto g(t)$ is a curve of C_l-metrics, for all sufficiently small t

$$\int_0^l g(t)(\dot\gamma_t(s),\dot\gamma_t(s))\,ds=l.$$

Thus, by differentiating at $t=0$ we get

$$\int_0^l [h(\dot\gamma_0(s),\dot\gamma_0(s))+2g(\dot\gamma_0(s),X_0(s))]\,ds=0.$$

But we now notice that $\int_0^l g(\dot{\gamma}_0(s), X_0(s))ds$ is also the value of the derivative at $t=0$ of $2\mathbb{E}_g(\gamma_t) = \int_0^l g(\dot{\gamma}_t(s), \dot{\gamma}_t(s))ds$, which is known to be zero by the first variation formula, see 1.109 (since γ_0 is a geodesic for g). The zero energy condition is established. □

5.87. Notice that in the preceeding proof we have not used the fact that γ_t is a geodesic for $g(t)$, but only the fact that γ_0 is a geodesic for g. This shows that the zero energy condition for the first jet of the variation of metrics $t \mapsto g(t)$ is also satisfied for any variation of a C_l-metric g which keeps a curve of energy l through each point. This suggests that we have not used the full strength of the condition that the variation of g was by C_l-metrics.

5.88. General comments on infinitesimal variations are in order. Firstly, the problem of studying infinitesimal variations is at the same time more and less general than the problem of studying finite variations satisfying some geometric property. Indeed, when an infinitesimal variation does exist, there remains the question of integrating it in a finite variation. Hence, infinitesimal variations can exist even when no finite variation exists. On the other hand, there might exist a flat finite variation (i.e., such that all of its jets at some point are zero).

Secondly, for any geometric property, say of a Riemannian metric g (we recall that, in differential geometry, *geometric* means invariant by diffeomorphisms), there are always infinitesimal variations of g which preserve this property and give rise to finite variations. Namely, the Lie derivatives of g. Indeed, if $h = \mathscr{L}_X g$ for some vector field X, then if (φ^t) denotes the flow of X, $(\varphi^{t*}(g))$ is a finite variation of g which must satisfy the same property, and it is clear that we should work modulo such variations.

5.89. Our infinitesimal version of Blaschke's conjecture is:

"on $\mathbb{K}P^n$, there does not exist any symmetric differential form satisfying the zero-energy condition for all geodesics of the canonical metric except Lie derivatives of the canonical metric."

We will prove this version for $\mathbb{R}P^d$; for $\mathbb{C}P^n$ we mention a partial result in this direction (see 5.93). The original proofs are due to R. Michel ([ML 1] and [ML 2]) and involve some lengthy computations of Laplacians on the unit tangent bundle, estimates of their eigenvalues and the Radon transform in its full generality. We will give a more conceptual proof.

5.90 Theorem. *The infinitesimal Blaschke's conjecture* (5.89) *is true for* $\mathbb{R}P^d$ $(d \geqslant 2)$.

Proof. The idea of the proof is to reduce the problem to the case $d=2$. We use the existence of totally geodesic two-dimensional projective spaces through any two-plane tangent to $\mathbb{R}P^d$, see 3.D. We also use a result which goes back to E. Calabi (see [CI] or [B. B-B-L]) and relates solutions of a certain system of partial differential equations and certain cohomology spaces.

Let π be a two-plane tangent to $\mathbb{R}P^d$ and let $i_\pi : \mathbb{R}P^2 \hookrightarrow \mathbb{R}P^d$ be a totally geodesic immersion of $\mathbb{R}P^2$ such that π is the image under Ti_π of the tangent space to $\mathbb{R}P^2$ at some point. Thus the symmetric differential two-form $i_\pi^*(h)$ also satisfies the zero-energy condition on $\mathbb{R}P^2$. We first prove the following

5.91 Lemma. *The infinitesimal Blaschke conjecture is true for* $\mathbb{R}P^2$.

Proof. Let k be a zero-energy symmetric differential two-form on $\mathbb{R}P^2$. Applying an infinitesimal version of the conformal representation theorem (see [BE-EB], theorem 5.2), we can write k as

$$k = \mathscr{L}_X g + f \cdot g$$

for some vector field X and some function f on $\mathbb{R}P^2$.

The part $\mathscr{L}_X g$ automatically satisfies the zero-energy condition, so that for all geodesic γ on $\mathbb{R}P^2$ we have $\int_\gamma f = 0$. This says precisely that the Radon transform of f (see 4.G) for definition) is zero. This proves that f is zero (for an elementary proof, see 4.58—4.62) the lemma is proved. □

5.92 *End of the Proof of Theorem* 5.90. From the preceding lemma, $i_\pi^* h$ is a Lie derivative of the canonical metric on $\mathbb{R}P^2$. Therefore, since the sectional curvature of $\mathbb{R}P^2$ is constant and equal to 4, the infinitesimal variation of the curvature deduced from $i^* h$ at each point of $i_\pi(\mathbb{R}P^2)$ is zero.

We now claim that the infinitesimal variation of the sectional curvature of the two-plane π deduced from h is also zero. This variation can be computed along any totally geodesic submanifold tangent to π. Since the two-plane π is arbitrary, this implies that the infinitesimal variation of the full curvature tensor deduced from h is also zero (by polarization, one recovers the full curvature tensor, see [G-K-M], p. 93).

But the result of Calabi quoted above asserts that any infinitesimal variation of the canonical metric on $\mathbb{R}P^d$ which induces no variation of the full curvature tensor is a Lie derivative of the metric (this is due to the fact that such variations modulo Lie derivatives are in one-to-one correspondance with $H^1(\mathbb{R}P^d, \Theta)$, where Θ denotes the sheaf of germs of infinitesimal isometries, and to the fact that $H^1(\mathbb{R}P^d, \Theta) \sim H^1(\mathbb{R}P^d, \mathbb{R}) \otimes \mathfrak{so}(d+1)$). □

5.93 Theorem (R. Michel, see [ML 2]). *Any infinitesimal variation of the canonical metric of* $\mathbb{C}P^n$, *which when restricted to any complex projective line is a Lie derivative of the canonical metric, is globally a Lie derivative of the canonical metric on* $\mathbb{C}P^n$.

For the proof, see [ML 2].

Chapter 6. Harmonic Manifolds

A. Introduction

6.1. Let M be a ROSS (see 3.16). The fact that its isometry group is transitive on UM or on pairs of equidistant points implies that a lot of things do not really depend on m and n in M but only on the distance between them $\varrho(m,n)$. We shall mainly consider two objects.

Firstly, fix a point m in a Riemannian manifold (M,g) and consider the function θ_m roughly defined as

6.2
$$n \mapsto \theta_m(n) = \frac{\mu_{\exp_m^* g}}{\mu_{g_m}}(\exp_m^{-1}(n)),$$

i.e., the quotient of the canonical measure of the Riemannian metric $\exp_m^* g$ on $T_m M$ (pull back of g by the map $\exp_m$) by the Lebesgue measure of the Euclidean structure g_m on $T_m M$. This quotient at the point $\exp_m^{-1}(n)$ is a real number and $n \mapsto \theta_m(n)$ is a well-defined function near m. For example, θ_m is identically 1 for every m in (M,g) when (M,g) is flat. The problem of determining where the map $(m,n) \mapsto \theta_m(n)$ is really defined causes us some trouble (it is certainly defined whenever $\varrho(m,n) < \operatorname{Inj}(g)$, see 5.11). We will circumvent this difficulty by using the following trick, suggested by the classical formula giving θ by means of Jacobi fields (see [B-G-M], G.V. 15, p. 137 for example).

6.3. Let v be a point in $\mathring{T}M$ and $\gamma: s \mapsto \exp_{p(v)}\left(\frac{s}{\|v\|} v\right)$ the geodesic with initial velocity vector $\frac{v}{\|v\|}$; let $\{Y_i\}_{i=2,\ldots,d}$ be Jacobi fields along γ such that $Y_i(0)=0$ for every i and such that $\{Y_i'(0)\}_{i=2,\ldots,d}$ is an orthonormal basis of $v^\perp$. We define a function $\theta: TM \to \mathbb{R}$ by $\theta(v)=1$ whenever v is a zero vector and

6.4
$$\theta(v) = \|v\|^{d-1} \det(Y_2(\|v\|), \ldots, Y_d(\|v\|))$$

otherwise, the determinant being understood with respect to the basis $\{Y_i'(0)\}$ of $v^\perp$. We notice (by [B-G-M], p. 137 already mentioned) that we have $\theta(v) = \theta_{p(v)}(\gamma(\|v\|))$ in the sense of (6.2) when $\|v\|$ is small enough. Here θ makes sense as soon as v belongs to the domain of definition of the exponential map and is in particular everywhere defined when (M,g) is complete.

6.5. We now remark that if (M, g) is a ROSS, then "$\theta(v)$ depends only on $\|v\|$". More precisely, there exists a function $\Theta: \mathbb{R}_+ \to \mathbb{R}$ such that $\theta(v) = \Theta(\|v\|)$ for every v in TM. Then we define a *globally harmonic* Riemannian manifold (M, g) to be one which is complete and for which the preceding condition holds (see 6.10).

6.6. We now suppose that the Riemannian manifold (M, g) is compact. Then it has a unique kernel for the heat equation $K: M \times M \times \mathbb{R}_+^* \to \mathbb{R}$ (cf. [B-G-M], p. 204 for example). If (M, g) is a ROSS, then the uniqueness and the invariance of this kernel implies that "K depends only on the distance". More precisely, there exists a map $\Xi: \mathbb{R}_+ \times \mathbb{R}_+^* \to \mathbb{R}$ such that $K(m, n, t) = \Xi(\varrho(m, n), t)$ for every m, n in M and t in $\mathbb{R}_+^*$. We define a *strongly harmonic* Riemannian manifold (M, g) to be one which is compact and for which the preceding condition holds (see 6.10).

6.7. We emphasize the fact that, besides ROSSes (see 3.16) or locally flat Riemannian manifolds and their discrete quotients, no other example of a globally harmonic manifold is known; no other strongly harmonic manifolds besides the CROSSes (see 3.16) are known. The content of this chapter will show that the study of harmonic manifolds, despite the simplicity of their definition, is a very hard one.

Why do we include this topic in the present book? Because Allamigeon's theorem (cf. 6.82) asserts that a simply connected, globally harmonic Riemannian manifold is either diffeomorphic to $\mathbb{R}^d$ or is a Blaschke manifold. This partially reduces the search for harmonic manifolds to Blaschke's conjecture (see 5.F).

A nearly complete reference on harmonic manifolds is [R-W-W], which is completed by [AN]; see also the recent references [WE] and [W-H].

6.8. The theory of harmonic manifolds originated as follows: in the Euclidean space $\mathbb{R}^d$ a nice solution of $\Delta f = 0$ is given by

$$f(x) = \|x\|^{2-d} \quad \text{if} \quad d > 2, \qquad f(x) = \log(\|x\|) \quad \text{if} \quad d = 2,$$

a solution which is a function only of the distance to the origin. In 1930 H. S. Ruse made an attempt to find, on any Riemannian manifold (M, g), a solution of $\Delta f = 0$ depending only on the distance $\varrho(m, .)$ for m in M. That attempt having failed, people defined a Riemannian manifold to be *harmonic* precisely when it admitted solutions of $\Delta f = 0$ of that type. The name harmonic is explained in 6.20.

Then mathematicians tried to find out what harmonic Riemannian manifolds are. They first found conditions on curvature (cf. Section C), in particular, harmonic manifolds are Einstein spaces. These curvature conditions are strong enough to conclude that harmonic Riemannian manifolds of dimension $\leqslant 4$ are symmetric spaces (cf. section E). However it is an open problem to conclude that these curvature conditions imply that the space is symmetric when the dimension is > 4.

Then Allamigeon's result (of a global nature) implies topological restrictions which are valid in any dimension via R. Bott's theorem, see 6.82, 5. E and 7.23 via 5.42.

The notion of a strongly harmonic manifold seems to be new and is studied in Section G. From it we are able to deduce quite startling consequences but no definitive conclusions.

6.9 Remark. In [R-W-W], p. 53ff, the reader will see that the notion of a harmonic manifold can be defined for manifolds endowed with a non elliptic metric. But then the situation is radically different: for any $d \geqslant 4$ one knows that there exist harmonic manifolds which are not isometric to a symmetric space (symmetric, of course, in the non elliptic sense). On the contrary, it is not too hard to prove, using the curvature conditions, that a Lorentzian harmonic manifold necessarily has constant curvature: see [R-W-W], p. 68.

In this chapter Riemannian manifolds are not necessarily complete.

B. Various Definitions, Equivalences

6.10. Definitions. A Riemannian manifold (M, g) is said to be *locally harmonic* (we will say that it is an LH-manifold) if for every m in M there exists a positive real number $\varepsilon(m)$ and a function $\Theta_m:[0, \varepsilon(m)[\to \mathbb{R}$ such that

$$\forall u \in T_m M, \quad \|u\| \leqslant \varepsilon(m): \quad \theta(u) = \Theta_m(\|u\|).$$

A Riemannian manifold (M, g) is said to be *globally harmonic* (and we shall call it a GH-manifold) if it is complete and if for every m in M there exists a function $\Theta_m: \mathbb{R}_+ \to \mathbb{R}$ such that

$$\forall u \in T_m M: \quad \theta(u) = \Theta_m(\|u\|).$$

A compact Riemannian manifold (M, g) is said to be *strongly harmonic* (we shall say SH-harmonic) if there exists a map $\Xi: \mathbb{R}_+ \times \mathbb{R}_+^* \to \mathbb{R}$ such that

$$\forall m, n \in M: \quad \forall t \in \mathbb{R}_+^*: \quad K(m, n, t) = \Xi(\varrho(m, n), t).$$

We will first prove that the functions Θ_m in the first two definitions do not in fact depend on m. This will unify the present definition with 6.5 and also simplify the notation.

6.11 Definition. The *canonical geodesic involution* i on the tangent bundle TM of a Riemannian manifold is defined as follows. Denote by Ω the domain of definition of the exponential map (see 1.52) and let v be in Ω. If γ denotes the geodesic $t \mapsto \exp(tv)$, then i is the map $\Omega \to \Omega$ defined by

$$i(v) = -\dot{\gamma}(1).$$

6.12 Lemma. *For every v in Ω, one has $\theta(i(v)) = \theta(v)$.*

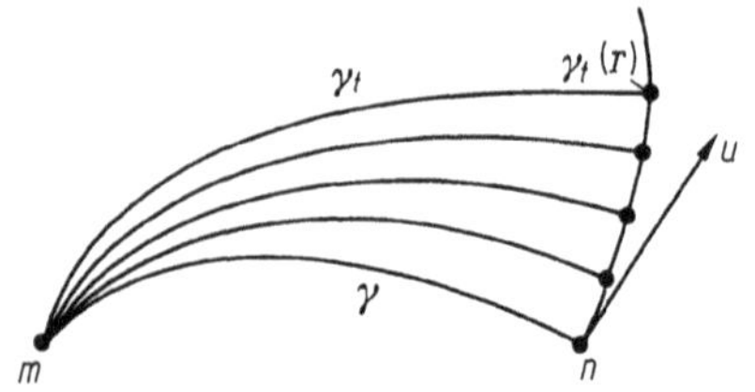

Fig. 6.12

We shall in fact show more. Pick any geodesic γ parametrized by arc length and identify tangent vectors along γ by parallel transport to vectors in a fixed Euclidean vector space V of dimension d (for example $V = T_{\gamma(0)}M$). The Jacobi field equation along γ (cf. 1.89) is

$$Y'' + R(Y) = 0,$$

where R is a symmetric endomorphism of V depending on s. As usual we consider the associated endomorphism-valued equation

6.13 $$A'' + R \circ A = 0$$

where A is a not necessarily symmetric endomorphism depending on s. By the Sturm trick, if $*$ denotes transposition of endomorphisms in the Euclidean space V, then the symmetry of R implies that

6.14 $$A'^{*}B - A^{*}B' = \text{constant}.$$

Fix s and now let A (resp. B) be the solution of (6.13) such that $A(0) = 0$ and $A'(0) = I$ (resp. $B(s) = 0$ and $B'(s) = -I$). Then by (6.14) we have

6.15 $$B(0) = A(s)^{*}.$$

In particular, (6.15) implies that

$$\theta(s\dot{\gamma}(0)) = s^{d-1}\det(A(s)) = s^{d-1}\det(B(0)) = \theta(-s\dot{\gamma}(s)). \quad \square$$

6.16 Proposition. *If (M, g) is a LH- or a GH-manifold, then there exists a function $\Theta : [0, \sup\{\varepsilon(m) : m \in M\}[\to \mathbb{R}$ such that for every m in $M : \Theta_m = \Theta \restriction [0, \varepsilon(m)[$.*

Proof. We shall prove the following equivalent assertion: for every real number r such that there exists m in M which satisfies $\Theta_m(r) \neq 0$, we have $\Theta_n(r) = \Theta_m(r)$ for every n in M.

Let r, m, be such that $\Theta_m(r) \neq 0$. By the continuity of θ there exists an open neighbourhood V of m on which $\Theta_n(r) \neq 0$ for every n in V. We shall prove that the function f defined on V by $n \mapsto \Theta_n(r)$ is constant. We shall then be done since θ is continuous and M connected. In fact, we shall prove that the differential of f is 0.

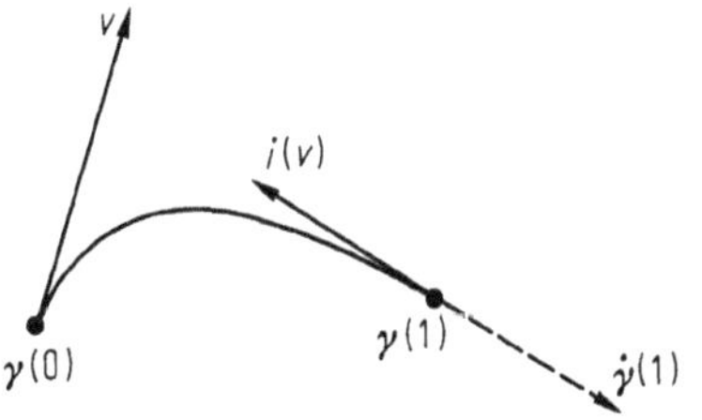

Fig. 6.16

Let n be in V and u in U_nM. To compute $df(u)$ we pick a geodesic γ through $m=\gamma(0)$ which is orthogonal to u (i.e., $(u,\dot\gamma(n))=0$) and we consider the point $n=\gamma(r)$. Since $\theta(r\dot\gamma(n))=\Theta_n(r)\neq 0$, the points m and n are not conjugate along γ. In particular, there exists a one-parameter family of geodesics γ_t with $\gamma_0=\gamma$, $\gamma_t(r)=m$ for every t and $\frac{\partial}{\partial t}(t\mapsto\gamma_t(r))(0)=u$. If we apply 6.12 to each of these geodesics we get $\Theta_{\gamma_t(r)}(r)=\Theta_m(r)$ for every t. Hence, $\frac{\partial}{\partial t}(\Theta_{\gamma_t(r)}(r))(0)=df(u)=0$. □

From now on we shall use the notation Θ instead of the Θ_m's.

Equivalences

Concerning equivalences between various definitions 6.10, we first remark that LH- and GH- are equivalent for a real analytic Riemannian manifold. It is an open problem to decide whether the same is true for C^∞ manifolds.

6.17 Theorem. *On the class of compact simply connected Riemannian manifolds, GH- and SH- are equivalent.*

Proof. The fact that SH- implies GH- is the content of 6.89. The converse, in the simply connected case, was recently proved by Dominique Michel (see [MI]). The proof is very technical and uses essentially Brownian motion techniques. □

6.18 Examples. The CROSSes are examples of SH-manifolds. Examples of GH-manifolds are the quotients of a ROSS, i.e., the complete locally symmetric spaces of rank one. Finally as an example of LH-manifolds we have any open submanifold of a ROSS. We again emphasize that no other examples of harmonic manifolds are known: see [R-W-W], p. 231.

6.19 Remarks. A geometric interpretation of local harmonicity is the following reformulation of 6.10: for small enough r's the geodesic spheres $S(m,r)$, equipped with the Riemannian structure induced by (M,g), are such that the action of the orthogonal group $O(T_mM)$ via $\exp_m$ on $S(m,r)$ preserves the volume element.

A geometric equivalence of LH- is the following: every sufficiently small geodesic sphere has constant mean curvature.

In [WE] the reader will find the notion of a k-harmonic manifold (the case $k=d-1$ being that of an LH-manifold). What is involved in the definition of these manifolds is the kth invariant of the $(d-1)$-square matrices in (6.15), the first one being the trace and the last one the determinant.

An Explanation of the Word "Harmonic"

6.20. We now give equivalent definitions of local harmonicity based on mean values of functions. This will explain the word "harmonic" in regard to the standard fact that in Euclidean spaces a harmonic function (i.e., one whose Laplacian Δ vanishes) has its mean value on every sphere equal to its value at the center of the sphere.

6.21 Proposition. *Given a point m in a Riemannian manifold (M,g), the following conditions are equivalent:*

(i) *(M,g) is locally harmonic at m (i.e., there exists a function Θ and a positive real number ε such that $\theta(u)=\Theta(\|u\|)$ for every u in T_mM with $\|u\|\leqslant\varepsilon$);*

(ii) *there exist a positive real number ε and a function $F:[0,\varepsilon[\to\mathbb{R}$ such that the function $f=\frac{1}{2}\varrho^2(m,.))$ satisfies $\Delta f=F(\varrho(m,.))$;*

(iii) *there exist a positive real number ε and a function $G:[0,\varepsilon[\to\mathbb{R}$ such that the function g defined on $B(m,\varepsilon)\setminus m$ by $n\mapsto G(\varrho(m,n))$ satisfies $\Delta g=0$;*

(iv) *there exists a positive real number ε such that every function $f:B(m,\varepsilon)\to\mathbb{R}$ with $\Delta f=0$ has a mean value on $S(m,r)$ equal to $f(m)$ for every r in $]0,\varepsilon[$.*

Proof. The idea is to use the classical formula (6.22) (see for example [B-G-M], p. 134) giving the Laplacian of a function of the form $G\circ\varrho(m,.)$ (i.e., "which depends only on the distance to m") as

6.22 $$\Delta g=-G''-G'\left(\frac{d-1}{\varrho(m,.)}+\frac{\theta'_m}{\theta_m}\right).$$

Here G' and G'' are the ordinary first and second derivatives of $G:[0,\varepsilon[\to\mathbb{R}$, $\theta_m=\theta\circ(\exp_m\restriction B(m,\varepsilon))^{-1}$ and θ'_m is the radial derivative of θ_m in T_mM.

Formula (6.22) immediately shows that (ii) implies (i) and that (iii) implies (i). Formula (6.22) yields the fact that (i) implies (iii) via the existence theorem for ordinary differential equations.

To prove equivalence with (iv) we denote by σ_r the canonical Riemannian measure on $(S(m,r), g\restriction S(m,r))$, by σ the canonical measure on the sphere U_mM (cf. 1.123) and by $\tilde\sigma_r$ the push-forward of σ on $S(m,r)$ by the diffeomorphism $U_mM\to S(m,r)$ given by $u\mapsto\exp_m(ru)$. The *mean value* $MV(f,r)$ of a function f on $S(m,r)$ is given by

$$MV(f,r)=\int_{S(m,r)} f r^{d-1}\theta d\tilde\sigma_r\Big/\int_{S(m,r)} r^{d-1}\theta d\tilde\sigma_r.$$

Now assume that (i) holds. Then θ is constant on $S=S(m,r)$ and

$$MV(f,r)=\int_S f d\tilde\sigma_r\Big/\int_S d\tilde\sigma_r.$$

Taking the derivative with respect to r (since $\tilde\sigma_r$ does not depend on r) and multiplying again by the constant $r^{d-1}\theta$ on S we get

$$\frac{d(MV(f,r))}{dr}=\int_S\frac{df}{dr}d\tilde\sigma_r\Big/\int_S d\tilde\sigma_r=\int_S\frac{df}{dr}d\sigma_r\Big/\int_S d\sigma_r.$$

But $\frac{df}{dr}$ is simply the normal derivative of f at the boundary $S(m,r)$ of the nice domain $B'(m,r)$. Therefore, by Green's formula since $\Delta f=0$ we get

$$\int_S\frac{df}{dr}d\sigma_r=\int_{B'(m,r)}\Delta f d\mu=0.$$

Let us now prove that (iv) implies (ii). Still with $S=S(m,r)$, we have

$$\frac{d}{dr}\left(\int_S f\,d\sigma_r\right)\Big/\int_S f\,d\sigma_r=\frac{d}{dr}\left(\int_S d\sigma_r\right)\Big/\int_S d\sigma_r .$$

But

$$\frac{d}{dr}\left(\int_S f\,d\sigma_r\right)=\frac{d}{dr}\left(\int_S f r^{d-1}\theta\,d\tilde{\sigma}_r\right)$$

$$=\int_S \frac{df}{dr}\,d\sigma_r+\int_S f\frac{d}{dr}(r^{d-1}\theta)\,d\tilde{\sigma}_r .$$

By Formula (6.22) $\frac{d}{dr}(r^{d-1}\theta)=r^{d-1}\theta\Delta(\varrho(m,.))$. Using Green's formula and the assumption that $\Delta f=0$, we get

$$\frac{d}{dr}\left(\int_S f\,d\sigma_r\right)=\int_S f\Delta r\,d\sigma_r .$$

Since $\frac{d}{dr}\left(\int_S d\sigma_r\right)\Big/\int_S d\sigma_r$ clearly does not depend on r, we call it $\zeta(r)$, so that

6.23 $$\int_{S(m,r)} f(\Delta r-\zeta(r))\,d\sigma_r=0 \text{ for every } f \text{ such that } \Delta f=0 .$$

By solving the Dirichlet problem on $B'(m,r)$, there exists a function f with $\Delta f=0$ on $B(m,r)$ and $f\upharpoonright S(m,r)=\Delta r-\zeta(r)$. Hence, (6.23) implies that $\Delta r-\zeta(r)=0$ and (iv) implies (ii). □

6.24 Remark. We leave as an exercise for the reader to prove that in a given harmonic manifold a function f is harmonic (i.e., $\Delta f=0$) if and only if $MV(f,r)=f(m)$ for every r in $]0,\varepsilon[$.

C. Infinitesimally Harmonic Manifolds, Curvature Conditions

6.25. Here the idea is to write, for an LH-manifold, the conditions obtained from Definition 6.10 by taking derivatives of any order at a point. More precisely, fix m in M, u in U_mM and define the *radial derivative of order* n of θ as

6.26 $$D_u^n\theta=\frac{d^n}{ds^n}(s\mapsto\theta(\exp_m(su)))(0).$$

6.27 Definition. A Riemannian manifold (M, g) is said to be *infinitesimally harmonic at m* if for every integer n there exists a real number $k_n(m)$ such that $D_u^n\theta = k_n(m)$ for every u in U_mM.

A Riemannian manifold is said to be *infinitesimal harmonic* (we will write IH-manifold) if for every integer n there exists a real number k_n such that $D_u^n\theta = k_n$ for every u in UM.

6.28 Remarks. In the real analytic case IH- and GH- are equivalent; in the C^∞ case, the equivalence is an open question.

We do not know if infinitesimally harmonic at every point implies infinitesimally harmonic, however, see 6.47.

6.29. Now by the general philosophy of normal coordinates (see for example [B-G-M], p. 97) the $D_u^n\theta$ can certainly be expressed in terms of covariant derivatives of the curvature tensor at m. Hence, we are sure that infinitesimally harmonic implies conditions on curvature. For the orders $n = 0, 1, 2, 3, 4, 5, 6$ these conditions can immediately be obtained by using the expansion of θ to order 6 given in [GY], p. 336. Here we shall use a nice trick due to Ledger (see [R-W-W], p. 61) which gives induction formulas for computing the $D_u^n\theta$ at any order, despite the fact that we shall only use the derived curvature conditions up to order 6.

6.30. Fix m in M and u in U_mM. Consider the geodesic $\gamma: s \mapsto \exp_m(su)$ through m with velocity vector u at m. Consider the endomorphism-valued function $s \mapsto A(s)$ defined as in (6.13) by the conditions $A'' + R \circ A = 0$, $A(0) = 0$, $A'(0) = I$. By (6.4) we know that

6.31 $$\det(A(s)) = s^{d-1}\theta(s),$$

where A and θ are functions of s taken along $s \mapsto \exp_m(su)$. By the classical formula giving the derivative of a determinant, we have

6.32 $$\operatorname{trace}(A'A^{-1}) = \frac{\theta'}{\theta} + \frac{d-1}{s}.$$

6.33. *Set* $C = sA'A^{-1}$ (a linear map depending on s). We take the derivative of C (taking (6.13) in account) to get

6.34 $$sC' = -s^2R - C^2 - C,$$

where R is as in (6.13) along γ. Now we take the $(n+1)$st derivative of (6.34), evaluate it at $s=0$ and obtain *Ledger's formulas:*

6.35 $$\begin{cases} \text{for every integer } n: \\ (n+1)C^{(n)}(0) = -n(n-1)R^{(n-1)}(0) - \sum\limits_{k=0}^{n} C^{(k)}(0)C^{(n-k)}(0). \end{cases}$$

Ledger's formulas allows us (at least theoretically) to compute the $C^{(n)}(0)$'s for any n. Here is the formula for $n=0, 1, \ldots, 6$:

6.36 $$C(0)=I,$$

6.37 $$C'(0)=0,$$

6.38 $$C''(0)=-\tfrac{2}{3}R(0),$$

6.39 $$C'''(0)=-\tfrac{3}{2}R'(0),$$

6.40 $$C^{(4)}(0)=-\tfrac{12}{5}R''(0)-\tfrac{8}{15}R(0)R(0),$$

6.41 $$C^{(5)}(0)=-\tfrac{10}{3}R'''(0)+\tfrac{5}{3}(R(0)R'(0)+R'(0)R(0)),$$

6.42 $$C^{(6)}(0)=-\tfrac{30}{7}R^{(4)}(0)-\tfrac{1}{7}(45R'(0)R'(0)+24R(0)R''(0)+24R''(0)R(0)+\tfrac{32}{3}R(0)R(0)R(0)),$$

where we have denoted the composition of endomorphisms by ordinary multiplication.

We now set:

6.43 $$C_u^{(n)}=C^{(n)}(0), \quad R_u: v\mapsto R(u,v)u, \quad R_u^{(n)}=\underbrace{D_u\ldots D_u}_{n\text{ times}}(R(u,.)u),$$

the objects being always endomorphisms of T_mM, associated with a given u in U_mM. From the formula

6.44 $$\operatorname{trace}(C)=s\frac{\theta'}{\theta}+d-1$$

(use (6.32)) and from Definition 6.27 we get the following

6.45 Lemma. *At a given m in M, if for every integer n the real number $D_u^n\theta$ does not depend on u in U_mM, then for every integer n the number* $\operatorname{trace}(C_u^{(n)})$ *also does not depend on u.*

6.46 Proposition. *If a Riemannian manifold (M,g) is infinitesimally harmonic, then there exist constants K, H, L such that for every u in UM one has*

$$\operatorname{trace}(R_u)=K, \quad \operatorname{trace}(R_u\circ R_u)=H, \quad \operatorname{trace}(32\,R_u\circ R_u\circ R_u-9\,R_u'\circ R_u')=L.$$

Proof. From the relation (6.38) we deduce that $\operatorname{trace}(R_u)=K$. If we now take derivatives of this equality along γ we get $\operatorname{trace}(R_u')=0$, $\operatorname{trace}(R_u'')=0$, $\operatorname{trace}(R_u''')=0$ and $\operatorname{trace}(R_u^{(4)})=0$. Then Relation (6.40) boils down to $\operatorname{trace}(R_u\circ R_u)=H$. Taking derivatives of this last relation, we obtain $\operatorname{trace}(R_u\circ R_u')=0$ and $\operatorname{trace}(R_u'\circ R_u')=-\operatorname{trace}(R_u\circ R_u'')$, and we have proved 6.46 via (6.42). □

6.47 Remark. If we suppose that (M, g) is infinitesimally harmonic at every point m in M, then we get as first relation

$$\operatorname{trace}(R_u) = K(m) \quad \text{for every } u \text{ in } U_m M.$$

Since $\operatorname{trace}(R_u) = \operatorname{Ric}(m)(u, u)$, we then have

$$\operatorname{Ric}(m) = K(m) g(m) \quad \text{at every } m \text{ in } M,$$

and Schur's theorem (see [K-N 1], p. 292, for example) implies that $K(m)$ does not depend on m. If we argue the same way for the second condition we get (using computations of the proof of 6.57)

$$\operatorname{trace}(R_u \circ R_u) = H(m) \quad \text{for every } u \text{ in } U_m M.$$

We shall see in 6.73 that this implies that $H(m)$ is constant. These remarks support the conjecture that infinitesimally harmonic at every point implies infinitesimally harmonic, cf. 6.28.

6.48. *Expression in Local Coordinates.* From 6.46 and by homogeneity we first get for every tangent vector u

6.49
$$\begin{cases} \operatorname{trace}(R_u) = K\|u\|^2, \quad \operatorname{trace}(R_u \circ R_u) = H\|u\|^4 \quad \text{and} \\ \operatorname{trace}(32 R_u \circ R_u \circ R_u - 9 R'_u \circ R'_u) = L\|u\|^6. \end{cases}$$

Then we pick any system of orthonormal coordinates and get from (6.49):

6.50 for every i, j, k, h:

$$\mathfrak{S}\Big(\sum_{a,b} R_{iajb} R_{kahb}\Big) = H \mathfrak{S}(\delta_{ij}\delta_{kh}),$$

6.51 for every i, j, k, h, l, m:

$$\mathfrak{S}\Big(32 \sum_{a,b,c} R_{iajb} R_{kbhc} R_{lcma} - 9 \sum_{a,b} D_i R_{jakb} D_h R_{lamb}\Big) = L \mathfrak{S}(\delta_{ij}\delta_{kh}\delta_{lm}),$$

where the sign $\mathfrak{S}$ means summation according to the action of the symmetric group in the four letters i, j, k, h for the first formula and that in the six letters i, j, k, h, l, m in the second formula.

D. Implications of Curvature Conditions

6.52 Proposition. *An IH-manifold is an Einstein space. In particular, the only IH-manifolds of dimension 2 or 3 have constant sectional curvature.*

Proof. We just need to apply 6.46 and [K-N 2], p. 293. □

From [BR 4], p. 45 and p. 51 we can get the

6.53 Corollary. *The canonical Riemannian structure on S^d is isolated among IH-structures on S^d. If a Kählerian metric on $\mathbb{C}P^n$ is IH- and has positive sectional curvature, then it is the canonical metric.*

An amusing consequence of 6.46 is the following

6.54 Proposition. *If a Riemannian manifold (M,g) satisfies $D_u^n\theta=0$ for every m in M, every u in U_mM and every $n=1,2,3,4$, then it is flat.*

Proof. From (6.44) and 6.46 we then have for all u in UM $\operatorname{trace}(R_u\circ R_u)=0$. Hence, $R_u=0$ for every u in UM and the curvature tensor is identically zero. □

Conditions of Order 4

6.55. We now play a little more with the condition of order 4. Let us introduce the symmetric bilinear differential form ξ defined for a Riemannian manifold by

6.56 $$\xi_{ij}=\sum_{u,v,w} R_{iuvw}R_{juvw} \quad \text{for every } i,j, \text{ in orthonormal coordinates.}$$

6.57 Proposition. *If an Einstein manifold (M,g) (with $\mathrm{Ric}=Kg$) satisfies the condition $\operatorname{trace}(R_u\circ R_u)=H$ for every u in UM and some constant H, then*

$$\xi=\tfrac{2}{3}((d+2)H-K^2)g.$$

In particular, the norm of the curvature tensor is constant, i.e.,

$$\|R\|^2=\frac{2d}{3}((d+2)H-K^2).$$

Proof. We work in orthonormal coordinates and set

$$\varphi(ijkh)=\sum_{p,q} R_{ipjq}R_{kphq}.$$

One always has $\varphi(ijkh)=\varphi(jihk)=\varphi(khij)$. If (M,g) satisfies $\mathrm{Ric}=Kg$, then using the symmetry properties of the curvature tensor we get

6.58 $$\sum_j \varphi(iijj)=K^2, \quad \sum_j \varphi(ijij)=\xi_{ii}, \quad \sum_j \varphi(ijji)=\tfrac{1}{2}\xi_{ii}$$

for every i,j. We now apply condition (6.50) with $i=j$ and $k=h$ and sum it over k. From (6.58) we get

$$\xi_{ii}=\tfrac{2}{3}((d+2)H-K^2).$$

We now use condition (6.50) with $i \neq j$, $k=h$ and sum it over k to get $\xi_{ij}=0$. Hence, the first assertion is true. The second follows from the fact that trace $\xi = \|R\|^2$. □

6.59. *Remarks on the Invariant* ξ. The form ξ in S^2M is a natural Riemannian invariant, algebraically the simplest one after the scalar and Ricci curvatures. However it does not seem to have received much attention in the literature. We quote a few facts concerning it.

If (M, g) is an Einstein space, then

6.60 $$\delta\xi = -\tfrac{1}{4} d(\|R\|^2) = -\tfrac{1}{4} d(\operatorname{trace}_g \xi).$$

6.61. From this one gets a Schur type consequence: if $\xi = fg$ for some function f and if $d > 4$, then f is constant.

On the other hand, for any 4-dimensional Riemannian manifold one always gets

6.62 $$\xi = \frac{\|R\|^2}{4} g.$$

Finally from (6.60) one gets in a straightforward way

6.63 $$\delta\delta\xi = -\tfrac{1}{4}\Delta(\|R\|^2).$$

Since the search for compact Einstein manifolds is still one of the big game in Riemannian geometry, we first suggest the search for a more restricted class of Riemannian manifolds, those which satisfy both $\mathrm{Ric} = kg$ and $\xi = hg$.

Conditions of Order 6

6.64. To derive consequences from the third condition in 6.46 we first recall that for an Einstein manifold (M, g) there are only *three* distinct orthogonal invariants of order 6 namely, in orthonormal coordinates:

$$|DR|^2 = \sum_{j,j,k,h,l} (D_l R_{ijkh})^2, \quad \hat{R} = \sum_{i,j,p,q,r,s} R_{ijpq} R_{pqrs} R_{rsij},$$

$$\text{and} \quad \mathring{R} = \sum_{i,j,p,q,r,s} R_{ipjq} R_{prqs} R_{risj}$$

(see for example [SI], p. 591 or [DI]). Our notation, a little different from that of [DI], is motivated by the following observation: if $\mathscr{R}: \Lambda^2 TM \to \Lambda^2 TM$ denotes the curvature operator acting in the second exterior power of TM, then $\hat{R} = \operatorname{trace}(\mathscr{R} \circ \mathscr{R} \circ \mathscr{R})$, which explains the symbol $\hat{}$. On the other hand, the symbol $\mathring{}$ is related to the action of the curvature tensor on the symmetric power S^2M.

From the classical computation of [LZ 2], p. 9–10, one gets (with our notation)

6.65 $$-\Delta(\|R\|^2) = 2K\|R\|^2 - \hat{R} - 4\mathring{R} + 2\|DR\|^2,$$

a formula which is valid for any Einstein manifold (M, g).

6.66. Notice that it is hopeless (in view of (6.63)) to get another relation by taking derivatives of relation (6.57) twice!

Now if (M, g) is an IH-manifold, consider Condition (6.51) in which we first assume that $i=j$, $k=h$ and $l=m$ and then sum over the three indices i, k, l. After a quite lengthy computation, where the only difficulty is playing with the symmetric group on six letters, one gets

6.67 $$32(dK^2+\tfrac{9}{2}K\|R\|^2+\tfrac{7}{2}\hat{R}-\mathring{R})-27\|DR\|^2=d(d^2+6d+8)L.$$

From (6.65) and (6.67) we then have the following

6.68 Proposition. *On an IH-manifold the following three functions are constant:*

$$\hat{R}+4\mathring{R}-2\|DR\|^2,\quad 112\hat{R}-32\mathring{R}-27\|DR\|^2 \quad \textit{and} \quad 197\hat{R}-172\mathring{R}.$$

E. Harmonic Manifolds of Dimension 4

6.69 Theorem (Lichnerowicz, Walker). *An IH-manifold of dimension 4 is flat or is locally isometric to a* ROSS.

References are [R-W-W], p. 142—150 and [WK].

The proof is quite lengthy. We start by recalling results from [S-T] on the curvature tensor of a 4-dimensional manifold. We fix m in M and work on $V=T_mM$, where we fix an orientation. Then on $\Lambda^2 V$ we have the Hodge operator

$$*:\Lambda^2V\to\Lambda^2V.$$

Explicitly, if $\Lambda^2 V$ is endowed with the standard Euclidean structure deduced from that of V, if $\mathscr{B}=\{e_i\}_{i=1,2,3,4}$ is a positive orthonormal basis of V then we denote by $\Lambda^2\mathscr{B}$ the basis

$$\Lambda^2\mathscr{B}=\{e_1\wedge e_2, e_1\wedge e_3, e_1\wedge e_4, e_3\wedge e_4, e_4\wedge e_2, e_2\wedge e_3\}$$

of $\Lambda^2 V$. Then the matrix of $*$ with respect to $\Lambda^2\mathscr{B}$ is $\begin{pmatrix}0 & I\\ I & 0\end{pmatrix}$ (where I is the 3×3-identity matrix).

For the curvature operator $\mathscr{R}:\Lambda^2V\to\Lambda^2V$ (already met in 6.64) the fact that (M, g) is an Einstein space is equivalent to the requirement that

6.70 $$*\circ\mathscr{R}=\mathscr{R}\circ *.$$

6.71. If (M, g) is an Einstein space, then there exists an orthonormal basis $\mathscr{B}$ of V, called a *Singer-Thorpe basis* (we shall write ST-basis), for which the matrix of $\mathscr{R}$

looks like

$$\begin{pmatrix} a & 0 & 0 & \alpha & 0 & 0 \\ 0 & b & 0 & 0 & \beta & 0 \\ 0 & 0 & c & 0 & 0 & \gamma \\ \alpha & 0 & 0 & a & 0 & 0 \\ 0 & \beta & 0 & 0 & b & 0 \\ 0 & 0 & \gamma & 0 & 0 & c \end{pmatrix} \quad \text{with} \quad \alpha+\beta+\gamma=0.$$

(the relation $\alpha+\beta+\gamma=0$ is simply the Bianchi identity $R_{1234}+R_{1342}+R_{1423}=0$). Moreover, $\mathscr{B}$ can be chosen with the following additional property: if σ denotes the sectional curvature at m, then as a function on the Grassmannian of two-planes we have

6.72 $$a=\sup\sigma \quad \text{and} \quad c=\inf\sigma.$$

We embark on the proof of a

6.73 Lemma. *If (M, g) is an IH-manifold of dimension 4, then for a suitably chosen orientation of V the curvature operator always satisfies the relation*

$$\left(\mathscr{R}-\frac{K}{3}I\right)\circ(I-*)=0$$

(where $\mathrm{Ric}=Kg$ *and I is the identity on $\Lambda^2 V$). Moreover, there exists a real number L'_m such that* trace $(R_u\circ R_u\circ R_u)=L'_m$ *for every u in U_mM.*

Proof. The identity between $\mathscr{R}, I, *$ being intrinsic, it can be proved in any basis. We take a ST-basis of course! Then

6.74 $$a+b+c=K.$$

In condition (6.50) we first take $i=j=k=h=1$ and get

6.75 $$a^2+b^2+c^2=H.$$

Then we take $i=j=1$ and $k=h=2,3,4$ to get

6.76 $$a^2+2bc+(\beta-\gamma)^2=b^2+2ca+(\gamma-\alpha)^2=c^2+2ab+(\alpha-\beta)^2=H.$$

Finally, with $i=1, j=2, k=3$ and $h=4$ we get

$$a(\beta-\gamma)+b(\gamma-\alpha)+c(\alpha-\beta)=0.$$

The last condition, coupled with $\alpha+\beta+\gamma=0$ yields

6.77 $$\frac{\alpha}{a-\dfrac{K}{3}}=\frac{\beta}{b-\dfrac{K}{3}}=\frac{\gamma}{c-\dfrac{K}{3}},$$

with the usual convention when some terms vanish. We *call* λ the common value of these quantities and when we substitute it in (6.76) we get

$$(\lambda^2-1)(a-b)\left(c-\frac{K}{3}\right)=0.$$

A first case is $a=b=c=\frac{K}{3}$ so that $\alpha=\beta=\gamma=0$ and we are done. There remains the case $\lambda=\pm 1$; but we can adjust the orientation so that we have the case $\lambda=1$. Then (6.77) is exactly the relation to prove when expressed in our basis.

We now prove the last assertion of the lemma. Let u be an element of U_mM, we complete it into an orthonormal basis $\mathscr{B}=\{e_1=u, e_2, e_3, e_4\}$ and we write the matrix of $\mathscr{R}$ with respect to $\Lambda^2\mathscr{B}$ as $\begin{pmatrix} A & B \\ C & D\end{pmatrix}$. Condition (6.70) and the fact that $\left(\mathscr{R}-\frac{K}{3}\right)\circ(I-*)=0$ imply that

$$\begin{pmatrix} A & B \\ C & D\end{pmatrix}=\begin{pmatrix} A & A-\frac{K}{3} \\ A-\frac{K}{3} & A\end{pmatrix}.$$

Notice that A is simply the matrix of R_u with respect to $\mathscr{B}$. Hence,

$$\operatorname{trace}(R_u\circ R_u\circ R_u)=\operatorname{trace}(A\circ A\circ A)=\tfrac{1}{8}\operatorname{trace}(\mathscr{R}\circ\mathscr{R}\circ\mathscr{R}).$$

6.78 Lemma. *In an IH-manifold, the three numbers, a, b, c appearing in an ST-basis with the additional condition* (6.72), *are constant on M. The number L'_m at the end of the Lemma* 6.73 *is also constant.*

Proof. From (6.74) and (6.75) we have

6.79 $$a+b+c=K, \quad a^2+b^2+c^2=H.$$

In our nice basis a direct computation of $\hat{R}$ and $\mathring{R}$ from 6.71 yields

$$\hat{R}=16(a^3+b^3+c^3+3a\alpha^2+3b\beta^2+3c\gamma^2),$$
$$\mathring{R}=12(2abc+a\beta\gamma+b\gamma\alpha+c\alpha\beta).$$

Using (6.79) we get

$$\hat{R}=192abc+\text{constant}, \qquad \mathring{R}=60abc+\text{constant}.$$

Then 6.68 implies that abc is constant, which yields the lemma via (6.79). □

6.80 Lemma. *The covariant derivative DR of the curvature tensor of an IH-manifold of dimension 4 vanishes identically.*

Pick any point m in M and an ST-basis of T_mM with the property (6.72). Extend this basis locally around m into a family of orthonormal frames (no longer ST in general!). By (6.72) we get in the neighbourhood $\mathscr{V}$ of m:

$$R_{1212}(n) \leqslant R_{1212}(m) = a \quad \text{for every } n \text{ in } \mathscr{V}.$$

Hence, we have

$$D_i R_{1212}(m) = 0 \quad \text{for every } i$$

(see [B-G-M], p. 40 for this type of argument). Since $R_{1414}(m) = c$ we similarly get

$$D_i R_{1414}(m) = 0 \quad \text{for every } i.$$

Since $\mathrm{Ric} = Kg$ we finally get

6.81 $$D_i R_{jkjk}(m) = 0 \quad \text{for every } i, j, k.$$

Now in any orthonormal basis $R_{1212} - R_{1234} = \dfrac{K}{3}$ because of 6.73, so (6.81) implies that

$$D_i R_{jkhl}(m) = 0 \quad \text{for every } i \text{ and every } j, k, h, l \text{ (all distinct)}.$$

We are left with the components of types $D_i R_{ijik}(m)$ and $D_i R_{ijkj}(m)$ (with i, j, k all distinct). Using the Bianchi identity and the fact that (M, g) is an Einstein space, one proves that $(ijk) = (jki) = (kij)$ if one *sets* $(ijk) = D_i R_{ijik}(m)$). Let us denote by $\textcircled{h}$ their common value, where h is an index such that i, j, k, h are all distinct. One also proves that $D_i R_{ijkj}(m) = \pm \textcircled{h}$.

Now by 6.46 and 6.73 there exists L''_m such that $\mathrm{trace}(D_u R_u \circ D_u R_u) = L''_m$ for every u in U_mM. Taking $u = e_1$ we get

$$L''_m = \sum_{a,b} (D_1 R_{1a1b}(m))^2 = \textcircled{2}^2 + \textcircled{3}^2 + \textcircled{4}^2.$$

By the same token, with $u = e_2, e_3, e_4$ we have

$$\begin{aligned}\textcircled{2}^2 + \textcircled{3}^2 + \textcircled{4}^2 = \textcircled{3}^2 + \textcircled{4}^2 + \textcircled{1}^2 = \textcircled{4}^2 + \textcircled{1}^2 + \textcircled{2}^2 \\ = \textcircled{1}^2 + \textcircled{2}^2 + \textcircled{3}^2,\end{aligned}$$

hence

$$\textcircled{1} = \pm \textcircled{2} = \pm \textcircled{3} = \pm \textcircled{4}.$$

Now apply (6.51) with $i = j = k = h = l = 1$ and $m = 2$. Using the various previously mentioned equalities relating the $D_i R_{ijik}(m)$'s and the $D_i R_{ijkj}(m)$'s, we get

$$\textcircled{1}\textcircled{2} = 0.$$

Therefore, ①=②=③=④=0 and $DR(m)=0$.

Proof of Theorem 6.69. Since $DR=0$, by (6.4) and (6.13) we see that along any geodesic $s\mapsto\exp(su)$ the function θ will be of the type $\prod_{i=1}^{d-1} f_i$, where every function f_i is one of the three following ones: s, $\frac{\sin\alpha_i s}{\alpha_i}$, and $\frac{\sinh\alpha_i s}{\alpha_i}$ for some real numbers α_i. Elementary calculus shows that if $\prod_{i=1}^{d-1} f_i=\prod_{i=1}^{d-1} g_i$ on some interval, then necessarily (up to permutation) $f_i=g_i$ for every i. Consequently, the endomorphisms R_u are all the same, up to similarity, when u runs through UM. Then, since (M,g) is a locally symmetric space by Elie Cartan's theorem (see [K-N 2], p. 244), the structure of symmetric spaces shows that the real numbers α_i's appearing in the f_i's above are precisely linear forms over the u's when u runs through a Cartan subalgebra of the symmetric space. Now such a non-zero linear form can be constant only when the Cartan subalgebra is one-dimensional, i.e., when the symmetric space has rank one. The preceding assertions use a good deal of the theory of symmetric spaces and we cannot refer the reader to an unique good reference. We used four basic facts: 1) the decomposition of our symmetric space into a product of irreducible ones; 2) the duality between the compact and the non-compact cases; 3) the structure theorem in the compact case: Cartan subalgebras and roots and 4) the formula giving the sectional curvature. See for example: [LS 1], Chapter IV, [K-N 2], p. 253—258, [LS 2], p. 58—60 and [K-N 2], p. 257. □

F. Globally Harmonic Manifolds: Allamigeon's Theorem

A historical break in the theory of harmonic manifolds was made by Allamigeon in [AN], p. 123 when he proved the following:

6.82. Theorem. *A simply connected GH-manifold is either diffeomorphic to $\mathbb{R}^d$ or is a Blaschke manifold.*

Proof. We pick some point m in M and consider formula (6.4):

$$\theta(su)=s^{d-1}\det(Y_2(s),\ldots,Y_d(s)) \quad \text{for } u \text{ in } U_mM.$$

The basic fact noted by Allamigeon is that along the geodesic $\gamma: s\mapsto\exp_m(su)$ the first conjugate point is determined by the first zero of the function $s\mapsto\theta(su)$. Moreover, the order of conjugacy is known as soon as $s\mapsto\theta(su)$ is known. Now suppose that (M,g) is globally harmonic. Then either there is no conjugate point on all geodesics issuing from m or (M,g) is an Allamigeon-Warner manifold at m by definition

5.23. In the first case M is diffeomorphic to $\mathbb{R}^d$ because $\exp_m$ is a covering map (see [K-N 2], p. 105). In the second case this holds at every m in M so that our (M,g) is a Blaschke manifold by the simply connectedness hypothesis and by Theorem 5.43. □

Since a Riemannian covering of a GH-manifold is again a GH-manifold we have the

6.83 Corollary. *If a Riemannian manifold (M,g) is a GH-manifold, then its universal Riemannian covering is either $\mathbb{R}^d$ or a Blaschke manifold.*

In particular, we have at our disposal the results of 5.E on the topology of M. We now make some remarks on the problem of deciding whether a GH-manifold is symmetric or not, the basic idea being to rely on 6.82.

6.84 Remarks. We now suppose that (M,g) is GH-manifold; then we are led to Blaschke's conjecture 5.F. See also Appendix D. Notice that here we know more. For example we know that our (M,g) is an Einstein space (and satisfies other curvature conditions). Even with all of this information we cannot go any further at the moment (a true scandal!). Here are some comments.

6.85. The first case to look at is $M=S^d$. Suppose that (M,g) is normalized with $\mathrm{Diam}(g)=\pi$. If d is even, then we know by Weinstein's Theorem 2.24, that

$$\mathrm{Vol}(S^d,g)=\beta(d).$$

From $\mathrm{Ric}=Kg$ we deduce by Myers's theorem (see [K-N 2], p. 88) and the fact that $\mathrm{Diam}(g)=\pi$, that $K\leqslant d-1$. The Bishop-Crittenden inequality (see [B-C], p. 256—257) then implies that $\mathrm{Vol}(g)\geqslant\left(\dfrac{K}{d-1}\right)^d\beta(d)$. So we cannot draw any conclusion (see also 5.78).

6.86. For a GH-metric g on S^d, normalized by $\mathrm{Diam}(g)=\pi$, we notice that every $S(m,\pi/2)$ is a minimal hypersurface. In fact, since the antipodal map is an isometry from 5.57, the function $\Theta:[0,\pi]\to\mathbb{R}$ is symmetrical about $\pi/2$ (i.e. $\Theta(\pi-s)=\Theta(s)$ for every s in $[0,\pi]$). In particular, $\Theta'(\pi/2)=0$, which implies the minimality of $S(m,\pi/2)$ (see 5.62).

Then a globally harmonic (S^d,g) carries many minimal hypersurfaces, in fact a d-parameter family. Notice that by the same token one can prove that in a compact, simply-connected, GH-Riemannian manifold normalized by $\mathrm{Diam}(g)=\pi/2$, every cut-locus $S(m,\pi/2)=\mathrm{Cut}(m)$ is a minimal submanifold.

6.88. Another case is that of a GH-structure g on the torus T^d:(T^d, g). This again leads (cf. 6.52 and 6.59) to the study of Einstein structures on tori. When $\mathrm{Ric} = Kg$ with a non-negative K, it is classical that g is flat (see [BE-EB], p. 391, for example) but for a non-positive K the corresponding question is open (it is of intrinsic interest even for a non harmonic (T^d, g)). Also notice that the harmonicity f (T^d, g) implies that (T^d, g) has no conjugate points. The search for metrics on T^d with no conjugate points is also an interesting open question (see [GN 1] and note that the proof of [BR 3], p. 275—276 is false).

G. Strongly Harmonic Manifolds

We recall (cf. 6.10) that a Riemannian manifold (M, g) is strongly harmonic (we shall write SH-harmonic) if it is compact and if there exists a map $\Xi:\mathbb{R}_+ \times \mathbb{R}_+^* \to \mathbb{R}$ with the property that the fundamental solution of the heat equation K on (M, g) can be written as $K(m, n, t) = \Xi(\varrho(m, n), t)$ for every m and n in M and t in $\mathbb{R}_+^*$.

We already quoted the fact that the SH- and GH-properties are equivalent in the simply connected case (cf. 6.17). Here we prove the

6.89 Proposition. *An SH-manifold is a GH-manifold.*

The proof is quite technical, but we shall on the way get results which are of independent interest.

For the end of this chapter we assume that (M, g) is strongly harmonic with the function Ξ being as in 6.10.

For an eigenvalue λ_α of the Laplacian Δ on (M, g) set

6.90 $$V_\alpha = \{f \mid \Delta f = \lambda_\alpha f\} \quad \text{and} \quad N_\alpha = \dim V_\alpha .$$

Denote an orthonormal basis of V_α for the global scalar product $\langle \varphi, \psi \rangle = \int_M \varphi\psi d\mu_g$ by $\{\varphi_i^\alpha\}$. Then we know (see [B-G-M], p. 205, for example) that

6.91 $$K(m, n, t) = \sum_\alpha e^{-\lambda_\alpha t} \left(\sum_{i=1}^{N_\alpha} \varphi_i^\alpha(m) \varphi_i^\alpha(n) \right) \quad \text{for every } m, n \text{ in } M \text{ and } t \text{ in } \mathbb{R}_+^* .$$

6.92. It is elementary to see that, as a function of t, the series $t \mapsto \sum_\alpha c_\alpha e^{-\lambda_\alpha t}$ (with all λ_α's different) is uniquely determined by the c_α's. Hence, from (6.91) we infer that

6.93
$$\begin{cases} \text{for every } \lambda_\alpha \text{ there exists a map } \Xi_\alpha : \mathbb{R}_+ \to \mathbb{R} \text{ with} \\ \sum_{i=1}^{N_\alpha} \varphi_i^\alpha(m)\varphi_i^\alpha(n) = \Xi_\alpha(\varrho(m,n)) \text{ for every } m, n \text{ in } M. \end{cases}$$

In particular, one has the

6.94 Lemma. *For every point m in M there exists a function $F:\mathbb{R}\to\mathbb{R}$ such that the function $f:M\to\mathbb{R}$ defined by $n\mapsto F(\varrho(m,n))$ is C^∞ and satisfies $\Delta f=\lambda f$ with $\lambda>0$.*

Proof. Fix any $\lambda=\lambda_\alpha$ and set $\Xi_\alpha=F$. Then $f=\sum_{i=1}^{N_\alpha}\varphi_i^\alpha(m)\varphi_i^\alpha$ is the desired function. □

6.95 Proposition. *Any SH-manifold is a Blaschke manifold.*

Proof. We prove that such a manifold is a Blaschke manifold at any given point m in M. Let n be a point in M such that

$$\delta=\varrho(m,n)=\sup\{\varrho(m,p):p\in M\}$$

and let γ be a geodesic segment with $\gamma(0)=m$ and $\gamma(\delta)=n$. We now pick functions f and F as in Lemma 6.94. Then for the arc length parameter s along γ we have $f(\gamma(s))=F(s)$. Hence, F is C^∞ on $[0,\delta]$ and we can apply Formula (6.22) to get

6.96
$$\Delta f=-F''-F'\left(\frac{d-1}{s}+\frac{\theta'_m}{\theta_m}\right)=\lambda f=\lambda F.$$

From (6.96) various facts follow: the first is that the zeros of F' on $[0,\delta]$ are isolated. The second is that $\theta(su)$ is certainly $\neq 0$ for every u in U_mM when $F'(s)\neq 0$. Using the connectedness of U_mM and the continuity of the cut-locus (cf. 5.17) we only have to show that the cut-locus of m is locally spherical about n. To do so we argue ab absurdo.

Consider geodesics γ', starting from m and close to γ. By the continuity of the cut-point, if γ' is close enough to γ, then its cut-point will appear at a distance $\mu(\dot\gamma'(0))=\delta'$ close to δ (see 5.18). Moreover, by the above and by the assumption of non-local sphericity of Cut(m) near n we can choose γ' such that $\theta(\delta'\dot\gamma'(0))\neq 0$. But $\theta(\delta'\dot\gamma'(0))\neq 0$ implies that $n'=\gamma'(\delta')$ is not conjugate to m along γ' (apply (6.4)) and then by 5.20 there are two distinct geodesic segments from m to n', which implies that the function $s\mapsto\varrho(m,\gamma'(s))$ cannot be differentiable at $s=\delta'$. But this function has to be differentiable at δ' because $s\mapsto f(\gamma'(s))=F(\varrho(m,\gamma'(s)))$ is differentiable at δ' (with $F(\delta')\neq 0$). □

Proof of 6.89. Since our manifold is a Blaschke manifold we have to prove that $\theta(u)$ depends only on $\|u\|$ for u in UM and $\|u\|$ in $[0, \mathrm{Diam}(g)]$. But this follows immediately from (6.96). □

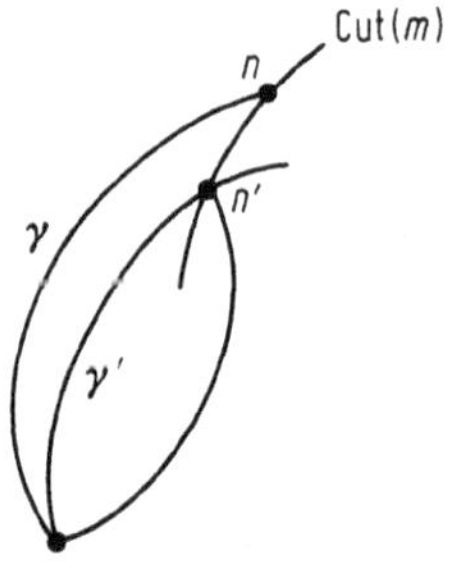

Fig. 6.96

6.97 Remark. In 6.68 we found that on an IH-manifold (M, g) the three invariants of order 6, namely $\hat{R}$, $\mathring{R}$ and $\|DR\|^2$ satisfy two independent linear relations with constant coefficients. This fact is not sufficient to prove that each of these invariants is a constant on M. But assume that (M, g) is an SH-manifold. Then by 6.10, all the coefficients in the Minakshisundaram-Pleijel asymptotic expansion.

$$K(m, m, t) = \sum_i e^{-\lambda_i t} \varphi_i^2(m)$$

are constant on M. If we use the values of these coefficients (cf. [B-G-M], p. 222 and 225) for the second and the third coefficients we find that the scalar curvature and the function $\|R\|^2$ are constant on M, which we already know by 6.52 and 6.57. But the fourth coefficient, computed in [SI], p. 598, yields the following relation

$$104\,\hat{R} + 224\,\mathring{R} - 189\,\|DR\|^2 = \text{constant}.$$

This new linear relation and the two relations of 6.68 are independent. Therefore we have

6.98 *for an SH-manifold, the functions* $\hat{R}$, $\mathring{R}$ *and* $\|DR\|^2$ *are constant.*

Nice Imbeddings

We now prove that an SH-manifold admits nice minimal isometric imbeddings into Euclidean spheres.

6.99 Theorem. *For an SH-manifold* (M, g) *fix any eigenvalue of the Laplacian (say* $\lambda = \lambda_\alpha$*) and set* $N = N_\alpha$, $V = V_\alpha$ *and* $\varphi_i = \varphi_i^\alpha$ $(i = 1, \ldots, N)$. *Then the map* $\Lambda = \Lambda_\lambda$ *from* M *to* $\mathbb{R}^N$ *defined by*

$$\Lambda(m) = (\varphi_1(m), \ldots, \varphi_N(m))$$

enjoys the following properties:

(i) *Λ is an immersion and the image $\bar{M}=\Lambda(M)$ is a d-dimensional submanifold of the sphere $S(0,R)$ in $\mathbb{R}^N$ of radius $R=(D/\mathrm{Vol}(g))^{1/2}$;*

(ii) *the submanifold $\bar{M}$ is minimal in $S(0,R)$;*

(iii) *the map Λ is a dilation of ratio $k=\dfrac{\lambda R^2}{d}$ between (M,g) and $(\bar{M},\bar{g})$, where $\bar{g}$ denotes the Riemannian structure induced on $\bar{M}$ by the Euclidean structure of R^N, i.e., $\Lambda^*\bar{g}=kg$;*

(iv) *Λ is an embedding except in the following case: the manifold M is diffeomorphic to S^d (say $M=S^d$) and $V=V_\lambda$ is made up of functions which are invariant under the action of the antipody σ of the Blaschke structure (S^d,g) (see 5.57); in this case the map $\Lambda: S^d\to\bar{M}$ is a two-sheet covering of $\bar{M}=\mathbb{R}P^d$.*

Proof. The basic idea of all what follows is Formula (6.100), which says roughly speaking that on $(\bar{M},\bar{g})$ the metric given by $\bar{g}$ is universally proportional to the metric induced by the canonical metric on $S(0,R)$.

From (6.93) and (6.94) we have for the scalar product on $\mathbb{R}^N$ and the function F of 6.94

6.100 $$(\Lambda(m),\Lambda(n))=\sum_{i=1}^{N}\varphi_i(m)\varphi_i(n)=F(\varrho(m,n))\quad\text{for every } m, n \text{ in } M\,.$$

First put $m=n$ and integrate over M to get

6.101 $$\begin{cases}\|\Lambda(m)\|^2=\sum\limits_{i=1}^{N}\varphi_i^2(m)=F(0)\,,\\ F(0)\,\mathrm{Vol}(g)=\sum\limits_{i=1}^{N}\int_M\varphi_i^2(m)\,d\mu_g=\sum\limits_{i=1}^{N}\|\varphi_i\|^2=N\end{cases}$$

since $\{\varphi_i\}$ is orthonormal. Then we have

$$\Lambda(M)\subset S(0,R)=(N/\mathrm{Vol}(g))^{1/2})\,.$$

Now differentiate (6.101) twice and get

6.102 $$\sum_{i=1}^{N}\varphi_i d\varphi_i=0\,,\quad \sum_{i=1}^{N}d\varphi_i\otimes d\varphi_i+\sum_{i=1}^{N}\varphi_i\,\mathrm{Hess}\,\varphi_i=0\,.$$

Differentiating (6.100) twice with respect to n, with m fixed, we get

$$\sum_{i=1}^{N}\varphi_i(m)\,\mathrm{Hess}\,\varphi_i=\mathrm{Hess}(F(\varrho(m,.)))\,.$$

Evaluating this at $m=n$ we obtain

$$\mathrm{Hess}(F(\varrho(m,.)))(m)=F''(0)g\,,\quad \sum_{i=1}^{N}\varphi_i\,\mathrm{Hess}\,\varphi_i=F''(0)g\,,$$

which implies by (6.102) that

$$\textbf{6.103} \qquad \sum_{i=1}^{N} d\varphi_i \otimes d\varphi_i = -F''(0) g\,.$$

But $\Lambda^* \bar{g} = \sum_{i=1}^{N} d\varphi_i \otimes d\varphi_i$. To compute the constant $F''(0) = -k$, we take traces:

$$\begin{aligned}\operatorname{trace}\left(\sum_{i=1}^{N} \varphi_i \operatorname{Hess} \varphi_i\right) &= -\sum_{i=1}^{N} \varphi_i \Delta \varphi_i = -\sum_{i=1}^{N} \lambda \varphi_i^2 \\ &= -\lambda R^2 = F''(0) \operatorname{trace}(g) = d\,F''(0)\,.\end{aligned}$$

This proves (iii).

From (6.103) our map is an immersion. Suppose now that $\Lambda(m) = \Lambda(n)$ for some points m, n in M with $m \neq n$. From (6.100) we deduce that $\Lambda(m) = \Lambda(n')$ when $\varrho(m,n) = \varrho(m,n')$. By 6.95 and 5.53 the set $\{n' \in M \mid \varrho(m,n) = \varrho(m,n')\}$ is a submanifold of M with positive dimension except when $M = S^d$ and when m, n are antipodal points (see 5.57). The fact that Λ is an immersion rules out the case in which the dimension is positive. If Λ is not an embedding, then $M = S^d$ and $\Lambda(m) = \Lambda(n)$ with $m \neq n$ implies that m and n are antipodal points for the Blaschke structure (S^d, g). Moreover, we have $\varphi_i(m) = \varphi_i(n)$ for every $i = 1, \ldots, N$ and every such pair. The proof of (i) and (iv) is finished.

The assertion (ii) is standard, see for example [K-N 2], p. 342. □

6.104 Corollary. *For an SH-manifold every eigenvalue of the Laplacian has multiplicity greater than or equal to $d+1$; if there exists an eigenvalue of multiplicity $d+1$, then necessarily (M,g) is isometric to (S^d, can).*

6.105 Remarks. If one wants to have an isometric immersion, one only has to replace Λ by $h\Lambda = (h\varphi_1, \ldots, h\varphi_N)$ with $h = \left(\frac{\lambda R^2}{d}\right)^{1/2}$.

If we could prove that for any Riemannian metric g on the sphere S^d the first eigenvalue always has multiplicity smaller than or equal to $d+1$, then 6.104 would imply that any SH-metric g on S^d is isometric to (S^d, can). This is open at the moment, except when $d=2$ (and then is of no use to us because of 6.52) (see [CG], Corollary 2.3).

Properties of the Nice Embeddings

We shall prove two properties that the nice embeddings 6.99 enjoy, namely 6.106 and 6.112 below

6.106 Proposition. *For any eigenvalue λ of the Laplacian of an SH-manifold (M,g) with multiplicity N there exists a function $\Gamma: [0, \mathrm{Diam}(g)] \to \mathbb{R}$ such that every*

$(N+1)$-tuple $(m_i)_{i=1,\dots,N+1}$ of points in M satisfies

$$\det(\Gamma(\varrho(m_i, m_j)))=0.$$

Proof. By [BM], p. 162, any $(N+1)$-tuple $(\bar{m}_i)_{i=1,\dots,N+1}$ of points on the sphere $S(0,R)$ in $\mathbb{R}^N$ satisfies

6.107 $$\det(R^{-1}(\bar{m}_i, \bar{m}_j)=0.$$

The proposition then follows with the function $\Gamma=R^{-1}(\cos\circ F)$. □

6.108 Notes. The universal metric relation (6.107) characterizes the metric space (S^d, can) (see [BM], p. 170 for a precise statement and a proof).

The relation of 6.106 is valid, in particular, for any CROSS. It would be nice to prove that this relation characterizes the CROSSes (compare with [SZ]).

Of course, we have universal metric relations for every eigenvalue λ. The most interesting one is expected to be that corresponding to the first eigenvalue λ_1 (because N_1 is probably the smallest of the N_α's and the corresponding Λ_1 the least "twisted"). But as already seen in 6.105, it appears difficult to characterize λ_1, at least in our context.

6.109 Lemma. *For a C^∞-curve $\gamma: I \to S^{N-1}$ (the unit sphere in $\mathbb{R}^N$) parametrized by arc length, the two following conditions are equivalent:*

(i) *there exists a function $\Omega: \mathbb{R}_+ \to \mathbb{R}$ such that for every close enough s and t in I one has $(\gamma(s), \gamma(t))=\Omega(|t-s|)$, i.e., the distance between points of γ, considered as points of S^{N-1}, is a universal function of the arc length of γ between these points;*

(ii) *there exists a one-parameter subgroup $(G(t))$ of the orthogonal group $O(N)$ and a real number t_0 in I such that $\gamma(t)=G(t-t_0)(\gamma(t_0))$ for every t in I.*

Moreover, such a curve is determined, by the function Ω, up to an element of $O(N)$.

Proof. The part "(ii) implies (i)" is easy. Since $O(N)$ leaves the scalar product invariant, we have

$$\begin{aligned}(\gamma(s), \gamma(t)) &= (G(t-t_0)(\gamma(t_0)), G(s-t_0)(\gamma(t_0)))\\ &= ((G(s-t_0))^{-1}G(t-t_0)(\gamma(t_0)), \gamma(t_0))\\ &= (G(t-s)(\gamma(t_0)), \gamma(t_0)).\end{aligned}$$

The last term seems to depend on $t-s$. In fact it depends only on $|t-s|$ because $G(t-s)=(G(s-t))^{-1}$ and $(G(t-s)u, u)=(u, (G(t-s)^{-1}(u))$ for every u (since $G(t-s)$ is in $O(N)$).

To prove the converse we need the theory of *curvatures* of curves in Euclidean spaces (see for example [SK 1], p. 1—60 or [DE 2], p. 323). The idea is that Condition (i) enables us to compute the curvatures of γ and to prove that they are all constant. For the convenience of the reader we shall briefly restate this result about curves together with the proof of the part "(i) implies (ii)".

Taking derivatives of the Ω-condition for γ with respect to t with fixed s we get that

$$\text{for every } k \text{ in } \mathbb{N} \text{ and every } s \text{ in } I\colon (\gamma(s), \gamma^{(k)}(s)) = \Omega^{(s)}(0) = c_k\,.$$

Notice that $c_k = 0$ when k is odd since Ω is an even function.

From $|\gamma(s)|^2 = 1$ and $|\gamma'(s)|^2 = 1$ for any s, an easy induction yields the formulas:

6.110
$$\begin{cases} \text{for every } s \text{ in } I \text{ and every } p, q \text{ in } \mathbb{N}: \\ (\gamma^{(p)}(s), \gamma^{(q)}(s)) = \begin{cases} (-1)^p c_{p+q} & \text{when } p+q \text{ is even} \\ 0 & \text{when } p+q \text{ is odd.} \end{cases} \end{cases}$$

Now recall that the Frenet frame $\mathscr{F}(s) = (f_1(s), \ldots, f_N(s))$ of γ at s is the orthonormalization of the N-tuple $(\gamma'(s), \gamma''(s), \ldots, \gamma^{(N)}(s))$. We can suppose that $\gamma'(s) \wedge \ldots \wedge \gamma^{(N)}(s) \neq 0$ for every s. In fact, if this were not true, then γ would be contained in a proper subspace of $\mathbb{R}^N$ and we could work in this subspace.

If we now consider the orthonormalization process and if we look at the same time at formulas (6.110), then we see that

$$\mathscr{F}'(s) = \mathscr{F}(s) \begin{pmatrix} 0 & k_1 & 0 & \ldots & 0 & 0 \\ -k_1 & 0 & k_2 & \ldots & 0 & 0 \\ 0 & -k_2 & 0 & \ldots & 0 & 0 \\ \vdots & \vdots & \vdots & \ldots & \vdots & \vdots \\ 0 & 0 & 0 & \ldots & 0 & k_{N-1} \\ 0 & 0 & 0 & \ldots & -k_{N-1} & 0 \end{pmatrix},$$

where the k_i are constants depending only on the c_k (and hence only on γ). In other words, we have

6.111 $\qquad \mathscr{F}^{-1}(s)\mathscr{F}'(s) = A \qquad$ for every s in I,

where A is a constant skew-symmetric matrix. Hence, the one-parameter subgroup $t \mapsto G(t) = \exp(tA)$ of $O(N)$ is the unique solution of the equation (6.111). We then have $\gamma(t) = G(t - t_0)\gamma(t_0)$ for every t in I by the very definition of $\mathscr{F}(t)$. $\square$

The lemma immediately yields the following:

6.112 Proposition. *Let $(\bar{M}, \bar{g})$ be the image of an SH-manifold (M, g) as in 6.99. Then for any two geodesics γ and δ of $(\bar{M}, \bar{g})$ there exists f in $O(N)$ such that $\delta = f \circ \gamma$, i.e., all geodesics in $(\bar{M}, \bar{g})$ are congruent curves in $\mathbb{R}^N$.*

Chapter 7. On the Topology of SC- and P-Manifolds

A. Introduction

7.1. The standard examples of SC-manifolds are the compact, rank-one Riemannian symmetric spaces (the CROSSes). But we described in Chapter 4 some examples of SC-manifolds which are not isometric to a CROSS, the so-called Zoll manifolds. Observe, however, that the underlying differentiable manifold in these examples is the standard sphere. In this chapter we will prove that, at least topologically, the SC-manifolds are not very different from CROSSes. The main result we prove is the following

7.2 Theorem (Bott-Samelson). *The integral cohomology ring of an SC-manifold is the same as that of a* CROSS.

One can look at 7.23 for a more precise statement.

Now R. Bott and H. Samelson gave two different proofs of this theorem, but both require only a much weaker assumption than the one quoted above: it is enough to assume that all the geodesics issuing from *one* point of the Riemannian manifold are closed with the same length (see 7.7, the definition of CL_l^m-manifold).

7.3. In Section B we begin with a study of various assumptions of "closedness" for the geodesics issuing from one point in a Riemannian manifold. Some of them will be well suited for the proofs which we know of the Bott-Samelson theorem and its generalization to P-manifolds. Other assumptions seem more natural, but turn out to be harder to work with. We also recall the definitions of SC- and P-manifolds and compare them with the pointed analogous definitions.

In Section C we discuss these definitions and give various examples and counterexamples in order to illustrate their relationships.

In Section D we prove Bott-Samelson theorem with the weakest hypothesis and the strongest conclusions we know.

In Section E we prove an analogous (but weaker) theorem for P-manifolds.

In Section F we add the hypothesis that the Riemannian manifold is homogeneous and prove that a homogeneous SC-manifold is isometric to a CROSS.

In Section G we give a list of questions we have not been able to answer in this chapter.

We conclude the chapter with historical notes in which we have tried to be fair.

Finally, in all of the chapter, we have used freely the vocabulary of algebraic topology (see [SR] for references).

B. Definitions

7.4. Throughout the chapter, we let (M, g) be a d-dimensional C^∞ connected Riemannian manifold, let m be a point in M and let l be some positive number. We will always consider geodesics as curves γ from some interval (usually $[0, l]$) into M, *parametrized by arc-length* (i.e., $|\dot{\gamma}| = 1$), so that the length of γ is l, and we will say that γ *issues from m* if $\gamma(0) = m$. Also, recall the following well-known

7.5 Definition. Let $\gamma: [0, l] \to M$ be a geodesic of length l issuing from m. Then the *index* of γ is the number of points t in $]0, l[$, counted with their multiplicities, such that $\gamma(t)$ is conjugate to m along γ.

We insist that $\gamma(l)$ *not* be counted, even in the case where γ is a closed curve of length l. (The "index" which is usually defined for *closed* geodesics in the literature on closed curves is different from the index defined above.)

We now define the notions of "closedness" we are interested in.

7.6 Definitions. (a) a *geodesic loop of length l issuing from m* is a geodesic $\gamma: [0, l] \to M$ such that $\gamma(0) = \gamma(l) = m$ and $\gamma(t) \neq m$ for each t in $]0, l[$ (i.e., l is the first length at which γ comes back to m);

(b) a geodesic $\gamma: [0, l] \to M$ is *periodic* with l as *a* period if $\gamma(0) = \gamma(l)$ *and* $\dot{\gamma}(0) = \dot{\gamma}(l)$ (or equivalently, if γ may be extended to a geodesic $\tilde{\gamma}:]-\infty, +\infty[\to M$ which is a periodic map with l as a period);

(c) a *closed geodesic of length l issuing from m* is a periodic geodesic from m whose *least* period is l;

(d) a *closed geodesic loop of length l issuing from m* is a geodesic loop of length l from m which is also a closed geodesic of length l from m;

(e) a geodesic loop of length l from m or a closed geodesic of length l from m is *simple* if it is one-to-one on $[0, l[$.

Notice that a simply closed geodesic of length l issuing from m is also a closed geodesic *loop* of length l from m.

Figures 7.6.1 to 7.6.5 below give some examples which illustrate the definitions 7.6.

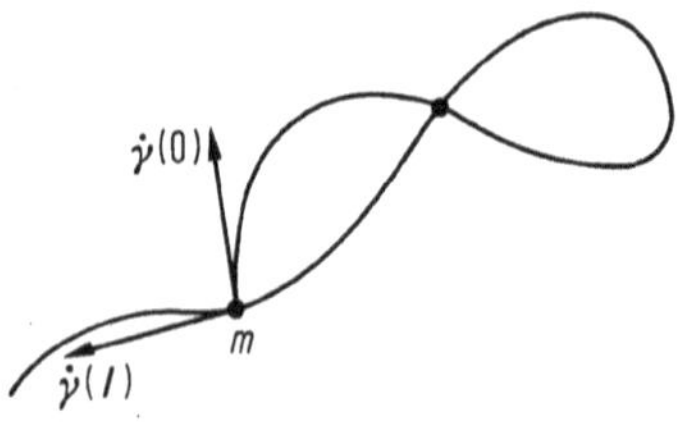

Fig. 7.6.1. A nonsimple nonclosed geodesic loop issuing from m

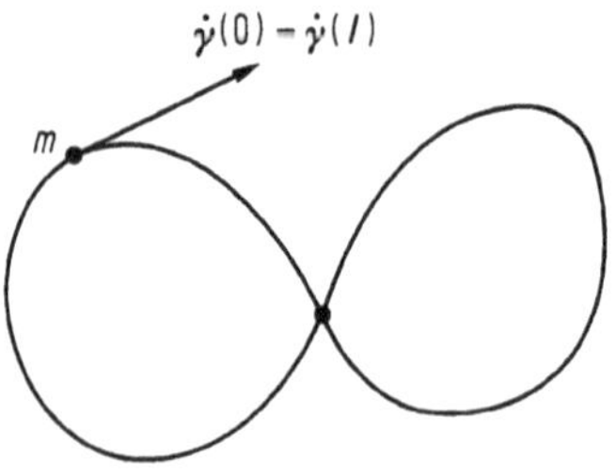

Fig. 7.6.2. A nonsimple closed geodesic issuing from m

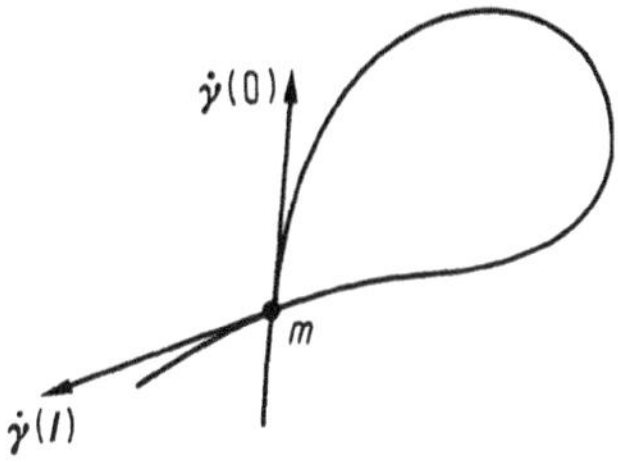

Fig. 7.6.3. A simple nonclosed geodesic loop issuing from m

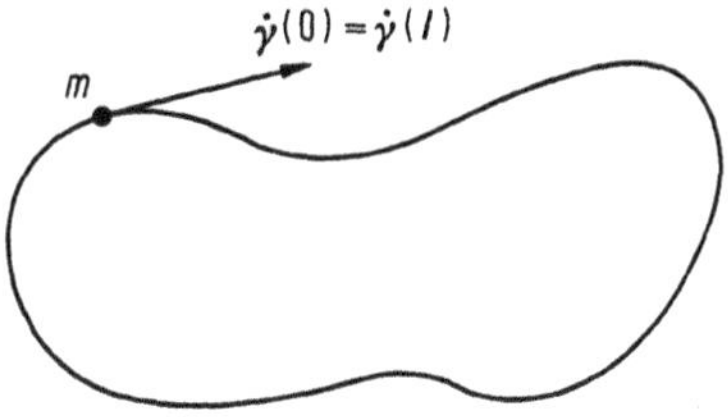

Fig. 7.6.4. A simply closed geodesic issuing from m

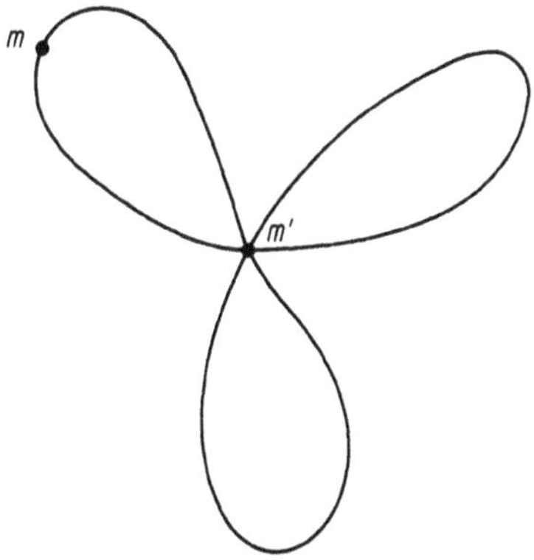

Fig. 7.6.5. This periodic geodesic with least period l is a closed geodesic loop with length l issuing from m, but it is *not* a geodesic loop with length l issuing from m'

We now give various definitions related to the closedness properties of geodesics issuing from one point.

7.7 Definitions. Let (M, g) be a Riemannian manifold, let m be a point in M and let l be a positive number.

Then (M, g) is a	if all the geodesics issued from m
(a) Z^m-manifold	come back to m;
(b) Q^m-manifold	are periodic;
(c) Y_l^m-manifold	come back to m at length l;
(d) P_l^m-manffold	are periodic with least *common* period l;
(e) C_l^m-manifold	are closed geodesics with length l;
(f) L_l^m-manifold	are geodesic loops with length l;
(g) CL_l^m-manifold	are closed geodesic loops with length l;
(h) SL_l^m-manifold	are simple geodesic loops with length l;
(i) SC_l^m-manifold	are simply closed geodesics with length l.

Remark. In the Y_l^m (respectively P_l^m) case, it may happen that some geodesic from m comes back to m before the length l. We only suppose that all these geodesics are back to m at length l (respectively admit the period l), and in what follows we shall take l to be the shortest such length (respectively period). We must confess that these Y- and P-definitions are rather technical, the natural ones being the Z- and Q-definitions but we have results in the Y- or P-cases that we have not been able to prove in the Z- or Q-cases (see the questions in G).

We now recall the global configurations we are interested in.

7.8 Definitions. (a) A Riemannian manifold (M,g) is an *SC_l-manifold* if it is an SC_l^m-manifold for each point m in M.

(b) A Riemannian manifold (M,g) is a *C_l-manifold* if it is a C_l^m-manifold for each point m in M.

(c) A Riemannian manifold (M,g) is a *P_l-manifold* if it is a P_l^m-manifold for each point m in M.

We shall say that (M,g) is an *SC- (resp. C-, P-)manifold* if there exists an l such that (M,g) is an SC_l-(resp. C_l-, P_l-)manifold.

We have the following result, which characterizes SC- and P-manifolds with weaker assumption than those in the definition, and says that there is no need to consider objects like L- or Z-manifolds.

7.9 Proposition. (a) *If (M,g) is an L_l^m-manifold for each m in M, then (M,g) is an SC_l-manifold.*

(b) *If (M,g) is a Z^m-manifold for each m in M, then (M,g) is a P-manifold.*

Proof of (b). We first see that all the geodesics of (M,g) are periodic. This is an easy consequence of the following elementary

7.10 Lemma. *Let $\gamma:\mathbb{R}\to M$ be a geodesic in a Riemannian manifold (M,g). If γ is* not *periodic, then the set of real numbers where γ is not one-to-one is countable.*

Proof of the lemma. For each integer n in $\mathbb{Z}$, let a_n be the infimum of the convexity radii of (M,g) at the points of the compact set $\gamma([n,n+1])$ in M. Now, if t belongs to $[n,n+1]$ and $\gamma(u)=\gamma(t)$, $u\neq t$, then $\gamma(u')\neq\gamma(t')$ for each t' in $[n,n+1]$ and each u' in $]u-a_n, u[\bigcup]u, u+a_n[$. Hence, the set $NI(n)$ of real numbers u such that $\gamma(u)=\gamma(t)$ for some t in $[n,n+1]$ and such that u does not belong to $[n,n+1]$ is countable. Then the set $NI=\bigcup_{n\in\mathbb{Z}} NI(n)$ of points where γ is not one-to-onc is also countable. □

If (M,g) has a non-periodic geodesic, then there exist a point m in M and a geodesic issuing from m which does not come back to m. This shows that if (M,g) is a Z^m-manifold for each m, then all of its geodesics are periodic. Now (b) will follow from the following

7.11 Lemma. *If all the geodesics of a Riemannian manifold are periodic, then they admit a common period.*

Proof of the lemma. Let us consider the bundle UM of unit tangent vectors to M, and the geodesic flow acting on UM. It is shown in 1.102 that there exists a Riemannian metric on UM such that the orbits of the geodesic flow are geodesics which project down to M to geodesics with the same arc-length parametrization and the same period (if they are periodic). All the geodesics of M are obtained in this way.

Hence the lemma [and the proof of (b)] follows from the following

7.12 Theorem (Wadsley). *If the orbits of a flow on a Riemannian manifold are periodic geodesics parametrized by arc-length, then the flow itself is periodic, so that the orbits have a common period.*

The theorem above is proved in §4 of [WY 2]. It shows that the following is a characterization of periodic flows:

7.13 A. W. Wadsley's theorem. *Let M be a C^r-manifold. Let $\mu:\mathbb{R}\times M\to M$ be a C^r-action ($3\leqslant r\leqslant\infty$) of the additive group of real numbers such that every orbit is a circle. Then there is a C^r-action $\varrho: S^1\times M\to M$ with the same orbits as μ if and only if there exists some Riemannian metric on M with respect to which the orbits of μ are embedded as totally geodesic submanifolds of M.*

For a proof, see [WY 2] or Appendix A. □

7.14 *Proof of* 7.9 (a): If (M, g) is an L_l^m-manifold for each point m, then it is a Z^m-manifold for each m and all its geodesics are periodic with a common period. It now suffices to show that all of these geodesics are simply closed and have the same length. Had we supposed that (M, g) was an L_l^m-manifold with the *same* l for each m in M, then the proof would have been easy. Indeed, let γ be a nonclosed geodesic loop issuing from m and let m' be some point on γ whose distance from m is smaller than the convexity radius of M at m. Then (with a shift of the origin) γ is also a geodesic issuing from m' and it cannot come back to m' at length l, which yields a contradiction (see Fig. 7.14 below).

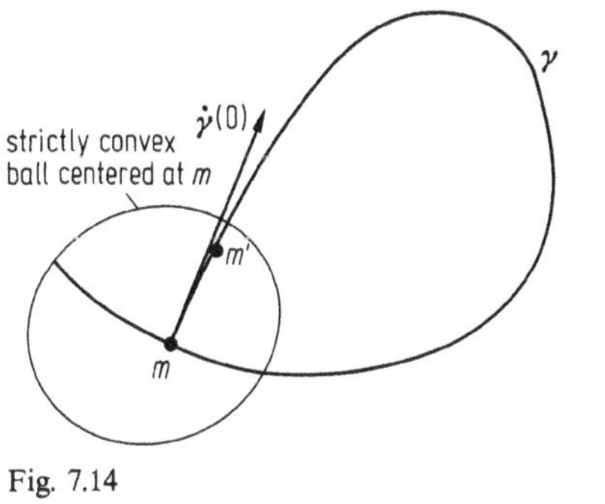

Fig. 7.14

7.15. The result is also valid without the assumption that l is constant. However, the proof requires more precise results on L_l^m-manifolds. We shall prove in 7.23 that in an L_l^m-manifold (M, g) the index k of a geodesic of length l issuing from m is a *topological* invariant of M. Hence, if (M, g) is an L_l^m-manifold for each m in M, then the indices k of all the geodesic loops in M are the same. Now let γ be a periodic geodesic and let m be a self-intersection point of γ. Under the L_l^m-assumption γ is the union of a certain number h of geodesic loops at m with the same length l and index k (see Fig. 7.15 below).

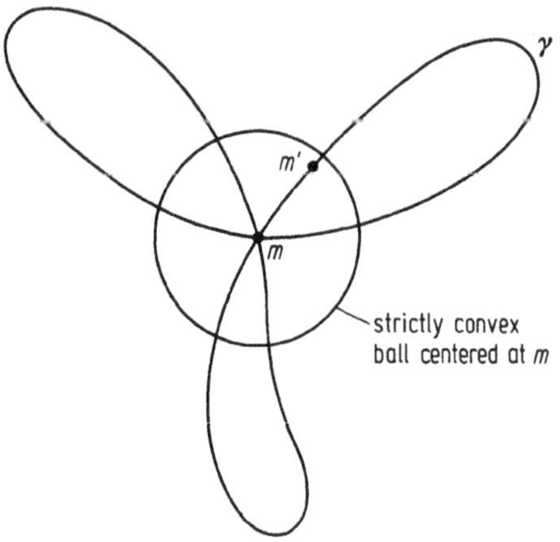

Fig. 7.15

Now let m' be a point on γ whose distance from m is smaller than the convexity radius of M at m. After a shift of the origin, γ becomes a closed geodesic issuing from m'. One easily computes (from Definition 7.5 and Paragraph 7.27) the index of this closed geodesic of length hl from m' and one finds that it equals $hk+(h-1)(d-1)$, number which is different from k if h is not 1. So (a) is proved. □

C. Examples and Counterexamples

7.16. We have the following diagram of implications, which are obvious from the definitions:

$$\begin{array}{ccccccccc} SC & \Longrightarrow & & & C & \Longrightarrow & P & & \\ \Downarrow & & & & \Downarrow & & \Downarrow & & \\ SC_l^m & \Longrightarrow & CL_l^m & \Longrightarrow & C_l^m & \Longrightarrow & P_l^m & \Longrightarrow & Q^m \\ \Downarrow & & \Downarrow & & & & \Downarrow & & \Downarrow \\ SL_l^m & \Longrightarrow & L_l^m & & \Longrightarrow & & Y_l^m & \Longrightarrow & Z^m \end{array}$$

7.17. In the preceding chapters we have already seen various examples of manifolds satisfying one of the definitions given in 7.7 or 7.8. Let us first recall examples of SC- and P-manifolds:

a) In Chapter 5 we saw that Blaschke manifolds are SC-manifolds (5.42);

b) In Chapter 3 we saw that compact, rank-one Riemannian symmetric spaces (CROSS'es) are Blaschke manifolds (3.35) and so are SC-manifolds;

c) It is easy to see that every quotient of an SC-manifold (and in particular of a CROSS) by a finite free group of isometries is a P-manifold (any geodesic is the image of a geodesic of the covering, and is hence periodic).

Let us recall that $\mathbb{R}P^{2n}$, $\mathbb{C}P^{2n}$, $\mathbb{H}P^n$ $(n \neq 1)$ and $\mathbb{C}\mathrm{a}P^2$ do not have any quotient (even topologically, see [BR-GB]) and that S^{2n} has only one *Riemannian* quotient, namely $\mathbb{R}P^{2n}$. Conversely, we saw in 5.32 that $\mathbb{C}P^{2n+1}$ has an (essentially unique) Riemannian $\mathbb{Z}_2$-quotient, and that S^{2n+1} has many quotients differing from $\mathbb{R}P^{2n+1}$ (among which are well-known lens spaces, see [WF 1] for their classification). These quotients are examples of P-manifolds which are not SC-manifolds.

Moreover, we will see as a corollary of the Bott-Samelson theorem that these P-manifolds cannot (for topological reasons) be SC-manifolds for any metric on them. So we may state the following

7.18 Proposition. *There exist P-manifolds which are not SC-manifolds (for any metric on them).* □

In Chapter 4, we studied other examples of SC-manifolds, the so-called Zoll surfaces or Zoll manifolds. And there (4.37) we saw that some of these examples are not Blaschke manifolds and, in particular, are not isometric to a CROSS. These Zoll manifolds are also examples of SC-manifolds which are not Blaschke manifolds at m for some points m.

We conclude these remarks on SC- and P-manifolds by remarking that we know of no example of a simply connected P-manifold which is not an SC-manifold.

The following construction allows us to give many more examples of manifolds satisfying one of the definitions 7.7.

7.19 Proposition (A. Weinstein). *Let $\xi(\pi: E \to B)$ be a vector bundle such that there exists a diffeomorphism σ of the associated sphere bundle $S(\xi)$ of ξ onto the standard sphere S^{d-1}. Let M be the differentiable manifold $D^d \bigcup_\sigma D(\xi)$ (which is the union of the associated disk bundle $D(\xi)$ of ξ and the standard disk D^d, glued along their boundaries by the diffeomorphism σ).*

Then there exists a Riemannian metric g on M such that (M, g) is a Blaschke manifold at m_0 (where m_0 is the centre of the disk D^d).

Notice that this is essentially the converse to the Allamigeon-Warner theorem (see 5.43). A proof is given in Appendix C, or see [WR 2]. □

Now Blaschke manifolds at m are SL_l^m-manifolds (by 5.40). There are many manifolds which can be written as $D^d \bigcup_\sigma D(\xi)$ for some (ξ, σ) as in the proposition, for example, any exotic sphere (for $d \neq 3, 4$, see [K-M]) and, more generally, any connected sum of an exotic sphere and a CROSS.

We have the following:

7.20 Proposition. *If $d \neq 3, 4$, then any connected sum of an exotic sphere and a CROSS (and in particular, exotic spheres themselves) admit metrics with the SL_l^m-property.*

In the non simply-connected case, one has more precise results. In dimension $d \geqslant 6$, one knows by differential and algebraic topology that the following statements are equivalent:

(a) $M = D^d \bigcup_\sigma D(\xi)$ and M is not simply-connected;

(b) M has the homotopy type of $\mathbb{R}P^d$ and its Browder-Livesay index is zero.

See [LO] for this result and the definition of the Browder-Livesay index (it is zero for even dimensions, has values in $\mathbb{Z}$ for dimension $4n-1$ and values in $\mathbb{Z}_2$ for dimension $4n+1$). There are also examples of manifolds satisfying (a) or (b) which are not diffeomorphic to $\mathbb{R}P^d$. We shall see in the Bott-Samelson Theorem 7.23 that a non simply-connected CL_l^m-manifold is diffeomorphic to $\mathbb{R}P^d$, so we have proved the following

7.21 Proposition. *Let M be homotopy equivalent to $\mathbb{R}P^d (d \geqslant 6)$ with zero Browder-Livesay index, and suppose that M is not diffeomorphic to $\mathbb{R}P^d$. Then there exist metrics on M which are Blaschke at m for some m (and in particular SL_l^m), but there does not exist any CL_l^m metric on M.*

Conversely, it is possible to put an SC_l^m-metric on certain exotic spheres (not diffeomorphic to the standard exotic sphere, see Appendix C, (C.19) or [B.B 1]).

Other well-known examples of $D^d \bigcup_\sigma D(\xi)$ are the Eells-Kuiper projective planes (see [E-K]). In particular, Eells and Kuiper found examples of manifolds which can be written as $D^d \bigcup_\sigma D(\xi)$, which have the same cohomology ring as $\mathbb{H}P^2$ or $\mathbb{C}a\,P^2$, but which do not have the same homotopy type. Hence, we have the following

7.22 Proposition. *There exist SL_l^m-manifolds (and even Blaschke manifolds at m) with the same cohomology ring as $\mathbb{H}P^2$ or $\mathbb{C}a P^2$, but not with the same homotopy type.*

Compare the proposition with the conclusions of the Bott-Samelson Theorem 7.23 for $k=0, 1$ or $d-1$. A detailed study of these examples may be found in [B.B 1].

D. Bott-Samelson Theorem (C-Manifolds)

In this paragraph we prove the main result of the chapter, namely:

7.23 Theorem. (Bott-Samelson [BT] [SN 2]). *Let (M, g) be an L_l^m-manifold of dimension $d \geqslant 2$. Let k be the index of one of the geodesics of length l issuing from m.*

Then M is compact, k is the same for all these geodesics and:

if $k > 0$, then M is simply-connected and the integral cohomology ring of M has exactly one generator. More precisely, one has only the following possibilities:

$k=1$, $d=2n$ and M has the homotopy type of $\mathbb{C}P^n$;

$k=3$, $d=4n$ and M has the integral cohomology ring of $\mathbb{H}P^n$;

$k=7$, $d=16$ and M has the integral cohomology ring of $\mathbb{C}a P^2$;

$k=d-1$, any d, and M has the homotopy type of S^d (so that M is homeomorphic to S^d if $d \neq 3, 4$).

if $k=0$, $\pi_1(M)=\mathbb{Z}_2$ and M has the homotopy type of $\mathbb{R}P^d$. Moreover, M is diffeomorphic to the union of the standard disk D^d and a (one-dimensional) disk bundle with boundary diffeomorphic to S^{d-1}, glued by this diffeomorphism (or, equivalently, if $d \geqslant 6$, then the Browder-Livesay index of M is zero). In fact, M is a Blaschke manifold at m.

Finally, if (M,g) is a CL_l^m-manifold and $k=0$, then M is diffeomorphic to $\mathbb{R}P^d$.

Remark. The k used in the present chapter differs by one from that used in Chapter 5 in the case of Blaschke manifolds.

7.24 *Proof.* The L_l^m-condition says that all the geodesics issuing from m come back to m at length l (and not before). By iteration of these geodesic loops, one sees that the exponential map at m, $\exp_m: T_m M \to M$, is well-defined on all of $T_m M$ and M is complete by the Hopf-Rinow theorem. Now every minimizing geodesic from m to any point p in M has length less than $l/2$, so that the diameter of M is less than l and M is compact (since it is complete and bounded). One also sees immediately that

7.25. *The point m is conjugate to itself with multiplicity $d-1$ along any geodesic from m to m (of any length).*

Now note that $d-1$ is the supremum of the possible multiplicities of a conjugate point. It is well known that for a (continuous) one-parameter family of geodesics issuing from some given point, the arc-length of the first, second, ..., n^{th} conjugate points to the given point (counting multiplicities) is a continuous function of the parameter (see [ME], p. 235, or 1.98). So, on a family of geodesics issuing from m, points conjugate to m cannot "cross" the values of the arc-lengths $l, 2l, \ldots, hl, \ldots$ for each integer h. In particular, k is constant.

One now needs a key lemma. Let us set (as in 5.8): $B(0_m, r)=\{v \in T_m M \mid \|v\|<r\}$ for the open ball of radius r around 0_m in $T_m M$; $B(m,r)=\{p \in M \mid \varrho(p,m)<r\}$ for the open ball of radius r around m in M; similarly $A(0_m, r_1, r_2)=\{v \in T_m \mid r_1<\|v\|<r_2\}$ for the annulus between the radii r_1 and r_2 around 0_m in $T_m M$; and finally B', A' for the closures of B, A.

7.26 Lemma. *Let (M,g) be an L_l^m-manifold. Then there exists a positive number ε such that:*

(a) $\exp_m$ *is a diffeomorphism of $B(0_m, \varepsilon)$ onto $B(m, \varepsilon)$;*

(b) $\exp_m$ *maps $A'(0_m, \varepsilon, l-\varepsilon)$ onto $M \setminus B(m, \varepsilon)$ (the complement of $B(m,\varepsilon)$ in M).*

Proof of the Lemma. (a) is well known (for any Riemannian manifold) as soon as ε is smaller than the injectivity radius of M at m. Let us suppose that (b) is false for every ε, $0<\varepsilon<\mathrm{Inj}_m(M)$. Then, for each integer n, there exist a unit tangent vector u_n in $T_m M$ and a length t_n with $1/n<t_n<l-1/n$, such that $\exp_m(t_n u_n)$ belongs to $B(m, 1/n)$. Because of the injectivity of $\exp_m$ on $B(0_m, \mathrm{Inj}_m(\mathrm{M}))$, one has in fact:

$\mathrm{Inj}_m(M) \leqslant t_n \leqslant l - \mathrm{Inj}_m(M)$ for each $n > 1/\mathrm{Inj}_m(M)$.

Let (t, u) be a limit point of the sequence (t_n, u_n) in the compact set $[\mathrm{Inj}_m(M), l - \mathrm{Inj}_m(M)] \times U_m M$. By continuity, $\exp_m(tu) = m$ with $\mathrm{Inj}_m(M) \leqslant t \leqslant l - \mathrm{Inj}_m(M)$ and this contradicts the condition L_l^m. □

7.27. Again by iteration of the geodesic loops at m, one now sees that, for each integer n in $\mathbb{N}$, $\exp_m$ maps $A(0_m, nl - \varepsilon, nl)$ diffeomorphically onto $B(m, \varepsilon)\setminus\{m\}$, $A(0_m, nl, nl + \varepsilon)$ diffeomorphically onto $B(m, \varepsilon)\setminus\{m\}$ and $A'(0_m, nl + \varepsilon, (n+1)l - \varepsilon)$ onto $M \setminus B(m, \varepsilon)$, see Figure 7.27. In particular, if m' belongs to $B(m, \varepsilon)\setminus\{m\}$, then m' is *not* conjugate to m along *any* geodesic from m to m'. Let $\gamma: \mathbb{R} \to M$ be the unique geodesic such that $\gamma(0) = m$ and $\gamma(\varrho(m, m')) = m'$. Then each geodesic from m to m' is the union of a certain number $n \geqslant 0$ of geodesic loops of length l issuing from m followed either by the segment $\gamma: [0, \varrho(m, m')] \to M$ or by the "opposite" geodesic $\bar{\gamma}: [0, l - \varrho(m, m')] \to M$ given by $\bar{\gamma}(t) = \gamma(l - t)$. The index of γ between m and m' is zero and that of $\bar{\gamma}$ between m and m' is k. Hence, the indices of the geodesics from m to m' are the numbers $n(k + d - 1)$ and $n(k + d - 1) + k$ for all integers $n \geqslant 0$.

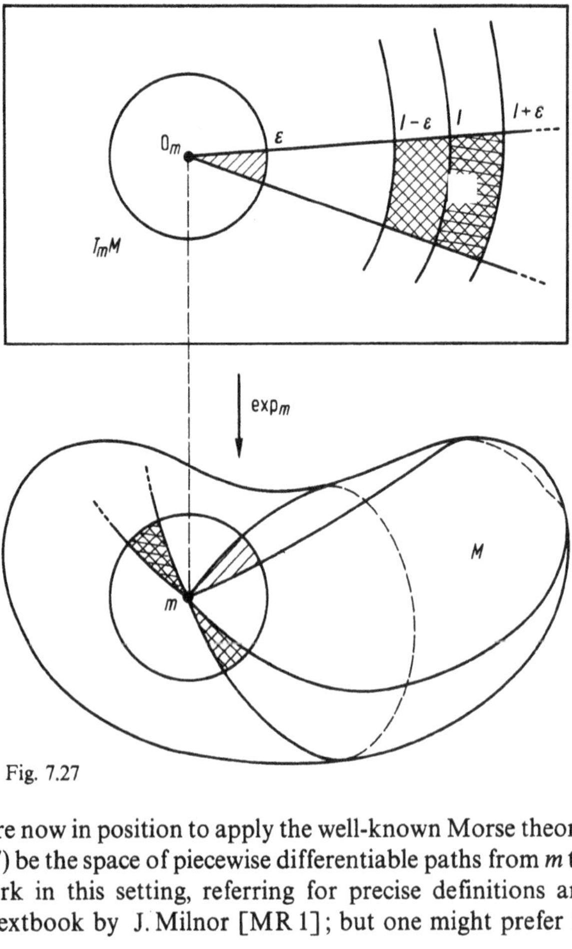

Fig. 7.27

7.28. We are now in position to apply the well-known Morse theory on path spaces. Let $\mathscr{C}(m, m')$ be the space of piecewise differentiable paths from m to m' (here we will always work in this setting, referring for precise definitions and results to the beautiful textbook by J. Milnor [MR 1]; but one might prefer to work with the space of paths of Sobolev class H^1 as in [F-K]). One knows that the energy $\mathbb{E}(\gamma) = 1/2 \int_\gamma \|\dot{\gamma}\|^2$ is a Morse function on $\mathscr{C}(m,m')$ whose critical points are exactly the geodesics from m to m'. These critical points are nondegenerate if and only if m' is not conjugate to m along the corresponding geodesic. The index of a critical point is exactly the index of the corresponding geodesic. Now $\mathscr{C}(m,m')$ has the homotopy type of a CW-complex K with one cell for each geodesic from m to m' (and the dimension of the cell is the index, that is, one of the numbers $n(k+d-1)$ or $n(k+d-1)+k, n \geqslant 0$). Let $\mathscr{C}(m,M)$ be the space of all (piecewise differentiable) paths issuing from m in M. The end-point map $w:\mathscr{C}(m,M) \to M$ is a Serre fibration with $\mathscr{C}(m,m')$ as fiber over m'. Since $\mathscr{C}(m,M)$ is contractible, the homotopy exact sequence of the fibration gives an isomorphism $\pi_{i+1}(M) = \pi_i(\mathscr{C}(m,m'))$ for all $i \geqslant 0$.

7.29. *First case:* $k>0$. Then K has only one 0-cell, hence is connected, and $\pi_1(M)=0$. Moreover, K has one k-cell and all the other cells have dimension greater than $k+d-1$, so that K has the same cohomology as S^k up to dimension $k+d-2$. In particular, $\pi_i(M)=0=H_i(M,\mathbb{Z})$ for $0<i<k+1$. Furthermore, the spectral sequence of the fibration gives a short Gysin exact seq ence. There exists a cohomology class U in $H^{k+1}(M,\mathbb{Z})$ such that the sequence

$$\ldots \to H^i(\mathscr{C}(m,M),\mathbb{Z}) \to H^{i-k}(M,\mathbb{Z}) \xrightarrow{\alpha} H^{i+1}(M,\mathbb{Z}) \to H^{i+1}(\mathscr{C}(m,M),\mathbb{Z}) \to \ldots$$

is exact for $i<k+d-1$, where α is the cup-product with U. Since $\mathscr{C}(m,M)$ is contractible, α is an isomorphism when $i<k+d-1$. Since M is compact, simply connected and d-dimensional, one has $H^d(M,\mathbb{Z})=\mathbb{Z}$ and $H^i(M,\mathbb{Z})=0$ for $i>d$. Hence we know the whole cohomology ring of M: one has $H^*(M,\mathbb{Z})=\mathbb{Z}(U)/U^n$, where the degree of U is $k+1$ and n satisfies $n(k+1)=d$.

7.30. Now, it is a difficult result in cohomology theory (involving cohomology operations) that if the integral cohomology ring of a manifold has only one generator, then the only possibilities are:

$$k=1, \quad d=2n\,;$$
$$k=3, \quad d=4n\,;$$
$$k=7, \quad d=16\,;$$

and

$$k=d-1, \quad \text{any} \quad d$$

(see [AD], [AS], [MR 2]).

If $k=d-1$, then M is a simply connected homology sphere, hence a homotopy sphere by Hurewicz' theorem and a topological sphere ($d \neq 3,4$) by Smale's theorem [SE].

If $k=1$ (i.e., when M has the same cohomology ring as $\mathbb{C}P^n$), then one can construct a map $f: M \to \mathbb{C}P^n$ inducing an isomorphism of cohomology rings, and hence a homotopy equivalence by Whitehead's theorem (see [KG 1] p. 537 for this construction). Note that we can have examples not homeomorphic to $\mathbb{C}P^n$ (see [B. B 1]).

For the two other cases, we have already seen in 7.22 that we can have different homotopy types.

7.31. *Second case:* $k=0$. Here K has two 0-cells and hence, at most two connected components. If $d \geqslant 3$, then K has no 1-cell, so it does indeed have two connected components and $\pi_1(M)=\mathbb{Z}_2$. If $d=2$, there are some 1-cells in K, but if $\pi_1(M)=0$, then M is S^2, and it is known that for any Riemannian metric on S^2 and any point, the cut locus always contains some conjugate point ([WN 1], [MS 1]). Hence, k cannot be zero for S^2 and we always have $\pi_1(M)=\mathbb{Z}_2$. In the unique non-trivial homotopy class of $\pi_1(M, m)$ one chooses a loop at m realizing the shortest length among all the (rectifiable) loops in the homotopy class. One easily sees that this loop must be a geodesic loop with length l issued from m. All of the geodesic loops with length l issued from m are in the same homotopy class since they are all homotopic by the following construction: if γ and γ' are two geodesic loops from m with length l, let γ_u be the geodesic issuing from m with unit tangent vector $v(u)$, where $v(u)$ is a one-parameter family of unit vectors at m with $v(0)=\dot{\gamma}(0)$ and $v(1)=\dot{\gamma}'(0)$. Then $(t, u)\mapsto\gamma_u(t)$ is a homotopy between γ and γ' with base point m.

7.32. Now let $\tilde{M}$ be the Riemannian universal covering of M and $\{m_1, m_2\}$ the fiber of m in $\tilde{M}$. From the preceding one immediately sees that $\tilde{M}$ is an $L^{m_1}_{2l}$-manifold, since any geodesic in M issuing from m_1 goes to m_2 at length l and then comes back to m_1 at length $2l$. Moreover, the index $\tilde{k}$ for $\tilde{M}$ is $d-1$. Applying results from the first case gives that M is a homotopy sphere, and it is known ([LO], p. 43) that any quotient of a homotopy sphere by a free differentiable $\mathbb{Z}_2$-action has the homotopy type of $\mathbb{R}P^d$.

For the last assertions of the theorem, one needs another lemma. Let us call V the set of points in $M\setminus\{m\}$ where two distinct geodesic loops of length l issuing from m intersect, or where such a geodesic loop intersects itself.

7.33 Lemma. *V is empty if $k=0$.*

Proof of the Lemma. First one sees that if ε is chosen as in Lemma 7.26 above, then $B(m, \varepsilon)\cap V=\emptyset$. Since $k=0$, m has no conjugate point before length l, so that $\exp_m$ is a local diffeomorphism on $B(0_m, l)$. If p is in V, then p is the image of at least four points in $B(0_m, l)$ under $\exp_m$, so that nearby points are also in V. Hence, V is open (see Fig. 7.33.1 below).

Now let γ be a geodesic loop with length l issuing from m which intersects V. Then $V\cap\gamma([0, l])$ is open in $\gamma([0, l])$. Let q be (one of) the point(s) of the closure of $V\cap\gamma([0, l])$ closest to m along γ. Let (q_n) be a sequence of points in $V\cap\gamma([0, l])$ which converges to q. Only finitely many of these points can be multiple points of γ (by the same kind of arguments as in Lemma 7.10). For the other points of the sequence, there exist a length t_n in $[\varepsilon, l-\varepsilon]$ and a unit vector u_n in T_mM (different from $\dot{\gamma}(0)$ and $-\dot{\gamma}(l)$) such that $\exp_m(t_n u_n)=q_n$. If (t, u) is some limit point of the sequence (t_n, u_n) in

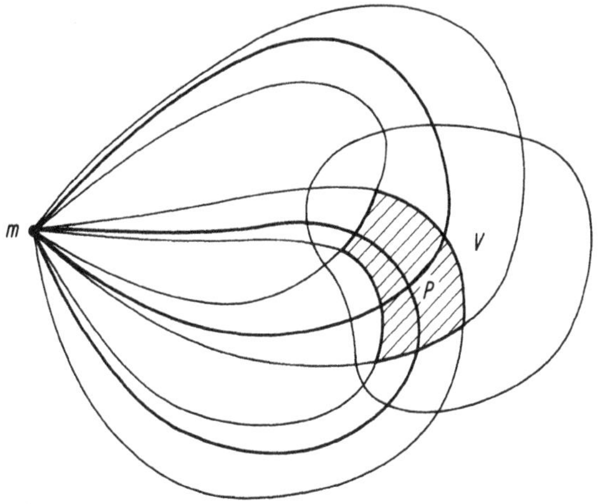

Fig. 7.33.1

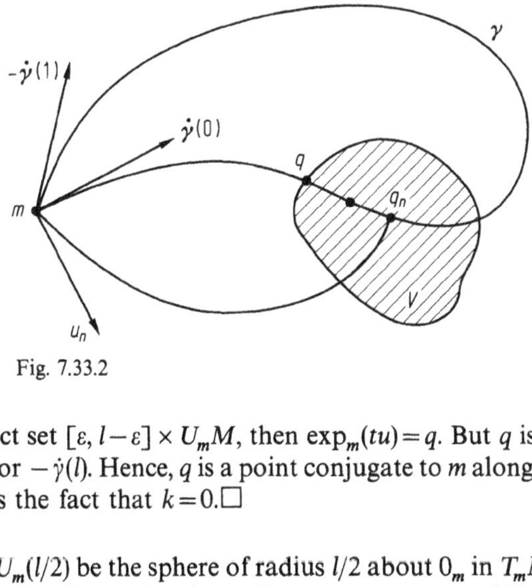

Fig. 7.33.2

the compact set $[\varepsilon, l-\varepsilon] \times U_m M$, then $\exp_m(tu)=q$. But q is not in V, so u must be either $\dot\gamma(0)$ or $-\dot\gamma(l)$. Hence, q is a point conjugate to m along γ or its opposite $\bar\gamma$. This contradicts the fact that $k=0$.□

7.34. Let $U_m(l/2)$ be the sphere of radius $l/2$ about 0_m in $T_m M$ and let $\mathrm{Cut}(m)$ be the cut-locus of m in M. An immediate consequence of Lemma 7.33 is that the restriction of $\exp_m$ to $U_m(l/2)$ is a two-fold covering of $U_m(l/2)$ onto $\mathrm{Cut}(m)$, so that $\mathrm{Cut}(m)$ is an imbedded submanifold of M. Moreover, (M, g) is a Blaschke manifold at m (also an Allamigeon-Warner manifold at m) and it has a spherical cut-locus at m (see Chapter 5 for the definitions and the relations between them). Now we apply the Allamigeon-Warner Theorem (5.29): M is the differentiable union of the disk $B'(m, l/4)$ and the submanifold $M \setminus B(m, l/4)$, which is diffeomorphic to the normal disk bundle to $\mathrm{Cut}(m)$ in M and whose boundary is the sphere $S(m, l/4)$ (the diffeomorphic image of $U_m(l/4)$ by $\exp_m$). Furthermore, $\mathrm{Cut}(m)$ is a $\mathbb{Z}_2$-quotient of a sphere, so that it has the homotopy type of $\mathbb{R}P^{d-1}$. Hence, M has the homotopy type of $\mathbb{R}P^d$ and the "desuspension theorem" of [LO] p. 13 says that (if $d \geqslant 6$) the Browder-Livesay index of M is zero.

7.35. Now let (M, g) be a CL_l^m-manifold with $k=0$. On $\mathbb{R}P^d$ let us choose a Riemannian metric g_l with constant curvature and whose geodesics are closed with length l. Let p be some point in $\mathbb{R}P^d$ and let f be a linear isometry from $T_p\mathbb{R}P^d$ to T_mM. Observe that the inverse image of a point q in $\mathbb{R}P^d \setminus \{p\}$ under $\exp_p$ is the set of points $(nl+\varrho(p,q))u$ or $((n+1)l-\varrho(p,q))u$ for all integers n (where u is a unit vector in $T_p\mathbb{R}P^d$ generating the unique geodesic from p to q).

But Lemma 7.33 says precisely that the same property is also true, with respect to the point m, for M satisfying the CL_l^m-condition. Hence, one may construct the following "Samelson map" $\varphi: \mathbb{R}P^d \to M$: first one sets $\varphi(p)=m$; then, if $q \neq p$ in $\mathbb{R}P^d$, we let u be the corresponding unit vector in $T_p\mathbb{R}P^d$ as above and set $\varphi(q) = \exp_m(f(\varrho(p,q)u))$. The function φ is well defined, one-to-one and onto. Now $\exp_p$ and $\exp_m$ are local diffeomorphisms near every vector whose norm is not an integral multiple of l, since k is zero for M and $\mathbb{R}P^d$. Therefore φ is a local diffeomorphism, and, being bijective, is a diffeomorphism. The theorem is proved. □

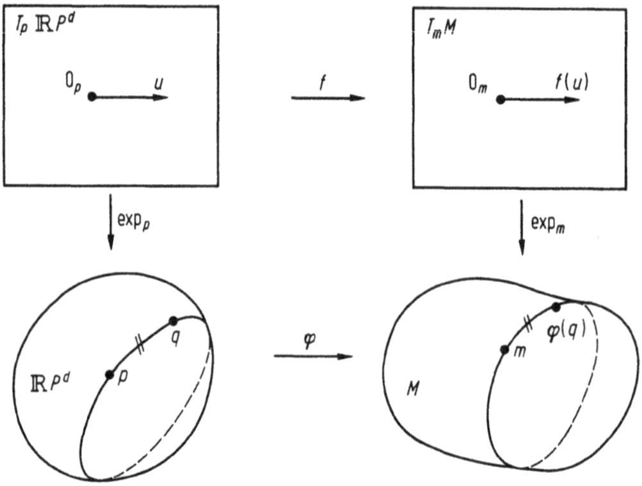

Fig. 7.35

7.36 Remark. One might as well construct such a Samelson map $\varphi: \mathbb{R}P^d \to M$ for each CL_l^m-manifold M (without the hypothesis $k=0$). Such a φ would be well defined, differentiable and onto, but not injective if $k \neq 0$ because then V is nonempty. Nevertheless, φ would be a local diffeomorphism near p thanks to Lemma 7.26. Hence, one gets a map of degree one from $\mathbb{R}P^d$ to M. This is the key point of H. Samelson's proof of the theorem in the CL_l^m-case.

E. P-Manifolds

We will prove the following weak form of the Bott-Samelson theorem for P-manifolds:

7.37 Theorem ([B.B 1]). *Let (M, g) be a Y_l^m-manifold. Then M is compact, $\pi_1(M)$ is finite and the* rational *cohomology ring of M has exactly one generator.*

Proof. All of the arguments in 7.24 carry over to Y_l^m-manifolds. Hence, M is compact, m is conjugate to itself with multiplicity $d-1$ along any geodesic of length l from m to m and all of these geodesics have the same index k.

But Lemma 7.26 is not true any more. We first look at the Riemannian universal covering of M.

7.38. With the same argument as at the end of 7.31 it is easy to see that any two geodesics with length l from m to m are homotopic as loops with (fixed) base point m. In particular, any such geodesic is homotopic to its opposite and any geodesic with length $2l$ issuing from m is homotopically trivial. Now, in each homotopy class in $\pi_1(M, m)$ there exists at least one geodesic from m to m which minimizes the length of the loops in the class, and the length of such a geodesic is trivially smaller than $2l$ (if not, it contains a geodesic of length $2l$ from m to m which is then homotopically trivial and may be removed without changing the homotopy class). Let $(\tilde{M}, \tilde{g})$ be the Riemannian universal covering of (M, g) and let F_m be the discrete fiber over m in M. The distances of a point $\tilde{m}$ in F_m to the other points in F_m are precisely the lengths of the minimizing geodesics from m to m in the various homotopy classes in $\pi_1(M, m)$. Hence, F_m is discrete and bounded. Therefore F_m is finite and $\pi_1(M, m)$, which has the same cardinality as F_m, is also finite. Moreover, any geodesic issuing from $\tilde{m}$ with length $2l$ comes back to $\tilde{m}$ because its projection onto M is a geodesic from m to m with length $2l$ and hence is homotopically trivial. Thus $\tilde{M}$ is a Y_{2l}^m-manifold.

7.39 Remark. The canonical real projective space is a Y_π^m-manifold whose Riemannian universal covering (which is the canonical sphere) is a $Y_{2\pi}^m$-manifold and *not* a Y_π^m-manifold.

One knows that if $\pi_1(M)$ is finite, then the rational cohomology ring of M has exactly one generator if and only if that of $\tilde{M}$ has the same property. Hence, from now on we will suppose that M is *simply connected*.

As a substitute for Lemma 7.26 we use the following well-known

7.40 Lemma. *Let (M, g) be a Riemannian manifold and m a point in M. Then there exists a point p in M such that p is not conjugate to m along any geodesic from m to p.*

Proof. This is a corollary of Sard's theorem applied to $\exp_m$. □

7.41. We now claim that there exist only a finite number $n+1$ of geodesics $\gamma_0, \gamma_1, \ldots, \gamma_n$ from m to p with length smaller than l. Indeed, if we let (γ_i) be an infinite family of distinct geodesics from m to p with lengths t_i smaller than l, then in the compact set $U_m M \times [0, l]$ there is a subsequence of the family $(\dot{\gamma}_i(0), t_i)$ which converges to a limit $(\dot{\gamma}(0), t)$. Then $\lim(\gamma_i(t_i)) = \gamma(t) = p$ is conjugate to m along the geodesic γ with initial vector $\dot{\gamma}(0)$, which yields a contradiction.

7.42. Let us call $k_0, k_1, \ldots, k_n$ the indices of the geodesics $\gamma_0, \gamma_1, \ldots, \gamma_n$ from m to p. One may as well suppose that $k_0 \leqslant k_1 \leqslant \ldots \leqslant k_n$. Now there exists at least one minimizing geodesic from m to p. Therefore $k_0 = 0$ and also $k_n = k$ by considering the opposite geodesic. (More generally, the same argument shows that $k_i + k_{n-i} = k$.)

7.43. Now we iterate these geodesics as in 7.27 and get easily the following description of all of the geodesics from m to p: such a geodesic is the union of a certain number N of geodesics from m to m with length l, followed by one of the γ_i's. Hence for each integer N of $\mathbb{N}$, there are exactly $n+1$ geodesics from m to p whose lengths are between Nl and $(N+1)l$, and their indices are the integers $k_i+N(d-1+k)$ because every Jacobi field along such a geodesic which is zero at the origin m is zero again every time the geodesic comes back to m at a length multiple of l.

7.44. We are now in position to apply Morse theory to the path space $\mathscr{C}(m, p)$ as in 7.28. For each integer j, the number of critical points with index j of the energy on $\mathscr{C}(m, p)$ is bounded by $n+1$. Hence, the (rational) Betti numbers of $\mathscr{C}(m, p)$ are also bounded by $n+1$, and the same is true for the space $\mathscr{C}(m, m)$ (because $\mathscr{C}(m, p)$ and $\mathscr{C}(m, m)$ are two fibers of the Serre fibration from $\mathscr{C}(m, M)$ to M given by the end point).

Now we may apply a theorem due to D. Sullivan:

7.45 Theorem. *The Betti numbers of the loop space $\mathscr{C}(m, m)$ of a simply-connected finite complex are bounded if and only if the rational cohomology ring of the complex is generated by one element.*

See [SL 1] p. 46. The proof uses Sullivan's theory of minimal models and the fact that $\mathscr{C}(m, m)$ is an H-space so Theorem 7.37 is proved. □

7.46 Remark. The proof works as well for any homology with coefficients in any field, except that there does not (yet?) exist an analog of Sullivan's theorem with coefficients in a field with non-zero characteristic.

F. Homogeneous SC-Manifolds

7.47. In this paragraph we examine the case of (Riemannian) homogeneous manifolds as an application of the Bott-Samelson theorem 7.23. That is, we will prove that a homogeneous SC-metric is *isometric* to that of a CROSS. To be precise, let us recall the following

7.48 Definition. The Riemannian manifold (M, g) *is homogeneous* if the group of isometries of (M, g) is transitive on M.

Then the metric g may be given by a scalar product on some invariant complement of the Lie algebra of an isotropy subgroup in the Lie algebra of the isometry group. In this setting, the Levi-Civita connection and the curvature of g can be algebraically described (see for example [K-N 2] Chapter X or [C-E] Chapter III). The geodesics are much more difficult to describe, and we have no direct proof that the SC-condition implies that M is a CROSS.

7.49. We shall use a less direct approach. We first use the Bott-Samelson theorem 7.23 and find which of the homogeneous spaces have the *integral* cohomology ring of a CROSS. This yields a rather short list of homogeneous spaces, each diffeomorphic to a CROSS. Then a case-by-case check settles the isometry question.

7.50 Theorem (Wang [WG 1], Borel [BL 1], Singh Varma [S.V]). *Let $M = G/H$ be a simply-connected homogeneous space with G and H compact and with G acting effectively on M. Then M has the integral cohomology ring of a* CROSS *if and only if M is diffeomorphic to a* CROSS. *Moreover G contains a simple compact subgroup G_0 which is transitive on M and such that $(G_0, H_0 = H \cap G_0, M)$ is one of the following:*

(a)	$G_0 = SO(n+1)$,	$H_0 = SO(n)$,	$M = S^n$;
(b)	$G_0 = SU(n+1)/\mathbb{Z}_{n+1}$,	$H_0 = S(U(1)U(n))/\mathbb{Z}_{n+1}$,	$M = \mathbb{C}P^n$;
(c)	$G_0 = Sp(n+1)/\mathbb{Z}_2$,	$H_0 = Sp(1)Sp(n)/\mathbb{Z}_2$,	$M = \mathbb{H}P^n$;
(d)	$G_0 = F_4$,	$H_0 = Spin(9)$,	$M = \mathbb{C}aP^2$;
(e)	$G_0 = G_2$	$H_0 = SU(3)$,	$M = S^6$;
(f)	$G_0 = Spin(7)$,	$H_0 = G_2$,	$M = S^7$;
(g)	$G_0 = SU(n+1)$,	$H_0 = SU(n)$,	$M = S^{2n+1}$;
(h)	$G_0 = Sp(n+1)$,	$H_0 = Sp(n)$,	$M = S^{4n+3}$;
(i)	$G_0 = Sp(n+1)/\mathbb{Z}_2$,	$H_0 = Sp(n)U(1)/\mathbb{Z}_2$,	$M = \mathbb{C}P^{2n+1}$;
(j)	$G_0 = Spin(9)$,	$H_0 = Spin(7)$,	$M = S^{15}$.

We will not prove this theorem, which involves precise calculations in the cohomology theory of homogeneous spaces (see the references given above for a complete proof). But we will indicate a brief

7.51 *Sketch of the Proof.* We first look at the *even*-dimensional case. Then the Euler characteristic of M is positive. This implies that the ranks of G and H are the same ([WG 1]). Now subgroups of maximal rank in a compact Lie groups are well known (see for example [B-S] or [WF 1] Chapter 8.10). One knows, in particular, that if G acts effectively on G/H, then G is semi-simple, and if $G_1, G_2, \ldots, G_k$ are the simple normal subgroups of G, then $G/H = \prod_{i=1}^{k} G_i/H \cap G_i$. But the cohomology ring of a product space has more than one generator, so that G must be simple. Now G (respectively H) has the same *real* cohomology ring as a product of odd-dimensional spheres with dimensions say $a_1, a_2, \ldots a_r$ (respectively $b_1, b_2, \ldots, b_r$), where r is the common rank of G and H, and the Hirsch formula says that the Poincaré polynomial of G/H is (see [BL 2], p. 450):

7.52 $$P(G/H, \mathbb{R}) = \prod_{i=1}^{r} (1 - t^{a_i+1})(1 - t^{b_i+1})^{-1}.$$

Using the list of subgroups of maximal rank in a simple Lie group and this formula, one can find all of the even-dimensional homogeneous spaces whose real cohomology ring has only one generator. We get the cases (a) (n even), (b), (c), (d), (e), (i) of the theorem, but also the cases $SO(2n+1)/SO(2)SO(2n-1)$, $G_2/SO(4)$, $G_2/U(2)$ (with two different embeddings of $U(2)$). Now the cohomology with $\mathbb{Z}_2$-coefficients of these spaces has more than one generator.

7.53. In the odd-dimensional case, one first sees that M has the homotopy type of a sphere, and so is homeomorphic to a sphere by Smale's theorem. Then one proves that H is totally non-homologous to zero in G and that rank $(G) = \text{rank}(H) + 1$ (see [SN 1]). The next step is to show that G contains a simple normal subgroup G_0 which is transitive on M [and G is G_0 or the product of G_0 with some subgroup of rank one: S^1, $SU(2)$ or $SO(3)$] (see [MA]). Now we have the equality for real cohomology:

7.54 $$H^*(G, \mathbb{R}) = H^*(H \times M, \mathbb{R}).$$

Knowing the real cohomology of simple Lie groups, one easily finds the *types* of possible H_0. Then one looks for the possible embeddings of these types, and one finds all of the odd-dimensional homogeneous spaces whose real cohomology ring has only one generator. We get the cases: (a) (n odd), (f), (g), (h) and (j) of the theorem, but also: $SO(2n+1)/SO(2n-1)$, $G_2/SU(2)$ (with two different embeddings of $SU(2)$), $G_2/SO(3)$ (with two different embeddings of $SO(3)$), $SU(3)/SO(3)$ and $Sp(2)/SU(2)$. And the $\mathbb{Z}_2$-cohomology of these spaces has more than one generator. □

We now prove the result that we announced at the beginning of the paragraph.

7.55 Theorem. *A homogeneous Riemannian manifold (M, g) which is an L_l^m-manifold for some point m in M and some length l is isometric to a* CROSS.

Proof. Notice first that by homogeneity, (M, g) is an L_l^m-manifold for each point m' in M, so that by Proposition 7.9(a), (M, g) is an SC-manifold. Secondly, it will suffice to consider the simply connected case. If M is not simply-connected, then by the Bott-Samelson Theorem 7.23, M is diffeomorphic to $\mathbb{R}P^d$ and a Riemannian metric on $\mathbb{R}P^d$ is symmetric if and only if its universal Riemannian covering is symmetric.

7.56. Let G be the isometry group of (M, g); M is compact since G is compact ([K-N 1], p. 239). The group G is effective and transitive on M by assumption. Applying Theorem 7.50 we see that G_0 and M are in the list (a) to (j). Let $\mathscr{G}_0$ be the Lie algebra of G_0 and let $\mathscr{H}_0$ be the sub-algebra corresponding to H_0. Let V be some $ad(H_0)$-invariant complement to $\mathscr{H}_0$ in $\mathscr{G}_0$. Then it is known (see [K-N 2], p. 200) that there is a natural one-to-one correspondence between the G_0-invariant Riemannian metrics on M and the $ad(H_0)$-invariant scalar products on V. This correspondence is given in the following way: if X, Y are two elements of V, one considers the vector fields X', Y' induced on M from X and Y by the action of G_0, and one sets

7.57 $$g(X', Y')_e = \langle X, Y \rangle.$$

Here $g(\cdot, \cdot)_e$ is the Riemannian metric at the point eH_0 in M, (e is the identity in G_0), and $\langle \cdot, \cdot \rangle$ is the scalar product on V. One then introduces the symmetric bilinear mapping $U: V \times V \to V$ defined by

7.58 $$2\langle U(X, Y), Z \rangle = \langle X, [Z, Y]_V \rangle + \langle [Z, X]_V, Y \rangle.$$

Now, if X', Y' are the vector fields associated with X, Y in V, then the element in V corresponding to $(D_{X'}Y')_{eH_0}$ (where D is the Levi-Civita connection of g) is:

7.59 $\quad -1/2[X,Y]_V+U(X,Y)$.

Let α_t be the one-parameter subgroup of G_0 generated by X. Then the curve $t\mapsto\alpha_t(eH_0)$ in M is an integral curve of X'. Moreover, X' is invariant by α_t and so $D_{X'}X'$ is zero on the whole curve as soon as it is zero at eH_0. We have proved the following

7.60 Lemma. *If $U(X,X)=0$, then the curve $t\mapsto\alpha_t(eH_0)$ of M (where α_t is the one-parameter subgroup of G_0 generated by X) is a geodesic.* □

We now apply all that to the cases (a) to (j) of the Theorem 7.50. Firstly, in the cases (a) to (f), the subgroup H_0 has irreducible representation on V, so that M has (up to a scalar) only one G_0-invariant metric, which is known to be symmetric in these cases (see for example [BR 1], p. 206 and 220 for (e) and (f)).

Therefore we have only to consider cases (g) to (j), in which there is more than one invariant metric.

We will treat the case (h). The other cases can be treated in exactly the same way.

7.61. Let V be the Killing orthogonal complement of $\mathscr{H}_0=\mathfrak{sp}(n)$ in $\mathscr{G}_0=\mathfrak{sp}(n+1)$. This complement has an orthogonal decomposition as $V=V_1\oplus V_2$, where V_1 is the centralizer of $\mathfrak{sp}(n)$ (and the Lie subalgebra of some $Sp(1)$) and V_2 is irreducible under the action of $ad(Sp(n))$. From this we deduce that V_1 and V_2 are orthogonal for any $ad(Sp(n))$-invariant scalar product on V, and that any two $ad(Sp(n))$-invariant scalar products on V_2 are proportional. Let us now consider the connected centralizer of $Sp(n)$ in $Sp(n+1)$, that is, the connected subgroup isomorphic to $Sp(1)$ generated by V_1. The orbit of the class H_0 in $M=G_0/H_0$ under $Sp(1)$ is a submanifold of M diffeomorphic to S^3.

7.62 Lemma. *This orbit is totally geodesic in M.*

Proof. One easily sees that $U(V_1,V_1)\subset V_1$, so if X, Y are in V_1 and X', Y' are the associated vector fields on M, then $D_{X'}Y'$ is tangent to S^3 along S^3. □

Note. For this technique of computation in homogeneous spaces, see [B.B 2].

Now we have the following.

7.63 Lemma. *Theorem 7.55 is true for S^3.*

Proof. For any scalar product on $\mathfrak{sp}(1)$, there exist at least three independent vectors X_i, $i=1,2,3$, such that $U(X_i,X_i)=0$. Indeed, it suffices to diagonalize the scalar product with respect to the Killing form. The three one-parameter sub-groups generated by the X_i's are *closed* geodesics since all one-parameter subgroups are compact. Then the total lengths of these three geodesics are the ratios of the norms of the X_i in the given scalar product and in the Killing form (up to a common factor).

Hence, these geodesics have the same length if and only if the scalar product is proportional to the Killing form, and this gives a symmetric metric on S^3. □

7.64. Thus the scalar product on V, restricted to V_1 or V_2, is proportional to the restriction of the Killing form of $\mathfrak{sp}(n+1)$ to these spaces. We observe that all of the vectors in V_2 generate closed geodesics in M, and that the length of such a geodesic is proportional to the ratio of the scalar product and of the Killing form on V_2. Moreover, there exists a scalar product which gives the standard metric with constant curvature on S^{4n+1} (with all geodesics closed). Now if another scalar product gives the SC-property, then it is proportional to the preceding one on V_1 and V_2, and with the *same* factor. These factors give the ratio of the lengths of the closed geodesics generated by vectors in V_1 or V_2 for these two metrics. Finally, the scalar product gives a metric which is proportional to the standard metric. □

7.65 Remark. Theorem 7.55 is probably also true for (simply connected) homogeneous P-manifolds, but the proof would be longer: firstly, there are a few more cases to consider (see the list in the proof of Theorem 7.50 or [OK]), and secondly, the last argument in the proof above is then no longer sufficient. The problem requires a more detailed study of geodesics.

G. Questions

7.66. In Paragraph C we gave all the examples and counterexamples we know for manifolds which satisfy one of the definitions in 7.7 or 7.8. Keeping in mind the Diagram 7.16 and the various results we have, we are led to some questions. For example, all of the examples of P-manifolds which were not SC-manifolds were also not simply connected; hence, we may ask

7.67 Question (a). Is a simply connected P-manifold always an SC-manifold?

More generally, we may ask

Question (a′). Is a simply-connected P_l^m- (or Y_l^m-)manifold always a CL_l^m- (or an L_l^m-)manifold?

In Definition 7.6 we carefully distinguished closed geodesics from simply closed geodesics. That was essentially because SC-manifolds are SC_l^m-manifolds for each m. However, the Bott-Samelson theorem requires only that the manifolds have the CL_l^m-property. But we know of no counterexample corresponding to

7.68 Question (b). Is an L_l^m- (or a CL_l^m-)manifold always an SL_l^m- (or an SC_l^m-) manifold?

We introduced C-manifolds because they appear naturally in Chapter 2. There is no example of a C-manifold which is not an SC-manifold, so we can ask the following

7.69 Question (c). Is a C-manifold always an SC-manifold?

More generally one may also ask

Question (c'). Is a C_l^m-manifold always an SC_l^m-manifold, or at least a CL_l^m-manifold?

One may even ask a more precise question by supposing that M is simply connected.

When we "provided" the notion of P-manifold with dots, we introduced the notions of Y_l^m-, P_l^m-, Z^m- and Q^m-manifolds, but we only used the Y_l^m-manifolds in Paragraph E. We confess that P_l^m-manifolds were defined mainly to keep some symmetry in Diagram 7.16. But it seems (at least to the author) that the Z^m- and Q^m-assumptions are in fact the most "natural" ones for the problem at hand. Thus, we are led to ask the following

7.70 Question (d). Is a Z^m- (or a Q^m-)manifold always a Y_l^m- (or a P_l^m-)manifold for some l?

Let us remark that Proposition 7.9 is the solution of the global problem corresponding to (*d*). Now the key point in Proposition 7.9 was to prove that if all the geodesics are periodic than they admit a common period. For this purpose we used Wadsley's Theorem 7.13. A glance at its proof shows that the difficult point is to show that these periods are bounded. Hence, we are led to the following (considered as a first step towards answering Question (d))

7.71 Question (d'). Is the length of the first return to m (or the least period) on a Z^m- (or Q^m-)manifold uniformly bounded for every unit vector from m?

There are also some questions of a topological nature. All of the examples we considered were of the type $D^d \bigcup_\sigma D(\xi)$ (or a finite quotient of such manifolds), see C. So we may ask the following

7.72 Question (e). Is a manifold satisfying one of the Definitions 7.7 or 7.8 always diffeomorphic to some $D^d \bigcup_\sigma D(\xi)$ (or to a finite quotient of such a manifold)?

For Question (e) a crucial case might be manifolds with the same homotopy type as $\mathbb{C}P^n$.

Finally, all the examples of SC- and P-manifolds we know are diffeomorphic to the CROSSes or their quotients by free groups of isometries. So we may ask

7.73 Question (f). Is an SC-manifold always diffeomorphic to a CROSS?

Question (f'). Is a P-manifold always diffeomorphic to a quotient of a CROSS by a free group of isometries?

One may even ask

Question (f"). Is a non simply-connected P-manifold *isometric* to the quotient of a CROSS by a free group of isometries?

To conclude let us ask an easier question. It is not too difficult (exercise!) to prove that if a manifold is the quotient of a CROSS by a finite free group of isometries, then all of its geodesics are simply closed curves (not all with the same length, of course). So we may ask

Question (f″). In a P-manifold, are the geodesics all simply closed?

H. Historical Note

7.74. The author has not found Proposition 7.9 in print anywhere. However, Part (a) seems to be part of the folklore; Part (b) requires Wadsley's theorem, which appeared only recently [WY 2] (and Appendix A) and was brought to the author's attention by D. Epstein. The Bott-Samelson theorem was proved by R. Bott in [BT] for SC-manifolds. R. Bott already noticed that his proof only requires the SC_l^m-condition. A very different proof (sketched in 7.36) was given shortly after that of R. Bott by H. Samelson for CL_l^m-manifolds [SN 2]. The proof we have given here in the L_l^m-case is taken with minor modifications from H. Nakagawa [NA]. It is a refinement of R. Bott's original proof. The conclusions of R. Bott were later strengthened, thanks to major advances in the theory of differentiable structures on manifolds (see 7.30). The last sentence of the theorem was proved by K. Sugahara [SA] in the SC_l^m-case. The proof which we give here in the CL_l^m-case comes from [B.B 1]. The theorem for P-manifolds and its proof are also taken from [B.B 1]. The use of Sullivan's theorem in this proof is based on a suggestion of D. Gromoll. Theorem 7.50 is the result of work by H. Wang [WG 1], A. Borel [BL 1] (in the case of spheres) and H. Singh Varma [S.V] (for the other cases). Theorem 7.55 seems new, but Lemma 7.63 was well known in mechanics (because the geodesics of a homogeneous metric on $SO(3)$ are the trajectories of a 3-dimensional solid body with fixed centre of mass, and these trajectories may be calculated with the help of elliptic functions, see [L-L], p. 160).

Chapter 8. The Spectrum of P-Manifolds

A. Summary

This chapter is mainly devoted to a theorem of J. Duistermaat and V. Guillemin (Theorem 8.9) which relates the spectrum of the manifold (M, g) to properties of its geodesic flow. The theorem with motivations is given in Section B; Sections C to E provide the basic ingredients of the proof (given itself in Section F).

In Section G we give a result of A. Weinstein which applies to the spectra of Zoll surfaces.

Finally, in Section H we give some results on the first nonzero eigenvalue of the Laplace operator on Blaschke manifolds.

B. Introduction

8.1. Let (M, g) be a C^∞ compact Riemannian manifold without boundary, whose volume element *is denoted* by μ_g (see 1.117). Given X in $\mathscr{C}M$ (a vector field on M), one defines a function, *denoted by* $\operatorname{div}_g X$ or $\operatorname{div} X$, as follows: given a volume form ω corresponding to a choice of a local orientation on M with $|\omega| = \mu_g$, one sets

$$(\operatorname{div} X)\cdot\omega = d(i_X\omega).$$

It is clear that $\operatorname{div} X$ does not depend on the particular choice of a local orientation. ([B-G-M] p. 120).

8.2. One *defines* an operator $\delta: \Omega^1 M \to \Omega^0 M$ by the following formula:

$$\forall \alpha \in \Omega^1 M, \qquad \delta\alpha = -\operatorname{div}(\#\alpha).$$

This operator turns out to be the formal adjoint of the exterior differential $d: \Omega^0 M \to \Omega^1 M$ with respect to the metrics induced by g on these spaces ([B-G-M] p. 120).

8.3 Definition. The *Laplace operator* Δ acting on C^∞-functions on M is defined by $\Delta = \delta d$ (i.e., for any f in $\Omega^0 M$ $\Delta f = \delta d f$).

8.4. Given a local coordinate system (x^i), let (g_{ij}) be the matrix with entries $g_{ij} = g\left(\frac{\partial}{\partial x^i}, \frac{\partial}{\partial x^j}\right)$; let (g^{ij}) denote the inverse matrix and put $\theta = \sqrt{\det(g_{ij})}$. Then Δ is locally given by

8.5 $$\Delta f = -\theta^{-1} \sum_{i,j=1}^{d} \frac{\partial}{\partial x^i}\left(\theta g^{ij} \frac{\partial f}{\partial x^j}\right) \quad \text{([B-G-M] p. 126).}$$

8.6 Example. On $(\mathbb{R}^d, \text{can})$, $\Delta f = -\sum_{j=1}^{d} \frac{\partial^2 f}{\partial x^{j2}}$.

8.7. The operator Δ is an elliptic, linear partial differential operator with principal symbol $q(\lambda) = g^{-1}(\lambda, \lambda)$, where g^{-1} denotes the norm induced by the Riemannian metric g on T^*M. In local coordinates, $q(x, \xi) = \sum_{i,j=1}^{d} g^{ij}(x)\xi_i\xi_j$ (see 1.23).

The operator Δ is formally self-adjoint and positive. It extends to a positive self-adjoint unbounded operator on $L^2(M, \mu_g)$ ([YA] Theorem 2, p. 317), and the extension Δ has a discrete spectrum (because its resolvent is compact: [YA] Theorem 1, p. 325). We will be interested in $\sqrt{\Delta}$ rather than in Δ itself, so we write the spectrum as

$$\text{Spec}(M): 0 = \mu_0^2 < \mu_1^2 \leqslant \mu_2^2 \leqslant \mu_3^2 \cdots,$$

where each eigenvalue occurs as many times as its multiplicity. We denote by (φ_j) an L^2-orthonormal basis of C^∞-real eigenfunctions associated with the μ_j's.

8.8. The following diagram is a first step towards general theorems. The entries are as follows: we consider the spectra of compact, rank 1, symmetric spaces (CROSSes). These manifolds have periodic geodesic flow (with least period T). Let β' denote the number of conjugate points, in the interval $]0, T[$ (counting multiplicities) along a geodesic and β the number of conjugate points in the interval $]0, T]$; then $\beta = \beta' + d - 1$. We let ν_k be the terms in the arithmetic progression

$$\nu_k = \frac{2\pi}{T}(k + \beta/4); \quad k \in \mathbb{N}.$$

M	$\dim_{\mathbb{R}} M$	μ_j^2 $k \geqslant 0$	$\mu_j \bmod \frac{1}{k}$ $k \geqslant 1$	T	β'	β	$\frac{2\pi}{T}(k+\beta/4) = \nu_k$ $k \geqslant 0$
S^n	n	$k(k+n-1)$	$k + \frac{n-1}{2}$	2π	$n-1$	$2(n-1)$	$k + \frac{n-1}{2}$
$\mathbb{R}P^n$	n	$2k(2k+n-1)$	$2\left(k + \frac{n-1}{4}\right)$	π	0	$n-1$	$2\left(k + \frac{n-1}{4}\right)$
$\mathbb{C}P^n$	$2n$	$4k(k+n)$	$2\left(k + \frac{n}{2}\right)$	π	1	$2n$	$2\left(k + \frac{n}{2}\right)$
$\mathbb{H}P^n$	$4n$	$4k(k+2n+1)$	$2\left(k + \frac{2n+1}{2}\right)$	π	3	$4n+2$	$2\left(k + \frac{2n+1}{2}\right)$
$\mathbb{C}aP^2$	16	$4k(k+11)$	$2(k+11/2)$	π	7	22	$2(k+11/2)$

For the eigenvalues given in the three first lines of this diagram see [B-G-M] p. 160—173. The results of the two last lines do not (to our knowledge) appear explicitely in the literature except in [AL] (see also [SH]).

The main result of this chapter is the following theorem (see [D-G] Theorem 3.1 p. 53 and Theorem 3.2 p. 55):

8.9 Theorem (J. Duistermaat and V. Guillemin). *Let $\varphi_t: \mathring{T}^*M \to \mathring{T}^*M$ be the geodesic flow of M.*

1) Assume that $\varphi_T = Id$ (i.e., M is a P_L-manifold and T is an integral multiple of L). Let β be the number of conjugate points in the interval $]0, T]$, and let $\nu_k = \frac{2\pi}{T}(k+\beta/4)$ for k in $\mathbb{N}$. Then for any s in $]0, 1/2[$, for any $C>0$ and for any ε in $]0, 1-2s[$,

$$\textbf{8.10} \qquad \frac{\operatorname{Card}\left\{J \mid \mu_j \leqslant \mu, \mu_j \notin \bigcup_{k=1}^{\infty} J_k\right\}}{\operatorname{Card}\{j \mid \mu_j \leqslant \mu\}} = 0(\mu^{-\varepsilon})$$

(here $J_k = [\nu_k - ck^{-s}, \nu_k + ck^{-s}]$ and Card *denotes cardinal).*

2) Let $\nu_k = ak+b$ be an arithmetic progression, and assume that there exist $s_0>0$, $C_0>0$ and $\varepsilon_0>0$ such that (8.10) *holds for $(s_0, C_0, \varepsilon_0)$. Then*

(i) $\varphi_{\frac{2\pi}{a}} = Id$

(ii) $\frac{4b}{a}$ *is an integer and is equal modulo 4 to the number of conjugate points on $\left]0, \frac{2\pi}{a}\right[$. Furthermore,* (8.10) *holds for any (s, C, ε) satisfying the condition of Assertion 1.*

8.11 Remarks. 1) The theorem given in [D-G] pp. 53—55 is actually much stronger. Roughly speaking, it says that the spectrum accumulates within intervals of the form $[\nu_k - ck^{-1}, \nu_k + ck^{-1}]$ for c large enough. On the other hand, the proof which we give for Theorem 8.9 is simpler than that of J. Duistermaat and V. Guillemin (op. cit). (It also applies to a simpler context!).

2) A well-known theorem of H. Weyl asserts that $\operatorname{Card}\{j \mid \mu_j \leqslant \nu\} \sim (2\pi)^{-d} \operatorname{Vol}(B^*M)\nu^d$ as ν tends to $+\infty$ (here B^*M is the unit ball in T^*M). In 1968 L. Hormander [HR3] proved that $\operatorname{Card}\{j \mid \mu_j \leqslant \nu\} = (2\pi)^{-d} \operatorname{Vol}(B^*M)\nu^d + O(\nu^{d-1})$ and it turns out that this is the best possible result in the sense that it cannot be improved for (S^d, can) (see [HR 3] p. 216), nor can it be improved for manifolds with periodic geodesic flow (see [D-G] p. 40).

C. Wave Front Sets and Sobolev Spaces

8.12. Let us first define *the wave front set of a distribution*. Let Ω be an open subset of $\mathbb{R}^d$. According to the Paley-Wiener Theorem ([HR 1] Theorem 1.7.7. p. 21) a distribution u in $\mathscr{D}'(\Omega)$ is C^∞ in a neighbourhood of a point x_0 in Ω if and only if there exists a C^∞-function with compact support ϕ, $\phi(x_0) \neq 0$, such that for any $N>0$

8.13. $|\widehat{\phi u}(\xi)| \leqslant C_N(1+\|\xi\|)^{-N}$ for any ξ in $\mathbb{R}^d$ (where $\widehat{\phi u}(\xi)=\langle e^{-i\langle\cdot,\xi\rangle}\phi, u\rangle$ is the Fourier transform of ϕu). This means that ϕu is in $C_0^\infty(\Omega)$. We define the *singular support of* u, Singsupp u, as the complement in Ω of the set of points in a neighbourhood of which u is C^∞.

8.14. Inequality (8.13) is required to hold for any ξ in $\mathbb{R}^d$. It turns out to be fruitful to look at those ξ-rays along which $\widehat{\phi u}$ is decreasing faster than any power of $\|\xi\|$ at infinity. For this reason we look at those $\xi_0 \neq 0$ such that for any $N>0$

8.15 $|\widehat{\phi u}(\tau\xi_0)| \leqslant C_N \tau^{-N}$ as τ tends to $+\infty$.

As usual, when some kind of regularity is concerned we want to have uniformity in some sense. Therefore we have the following

8.16 Definition. The *wave front set* $WF(u)$ of u in $\mathscr{D}'(\Omega)$ is the complement in $\Omega \times (\mathbb{R}^d \setminus \{0\})$ of the set of points (x_0, ξ_0) for which there exist a function ϕ in $C_0^\infty(\Omega)$, with $\phi(x_0) \neq 0$, and a neighborhood V of ξ_0 in $\mathbb{R}^d \setminus \{0\}$ such that for any N,

8.17. $|\widehat{\phi u}(\tau\xi)| \leqslant C_N \tau^{-N}$ as τ tends to $+\infty$, uniformly for ξ in V.

8.18. We can restate this as follows: $\langle e^{-i\tau\langle\cdot,\xi\rangle}\phi, u\rangle = O(\tau^{-N})$ uniformly for ξ in V as τ tends to $+\infty$. This means that we have tested u by the function $e^{-i\tau\langle x,\xi\rangle}\phi(x)$ which is rapidly oscillating as τ tends to $+\infty$. The phase $\langle x, \xi\rangle$ depends on the parameter ξ in V and (8.18) is required to hold uniformly in V. We notice that ξ is the normal vector to the hypersurface of constant phase $\langle x, \xi\rangle = $ constant. Then it is natural to test u by functions of the form $e^{-i\tau\psi(x,a)}\phi(x)$, where x is in $\mathbb{R}^d$, and a in some open set A of $\mathbb{R}^p$, and where ψ is in $C^\infty(\mathbb{R}^d \times A)$ and takes real values, and ϕ is in $C_0^\infty(\mathbb{R}^d)$. As before, we are interested in the normal vectors to the hypersurfaces of constant phase $\psi(x, a) = ct$ [i.e., in the vector gradient$_x$ $\psi(x, a)$]. However, the Euclidean structure is not needed any more and therefore it is more natural to consider the conormal vectors $d_x\psi(x, a)$.

8.19 Proposition. *Let Ω be an open set in $\mathbb{R}^d$ and let p_Ω be the projection onto the first factor $p_\Omega: \Omega \times \mathbb{R}^{d*} \setminus \{0\} \to \Omega$ (* means dual).*

1) Given u in $\mathscr{D}'(\Omega)$, $WF(u)$ is a closed cone in $\Omega \times \mathbb{R}^{d*} \setminus \{0\}$ and $p_\Omega(WF(u)) = singsupp(u)$.

2) The point (x_0, ξ_0) does not belong to $WF(u)$ if and only if for any real valued C^∞-function $\psi(x, a)$ of (x, a) in $\mathbb{R}^d \times A$ with $d_x\psi(x_0, \alpha_0) = \xi_0$ there exist a function ϕ in $C_0^\infty(\Omega)$ and an open neighbourhood A_0 of a_0 in A such that for any N, $|\langle e^{-i\tau\psi(\cdot,a)}\phi, u\rangle| \leqslant C_N \tau^{-N}$ as τ tends to $+\infty$, uniformly for a in A.

3) The point (x_0, ξ_0) does not belong the $WF(u)$ if and only if there exists a properly supported pseudo-differential operator E of degree 0, with principal symbol $\sigma(E)$ (x, ξ) such that $\sigma(E)$ $(x_0, \xi_0) \neq 0$ and Eu is in $C^\infty(\mathbb{R}^d)$.

Proof. 1) is clear; 3) follows from standard methods in the theory of pseudo-differential operators (see for example [NG] p. 41 and [HR 4] pp. 120 ff.). 2) The if part is clear. Now assume that (x_0, ξ_0) does not belong to $WF(u)$ and let χ in $C_0^\infty(\Omega)$

be equal to 1 on Suppϕ. Using Passeval's indentity one can write

$$\langle e^{-i\tau\psi(\cdot,a)}\phi, u\rangle = \langle \mathscr{F}^{-1}(e^{-i\tau\psi(\cdot,a)}\chi), \mathscr{F}(\phi u)\rangle$$

(where $\mathscr{F}$ denotes Fourier transform). Then one can write

$$\begin{aligned}\langle e^{-i\tau\psi(\cdot,a)}\phi, u\rangle &= (2\pi)^{-d} \iint e^{i\langle x,\xi\rangle} e^{-i\tau\psi(x,a)}\chi(x)\mathscr{F}(\phi u)(\xi)dx d\xi \\ &= (2\pi)^{-d}\tau^d \iint e^{i\tau[\langle x,\xi\rangle - \psi(x,a)]}\chi(x)\mathscr{F}(\phi u)(\tau\xi)dx d\xi .\end{aligned}$$

Now consider the function $F(\xi,\tau,a) = \int e^{i\tau[\langle x,\xi\rangle - \psi(x,a)]}\chi(x)dx$. Since $d_x\psi(x_0,a_0) = \xi_0$, we can choose suppϕ and A_0 so small that $|d_x\psi(x,a) - \xi|$ is bounded away from zero when (x,a) is in (Suppϕ) $\times A_0$ and ξ does not belong to a neighbourhood V of ξ_0 (see Definition 8.16). Then taking a in A_0, one can show (using integration by parts) that for any N $|F(\xi,\tau,a)| \leqslant C_N\tau^{-N}(1+|\xi|)^{-N}$ when ξ is not in V and $\tau \geqslant 1$. On the other hand, the Paley-Wiener Theorem gives the estimate $|\mathscr{F}(\phi u)(\tau\xi)| \leqslant C(1+\tau|\xi|)^{N_0}$ for some N_0, and the fact that (x_0,ξ_0) does not belong to $WF(u)$ gives estimates of the form $|\mathscr{F}(\phi u)(\tau\xi)| \leqslant C_L\tau^{-L}$ for any L when Suppϕ is small enough, ξ in V and $\tau \geqslant 1$. Piecing these facts together yields the proof. □

8.20. It follows from Proposition 8.19 that $WF(u)$ does not depend on a choice of coordinates and that $WF(u) \subset \mathring{T}^*\Omega$. Hence, one can define $WF(u)$ for any u in $\mathscr{D}'(M)$ (where M is a manifold) and $WF(u)$ is a closed cone in $\mathring{T}^*M$.

8.21. Let us now define the Sobolev spaces. Let k be an integer. The *Sobolev space* $H^k(\mathbb{R}^d)$ is defined as the set of tempered distributions u whose derivatives up to order k are in $L^2(\mathbb{R}^d)$.

8.22 Proposition. $H^k(\mathbb{R}^d) = \{u \in \mathscr{S}'(\mathbb{R}^d) | \hat{u}(\xi)\,(1+|\xi|^2)^{k/2} \in L^2(\mathbb{R}^d)\}$.

This proposition leads to the definition of the Sobolev spaces for any real order s: for s in $\mathbb{R}$ $H^s(\mathbb{R}^d)$ is the set of tempered distributions u in $\mathscr{S}'(\mathbb{R}^d)$ such that $\hat{u}(\xi)(1+|\xi|^2)^{s/2}$ is in $L^2(\mathbb{R}^d)$.

One can localize these notions. Let Ω be an open subset of $\mathbb{R}^d$, $H^s_{\text{loc}}(\Omega) = \{u \in \mathscr{D}'(\Omega) | \forall \varphi \in C_0^\infty(\Omega),\ \varphi u \in H^s(\mathbb{R}^d)\}$. It is clear that for k in $\mathbb{N}$ $H^k_{\text{loc}}(\Omega)$ is invariant under changes of coordinates since $L^2_{\text{loc}}(\Omega)$ is. In fact, one can prove that any pseudo-differential operator P of order m on Ω defines a continuous map

$$P: H^s(\mathbb{R}^d) \cap \mathscr{E}'(\Omega) \to H^{s-m}_{\text{loc}}(\Omega) .$$

Then one can define the Sobolev spaces on manifolds by making use of elliptic pseudo-differential operators (see e.g. [HR 2] p. 169 ff.).

8.23. On a compact Riemannian manifold (M,g) we can give a more convenient definition of Sobolev spaces (at least a posteriori). Using Seeley's theorem on complex powers of elliptic operators ([SY] p. 289) we define $H^s(M)$ for any real s as

$$H^s(M) = \{u \in \mathscr{D}'(M) | (1+\Delta)^{s/2} u \in L^2(M,\mu_g)\} .$$

This definition agrees with 8.21 when s is an even integer. Furthermore, it provides $H^s(M)$ with a natural norm.

8.24. Let us now turn to the relationship between wave front sets and Sobolev spaces. We can refine Definition 8.16. as follows: let Ω be an open set in $\mathbb{R}^d$ and let u be an element of $\mathscr{D}'(\Omega)$. We will say that (x_0, ξ_0) is not in $WF_s(u)$ if there exist a function ϕ in $C_0^\infty(\Omega)$, with $\phi(x_0) \neq 0$, and an open cone Γ_{ξ_0}, containing ξ_0, in $\mathbb{R}^d \setminus \{0\}$ such that $(1+\|\xi\|^2)^{s/2}\widehat{\phi u}(\xi)$ is in $L^2(\Gamma_{\xi_0})$. Equivalently, one could say that (x_0, ξ_0) does not belong to $WF_s(u)$ if there exists a pseudo differential operator A of order 0 with principal symbol $\sigma(A)$, such that $\sigma(A)(x_0, \xi_0) \neq 0$ and such that Au is in $H^s_{\mathrm{loc}}(\Omega)$. This clearly extends to manifolds. Using standard techniques in the theory of pseudo-differential operators (e.g., [D-H], p. 202 or [NG], p. 41 ff.) one can prove the following

8.25 Proposition. *The point (x_0, ξ_0) is not in $WF_s(u) \subsetneq \mathring{T}^*M$ if and only if there exist u_1 in $H^s(M)$ and u_2 in $\mathscr{D}'(M)$ such that $u = u_1 + u_2$ and (x_0, ξ_0) does not belong to $WF(u_2)$.*

D. Harmonic Analysis on Riemannian Manifolds

8.26. Let (M, g) be a d-dimensional compact Riemannian manifold with volume element μ_g. We know that (μ_j^2, φ_j) is an orthonormal basis of C^∞ real functions in $L^2(M, \mu_g)$ (see 8.7).

Given any f in $L^2(M)$, one can write $f = \sum_{j=0}^{\infty} a_j \varphi_j$ (with equality in the L^2-sense), where $a_j = \int_M f \varphi_j \mu_g$. The series (a_j) is called the Fourier series of f. One can read off from this series the regularity properties of f.

8.27 Proposition. a) *The distribution f is in $H^s(M)$ if and only if $\sum_{j=1}^{+\infty} |a_j|^2 \mu_j^{2s} < +\infty$.*

b) *This distribution f is in $C^\infty(M)$ if and only if (a_j) is a rapidly decreasing sequence.*

Proof. a) Recall that f is in $H^s(M)$ if and only if $(1+\Delta)^{s/2} f$ is in $L^2(M)$, i.e., if and only if $\sum_{j=0}^{+\infty} |a_j|^2 (1+\mu_j^2)^s < +\infty$.

b) Using Sobolev's lemma and the compactness of M one has the equality $\bigcap_{s \in \mathbb{R}} H^s(M) = C^\infty(M)$.

8.28. Recall (cf. 8.11.2)) that $\operatorname{Card}\{j \mid \mu_j \leqslant \lambda\} \sim (2\pi)^{-d} C(d) \operatorname{Vol}(M) \lambda^d$, with

$$C(d) = \int_{\{x \in \mathbb{R}^d \mid |x| \leqslant 1\}} dx \quad \text{and} \quad \operatorname{Vol}(M) = \int_M \mu_g .$$

From this we deduce that

$$\mu_j \sim 2\pi \left(\frac{j}{C(d)\operatorname{Vol}(M)}\right)^{1/d}.$$

Assume that $\sum_{j=1}^{+\infty} a_j^2 \mu_j^{2s} < +\infty$ for any s. From the estimate for μ_j we deduce that $\sum_{j=1}^{+\infty} |a_j|^2 j^{2\alpha} < +\infty$ for any α and hence, that the sequence (a_j) is rapidly decreasing. □

Given a distribution u in $\mathscr{D}'(M)$ we define its Fourier coefficients by $a_j(u) = \langle u, \varphi_j \rangle$. We have the following

8.29 Proposition. *The mapping $u \mapsto (a_j(u))$ is an isomorphism from $\mathscr{D}'(M)$ onto $\mathscr{S}'(\mathbb{N})$ (slowly increasing sequences). Furthermore, if u belongs to $\mathscr{D}'(M)$ and f is in $C^\infty(M)$, then $\langle u, f \rangle = \sum_{j=0}^{+\infty} a_j(u)\overline{a_j(f)}$.*

Notice that these results apply in particular to the manifold $M \times M$ with $\mu_{ij}^2 = \mu_i^2 + \mu_j^2$, and $\varphi_{ij}(x, y) = \varphi_i(x) \otimes \varphi_j(y)$.

8.30. We now turn to the study of the basic objects, namely the wave equation and the half-wave equation. We look for a solution u of the wave equation on M, i.e., we look for, say a C^∞ function u on $M \times \mathbb{R}$ such that $\frac{\partial^2 u}{\partial t^2} + \Delta u = 0$.

One can try to write $u(x,t)$ as $u(x,t) = \sum_{j=0}^{\infty} a_j(t)\varphi_j(x)$. This leads to the equations $a_j''(t) + \mu_j^2 a_j(t) = 0$, $j \geqslant 0$, and hence, to $u(x,t) = a_0 + b_0 t + u_+(x,t) + u_-(x,t)$ with $u_+(x,t) = \sum_{j=1}^{\infty} a_j e^{-i\mu_j t}\varphi_j(x)$ and $u_-(x,t) = \sum_{j=1}^{\infty} b_j e^{i\mu_j t}\varphi_j(x)$.

8.31. Let $\sqrt{\Delta}$ be the operator defined by $\sqrt{\Delta}\varphi_j = \mu_j \varphi_j$. This operator is $\Delta^{1/2}$ in the sense of spectral theory. Then according to Seeley's theorem $\sqrt{\Delta}$ is a pseudo-differential operator of degree 1. (see [SY]).

Clearly $u_\pm$ satisfies the equation $P_\pm u_\pm = 0$, where we let $P_\pm = \frac{1}{i}\frac{\partial}{\partial t} \pm \sqrt{\Delta}$, *(half-wave equations)*. Notice that u_+ and u_- are solutions of the wave equation $\left(P_+ P_- = P_- P_+ = \frac{\partial^2}{\partial t^2} + \Delta\right)$.

8.32. Clearly, $u_+(x,t)$ is determined by $u_+(x,0) = f(x)$. If $f(x) = \sum_{j=1}^{\infty} a_j \varphi_j(x)$, then $u_+(x,t) = \sum_{j=1}^{\infty} a_j e^{-it\mu_j}\varphi_j(x)$. We shall use the following notation: $u_+(x,t) = (e^{-it\sqrt{\Delta}} f)(x)$. This means that we have a well-defined operator $e^{-it\sqrt{\Delta}}$ which maps $H^s(M)$ into

$H^s(M)$ (see Proposition 8.27). Let us denote by $K_t(x, y)$ the distribution kernel in $\mathscr{D}'(M \times M)$ of the operator $e^{-it\sqrt{\Delta}}$ (t is fixed), i.e., we have

$$\langle K_t(x, y), u(x) \otimes v(y) \rangle = \int_M (e^{-it\sqrt{\Delta}} u) \bar{v} \mu_g,$$

for any u and v in $C^\infty(M)$. If we take $u(x) = \varphi_j(x)$ and $v(y) = \varphi_k(y)$, then it follows that $K_t(x, y) = \sum_{j=1}^{\infty} e^{-it\mu_j} \varphi_j(x) \otimes \varphi_j(y)$.

E. Propagation of Singularities

8.33. Some more facts about wave front sets: let X and Y be C^∞ manifolds and let $A: C_0^\infty(Y) \to \mathscr{D}'(X)$ be a continuous operator with distribution kernel K_A in $\mathscr{D}'(X \times Y)$. In [HR 4], p. 132, L. Hörmander proves the following

8.34 Proposition. *Assume that $A: C_0^\infty(Y) \to \mathscr{D}'(X)$ is a continuous linear operator and that $WF(K_A) \subset \mathring{T}^*X \times \mathring{T}^*Y$. Then A extends to $\mathscr{E}'(Y)$ and*

$$\begin{aligned} WF(Au) &\subset WF'(A) \circ WF(u) \\ &= \{\lambda \in \mathring{T}^*X \mid \exists \mu \in WF(u), (\lambda, \mu) \in WF'(A)\}. \end{aligned}$$

Here $WF'(A)$ is defined as

$$WF'(A) = \{(\lambda, -\mu) \in \mathring{T}^*X \times \mathring{T}^*Y \mid (\lambda, \mu) \in WF(K_A)\}.$$

8.35. According to [D-G], p. 43, $K_t(x, y)$, the distribution kernel of $e^{-it\sqrt{\Delta}}$ is a Fourier integral and $WF'(e^{-it\sqrt{\Delta}}) = \operatorname{graph}(\varphi_t)$, where φ_t is the geodesic flow on (M, g). If u_t is the value at time t of the solution of the Cauchy problem

$$\left(\frac{1}{i}\frac{\partial}{\partial t} + \sqrt{\Delta}\right) u(t, x) = 0, \qquad u(0, x) = u_0(x),$$

then $u_t = e^{-it\sqrt{\Delta}} u_0$. Propositions 8.34 and 8.32 imply that

$$WF(u_t) \subset \varphi_t(WF(u_0)) \subset \varphi_t(\varphi_{-t}(WF(u_t)))$$

and hence,

8.36 $$WF(u_t) = \varphi_t(WF(u_0)).$$

Notice that (8.36) is a result on the *propagation of singularities*. Indeed, projecting (8.36) down onto M shows that the singular support of u_t is the set of points x in M

such that there exist a point y in Singsupp(u_0), and a geodesic γ of length $|t|$, such that $\gamma(0)=x$ and $\gamma(t)=y$. However, it should be noted that one cannot quite apply the theorem on propagation of singularities of [D-H] (Theorem 6.1.1, p. 196) since $\frac{1}{i}\frac{\partial}{\partial t}+\sqrt{\Delta}$ is not a pseudo-differential operator ([NG], p. 49).

One could prove more precise results. Namely,

8.37 $\quad \forall s \geqslant -d/2, \quad WF_s'(e^{-it\sqrt{\Delta}})=\operatorname{graph}(\varphi_t)$

and

8.38 $\quad \forall s < -d/2, \quad WF_s'(e^{-it\sqrt{\Delta}})=\phi, \quad$ i.e., $K_t \in H^s$.

These statements follow from the fact that for a fixed value of t, $e^{-it\sqrt{\Delta}}$ is a Fourier integral operator of degree 0 and hence, that its H^s-regularity is similar to that of a pseudo-differential operator of degree 0, e.g., the identity.

8.39 Remark. The statement on the propagation of singularities can be compared to the results on the propagation of light in $\mathbb{R}^d$.

F. Proof of the Theorem 8.9 (J. Duistermaat and V. Guillemin)

8.40 *Proof of Statement* 1. If $\varphi_T = Id$, then (8.36) implies that

$$WF'(e^{-iT\sqrt{\Delta}})=WF'(Id)=\operatorname{Graph}(Id)=\operatorname{Diag}(\mathring{T}^*M).$$

More precisely, the symbolic calculus of Fourier integral operators implies that $e^{-iT\sqrt{\Delta}}=i^{-\beta}Id+R$, where R is an operator of degree -1 and β is as in 8.8.

8.41 Remark. The factor $i^{-\beta}$ conveys the well-known fact of geometric optics that crossing a generic caustic induces a shift of $-\frac{\pi}{2}$ for the phase of light beams. Here, going through a conjugate point of index m induces a shift of $-m\cdot\frac{\pi}{2}$ in the phase of the singular part of the singularity of u_t. One might view this as resulting from interference with infinitely close light beams.

The H^s-regularity results (8.37) and (8.38) extend to Fourier integral operators of order -1, implying that $K_T - i^{-\beta}K_0$ is in $H^{-\frac{d}{2}+1-\varepsilon'}$ for any $\varepsilon'>0$. Hence, using Proposition 8.27 and Section 9, we have

$$\sum_{j=1}^{\infty}\frac{|1-\exp(-i\mu_j T + i\beta\frac{\pi}{2})|^2}{\mu_j^{d-2+2\varepsilon'}}<+\infty.$$

We *denote by* I_k the interval $I_k=\left[\frac{\nu_{k-1}+\nu_k}{2},\frac{\nu_k+\nu_{k+1}}{2}\right[$ *and by* α_k the number Card$\{j \mid \mu_j \in I_k \setminus J_k\}$ (see 8.8 and 8.10):

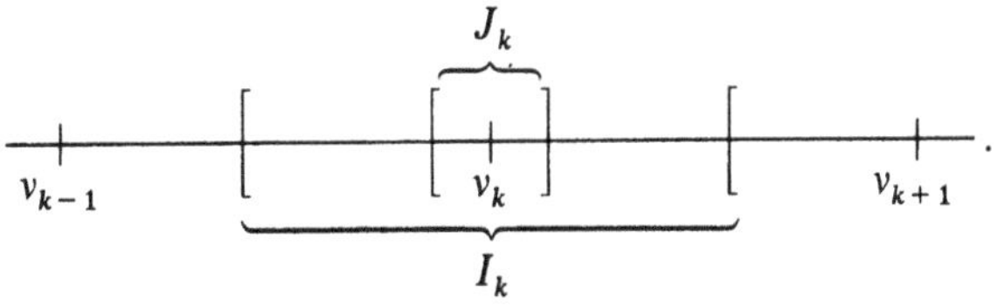

Then $\sum_{k \geq 1} \frac{\alpha_k}{k^{d-2+2s+2\varepsilon'}} < +\infty$.

Choosing ε' small enough we find that $\alpha_k = O(k^{d-1-\varepsilon})$, and therefore that $\operatorname{Card}\{j|\mu_j \leq \mu, \mu_j \notin \bigcup_k J_k\} = O(\mu^{d-\varepsilon})$. The theorem of H. Weyl (see 8.28) finishes the proof. □

8.42 *Proof of Statement* 2. Assume that we can prove that $K_0 - i^{\beta} K_T$ is of class $H^{-d/2}$. Then it follows that $WF'_{-d/2}(K_0) = WF'_{-d/2}(K_T)$ and hence, that $\operatorname{Graph}(\varphi_T) = \operatorname{Graph}(Id)$ (i.e., $\varphi_T = Id$), and we are done. The assertion follows if we can prove that

$$\sum_{j=1}^{\infty} \frac{|1 - \exp(i\frac{\pi}{2}\beta - i\mu_j T)|^2}{\mu_j^d} < +\infty .$$

Define β_k by $\beta_k = \operatorname{Card}\{j|\mu_j \in J_k\}$. Using the proof of Statement 1, we see that it suffices to prove that

$$\text{(i) } \sum_k \frac{\beta_k}{k^{d+s_0}} < +\infty \quad \text{and} \quad \text{(ii) } \sum_k \frac{\alpha_k}{k^d} < +\infty .$$

By H. Weyl's theorem (8.28), we have $\sum_{k=0}^{K} \beta_k \lesssim C_1 K^d$, and the assumption in 2) of the theorem implies that $\sum_{k=0}^{K} \alpha_k \lesssim C_2 K^{d-\varepsilon_0}$ for some real numbers C_1 and C_2. The assertions (i) and (ii) follow by using Abel's transformation formula [AEL]. □

G. A. Weinstein's Result

8.43 Theorem ([WN 4] Theorem 5.1). *Let g_0 and g_1 be two P_T-metrics on a manifold M. Assume that there exists a C^∞ one parameter family $g(t)$ of P_T-metrics on M such that $g(0) = g_0$ and $g(1) = g_1$. Then $|\mu_j^2(g_0) - \mu_j^2(g_1)| = O(1)$ as j tends to infinity. In particular, this applies to Zoll surfaces (see Chapter* IV.F*). Let $(\mu_{0,j}^2)$ be the eigenvalues for $(S^2, g_0 = \text{can})$, $\mu_{0,j}^2 = k_j(k_j+1)$ and let $(\mu_{1,j}^2)$ be the eigenvalues for (S^2, g_1) where g_1 is a Zoll metric (see* 4.F*). Then $\mu_{1,j}^2 = k_j(k_j+1) + 0(1)$ and hence,* $\mu_{1,j} = k_j + \frac{1}{2} + O\left(\frac{1}{k_j}\right)$.

8.44. The idea of the proof is the following: there exists a homogeneous canonical transformation $\chi: \mathring{T}^*S^2 \to \mathring{T}^*S^2$ such that $q_1 \circ \chi = q_0$, where the q's are the norms associated with g_1 and g_0 (see 4.F). With this canonical transformation A. Weinstein associates an isometric Fourier integral operator A:

$$A : L^2(S^2, \text{can}) \to L^2(S^2, g_1)$$

such that $\Delta_1 \circ A - A \circ \Delta_{\text{can}}$ is a bounded operator (i.e., up to a bounded operator, A intertwines the Laplace operators). The result follows by using the min-max characterization of eigenvalues ([C-H], p. 161).

H. On the First Eigenvalue $\lambda_1 = \mu_1^2$

8.45. Let (M, g) be a compact Riemannian manifold and let σ be an isometry of M such that $\sigma^2 = Id$. The Blaschke manifolds described in Chapter V admit such an isometry. One can define odd and even functions with respect to σ. One denotes by C_+ the set of smooth functions f such that $f \circ \sigma = f$ and by C_- the set of smooth functions f such that $f \circ \sigma = -f$. Let V_λ be the kernel of $\Delta - \lambda Id$ in $C^\infty(M)$ and let $V_\lambda^\pm$ be the intersection of V_λ with $C_\pm$. In $L^2(M, \mu_g)$, V_λ is the orthogonal direct sum of V_λ^+ and V_λ^-. In analogy to eigenvalues one can define:

$$\lambda_1^+ = \inf\{\lambda > 0 \mid V_\lambda^+ \neq \{0\}\} \quad \text{and} \quad \lambda_1^- = \inf\{\lambda > 0 \mid V_\lambda^- \neq \{0\}\}.$$

8.46. Now consider the standard differentiable d-dimensional sphere S^d and let σ be its antipodal map. Then for the standard Riemannian metric can on S^d, we have (see [B-G-M], p. 160)

$$\lambda_1(\text{can}) = \lambda_1^-(\text{can}) < \lambda_1^+(\text{can}).$$

Now if g is any Riemannian metric on S^d which is invariant under σ, i.e., $\sigma^* g = g$, then we can consider $\lambda_1^-(g)$ and $\lambda_1^+(g)$. By the result above we have

$$\lambda_1(g) = \lambda_1^-(g) < \lambda_1^+(g)$$

for every g close enough to the metric can.

Problem. Investigate whether $\lambda_1(g) = \lambda_1^-(g)$ for every Riemannian metric g on S^d which is invariant under σ when $d \geqslant 3$.

For $d = 2$ we have the following

8.47 Proposition. *For every Riemannian metric on S^2 which is invariant by σ one has $\lambda_1(g) = \lambda_1^-(g) < \lambda_1^+(g)$.*

Proof. Suppose that $\lambda_1(g) = \lambda_1^+(g)$ and let f be such that $\Delta f = \lambda_1^+(g) f$. According to a result of Cheng ([CG], p. 52), the set $\gamma = f^{-1}(0)$ is a simple C^∞ closed curve in S^2. Since f is invariant by σ the curve γ is also invariant by σ. Being connected, this simply closed curved has to join two antipodal points and hence, it projects down to a curve $\underline{\gamma}$ in $\mathbb{R}P^2$ which is not homotopic to zero. Therefore, $\mathbb{R}P^2 \setminus \underline{\gamma}$ is connected, and the projected function $\underline{f}$ of f on $\mathbb{R}P^2$ has the same sign everywhere. But f itself would then have a constant sign, contradicting $\int_{S^2} f \, d\mu_g = 0$. $\square$

8.48. Now consider a Riemannian metric g on the d-dimensional sphere S^d which makes (S^d, g) into a Blaschke manifold. By 5.57 we can consider S^d as the standard differentiable sphere, endowed with its antipodal map σ. In particular, by the result above we can define the eigenvalue $\lambda_1^-(g)$. We then have

8.49 Theorem ([BR 6], p. 146). *For a Blaschke Riemannian metric* (S^d, g), *normalized by taking the diameter equal to* π, *one always has* $\lambda_1^-(g) \geq d$. *Furthermore, if* $\lambda_1^-(g) = d$, *then* (S^d, g) *is isometric to* (S^d, can).

Proof. We set $S^d = M$ for convenience of notation. First remark that $\int_M f d\mu_g = 0$ for every f in C_-. Now a direct extension of the minimum principle (see for example [B-G-M], p. 186) to our situation shows that

$$\textbf{8.50} \qquad \lambda_1^- = \inf\left\{\frac{\|df\|^2}{\|f\|^2} \;\middle|\; f \in C_-\right\}.$$

Hence, we will be done if we can show that

$$\frac{\|df\|^2}{\|f\|^2} \geq d \qquad \text{for every } f \text{ in } C_-.$$

Let u be in UM and let γ_u be the geodesic $t \mapsto \exp_{p(u)}(tu)$ generated by u. For any function $f: M \to \mathbb{R}$ denote by f_u the real valued function $f_u = f \circ \gamma_u$. If $f \in C_-$, we have, in particular, $f_u(t+\pi) = -f_u(t)$ for every t in $\mathbb{R}$ and every u in UM. Therefore, the mean value $\int_0^\pi f_u(t)dt$ is equal to zero. Hence, by Wirtinger's inequality (see for example [BE-GO], p. 345) we have:

$$\textbf{8.51} \qquad \int_0^\pi (f_u'(t))^2 dt \geq \int_0^\pi f_u^2(t)dt.$$

The idea of the proof is now to repeat the proof of 5.64 with (5.65) replaced by (8.51), i.e., to integrate (8.51) over the unit bundle UM. In the following inequalities and equalities we only use Fubini's theorem (as in the proof of 5.64), Liouville's theorem on the geodesic flow via 1.125 and the formula given at the end of the proof of 5.64 concerning the integral of a quadratic form over a Euclidean sphere:

$$\int_{UM}\left(\int_0^\pi (f_u'(t))^2 dt\right) d\mu_1 \geq \int_{UM}\left(\int_0^\pi f_u^2(t)dt\right) d\mu_1.$$

But $f_u'(t) = df(\zeta^t(u))$. Hence,

$$\begin{aligned}
\int_{UM}\left(\int_0^\pi (f_u'(t))^2 dt\right) d\mu_1 &= \int_{UM \times [0,\pi]} (df(\zeta^t(u)))^2 d\mu_1 \otimes dt \\
&= \int_0^\pi \left(\int_{UM} (df(u))^2 d\mu_1\right) dt = \pi \int_{UM} (df(u))^2 d\mu_1 \\
&= \pi \int_{m \in M}\left(\int_{u \in U_m M} (df(u))^2 d\sigma\right) d\mu_g.
\end{aligned}$$

Since

$$\int_{U_m M} (df(u))^2 = \frac{1}{d}\beta(d-1)\,\mathrm{trace}_g((df \otimes df)(m))$$
$$= \frac{\beta(d-1)}{d}|df(m)|^2,$$

we finally have

$$\int_{UM}\left(\int_0^\pi (f_u'(t))^2 dt\right) d\mu_1 = \frac{\beta(d-1)}{d}\|df\|^2.$$

For the projection $p: UM \to M$ we have:

$$\int_{UM}\left(\int_0^\pi f_u^2(t)dt\right) d\mu_1 = \int_{UM\times[0,\pi]} (f(p(\zeta^t(u))))^2 d\mu_1 \otimes dt$$
$$= \pi \int_{UM} (f(p(u)))^2 d\mu_1 = \pi\beta(d-1)\|f\|^2.$$

Hence, as claimed, $\|df\|^2 \geqslant d\|f\|^2$ for every f in C_-.

If $\lambda_1^- = d$ we necessarily have equality in Wirtinger's inequality (8.51) for every u in UM. It is known that this implies the existence for every u in UM of real numbers a and b such that $f_u(t) = a\cos t + b\sin t$ for every t in $\mathbb{R}$. Then we argue as in the proof of Obata's theorem, see [B-G-M], pp. 180—182. □

8.52 Corollary. *For a Blaschke manifold (S^2, g) one always has $\lambda_1(g) \geqslant 2$.*

Appendix A. Foliations by Geodesic Circles

By D. B. A. Epstein

I. A.W. Wadsley's Theorem

a. Introduction

A.1. Let M be a C^∞-manifold with a C^∞-foliation by circles. We prove the following theorem of A.W. Wadsley [WY 2]:

A.2 Theorem. *The following conditions are equivalent:*

1) *There is a C^∞ Riemannian metric with respect to which all the circles are geodesic.*

2) *For any compact subset K of M, the circles meeting K have bounded length (the bound depending on K) with respect to some (and hence any) C^∞ Riemannian metric.*

3) *Let $\tilde{M}$ be the double cover of M obtained by taking the two different possible local orientations of the leaves (so that we can put a coherent set of arrows on the leaves in $\tilde{M}$). There is a C^∞-action of the orthogonal group $O(2)$ on $\tilde{M}$ and the non-trivial covering translation $\sigma: \tilde{M} \to \tilde{M}$ is an element of the non-trivial component of $O(2)$. Each orbit under the $O(2)$-action consists of two components and each component is mapped diffeomorphically onto a leaf of M by the projection $\tilde{M} \to M$.*

A.3 Remark. The Klein bottle is fibred over the circle by circles. In this case the double cover referred to in 3) is the torus. By multiplying the whole situation with a non-orientable manifold, we see that $\tilde{M}$ does not need to be orientable. If one can already put a coherent set of arrows on the foliation of M, then $\tilde{M}$ will be disconnected and Condition 3) reduces to the existence of a C^∞-action of $SO(2) = S^1$ on M, such that each orbit is a leaf.

A.4 Corollary. *Let $\mu: M \times \mathbb{R} \to M$ be a C^∞-action of the additive group of real numbers, such that each orbit is a circle. Then there is a C^∞-action of the circle on M with the same orbits as μ if and only if M has a Riemannian metric with respect to which each orbit is a geodesic.*

A.5 Remark. The differentiability assumptions can be improved. For example, one only needs to suppose that the foliation is C^2 rather than C^∞. To avoid tedious complications, we do not strive for utmost generality.

A.6. It is known (see, for example, Epstein [EPN 3]) that the second condition of the theorem above is equivalent to many other conditions, for example that the quotient space obtained by pinching each leaf to a point is Hausdorff, or that for each of the circles, the Poincaré map (see A.14) on a transverse disk has finite order. The last condition is useful for giving a geometric picture of what is going on. For example, for oriented manifolds in dimension three, it is known (see [EPN 1]) that the foliation is locally diffeomorphic to the foliation given by the orbits of the following flow on $D^2 \times S^1$ ($\subset \mathbb{C} \times \mathbb{C}$):

A.7 $$(z_1, z_2)_t = (z_1 \exp(2\pi i p t/q), z_2 \exp(2\pi i t))$$

where $|z_1| \leqslant 1$ and $|z_2| = 1$. This example shows that the length of the circles is not a continuous function.

A.8. There is an example due to Sullivan [SL 2] of a flow on a compact 5-manifold with every orbit a circle, and such that the circles have unbounded length. Thurston (see § II or [SL 2]) has a similar example which is real analytic. Epstein [EPN 1] has shown that a compact 3-manifold foliated by circles satisfies Condition 2) of Wadsley's theorem. This is false for certain foliations of non-compact 3-manifolds. Epstein and Vogt have recently found a real analytic example on a compact 4-manifold foliated by circles of unbounded length.

A.9. There is of course no need to restrict consideration of the problem to leaf dimension one, though this is perhaps the most interesting case. For higher dimensions we ask about a manifold M, foliated with each leaf compact. (See Epstein [EPN 3] for an expository account.)

A.10 Problem. Find a necessary and sufficient condition of a differential geometric nature which ensures that the volume of the leaves is locally bounded. Such a condition will only be an acceptable solution to the problem if it makes sense separately for each embedded leaf, without reference to the other leaves—e.g., the condition that each leaf should be a geodesic circle. Wadsley has partial results in his theses [WY 1].

b. Killing Vector Fields

A.11. Let M be a manifold with C^∞ Riemannian metric g. A C^∞ vector field X on M is said to be a *Killing vector field* if $\mathscr{L}_X g = 0$. In this case, the 1-parameter group of diffeomorphisms of M generated by X consists of isometries.

A.12 Lemma. *Let M be a Riemannian manifold with a Killing vector field X and suppose that $g(X,X) = 1$. Then the orbits of X are geodesics.*

Proof. Let $\varepsilon > 0$ be small and let $\alpha : (-\varepsilon, \varepsilon) \to M$ be a solution curve. Let D be a small open disk of codimension one, transverse to X, meeting α at $\alpha(0)$ and orthogonal to α at this point. Let φ_t be the one-parameter family of diffeomorphisms generated by X, and suppose that $\varphi : D \times (-\varepsilon, \varepsilon) \to M$ is a diffeomorphism onto an open subset U of M.

Let Y be any vector orthogonal to X at $\alpha(0)$. We extend Y to a vector field on D and then extend it over U by defining $Y_{\varphi(x,t)} = \varphi_{t*} Y_*$ for $x \in D$ and $+\varepsilon < t < \varepsilon$. Then

$[X,Y]=\mathscr{L}_X Y=\frac{d}{dt}(\varphi_{-t})_* Y=0$ on U. It follows that if D is the Riemannian connection, then $D_X Y=D_Y X$.

Since φ_t is an isometry, Y is orthogonal to X at each point of α. It follows that along α we have

$$\begin{aligned} g(D_X X, Y) &= X\cdot g(X,Y)-g(X,D_X Y)\\ &= -g(X,D_X Y)\\ &= -g(X,D_Y X)\\ &= -(1/2)\,Y\cdot g(X,X)\\ &=0. \end{aligned}$$

Since $Y_{\alpha(0)}$ was arbitrary, we see that the component of $D_X X$ orthogonal to X is zero. But one also has

$$0=X\cdot g(X,X)=2g(D_X X,X).$$

Hence $D_X X=0$ and the proof is complete. □

A.13 Lemma. *Let M be a manifold with a C^∞ metric g and let X be a never zero Killing vector field. Then there is a unique conformally related metric g' such that $g'(X,X)=1$. X is a Killing vector field with respect to g' (so that by Lemma A.12, the solution curves are geodesic with respect to g').*

Proof. We have $g'(Y,Z)=\dfrac{g(Y,Z)}{g(X,X)}$.

Since $\mathscr{L}_X g=0$, we have

$$\mathscr{L}_X(g(X,X))=2g([X,X],X)=0.$$

Therefore, multiplication by $g(X,X)$ commutes with the derivation $\mathscr{L}_X$. It follows that

$$\mathscr{L}_X g'=(1/g(X,X))\mathscr{L}_X g=0. \quad \square$$

c. Local Analysis

A.14. Let X be a C^∞ vector field on a manifold M', with a periodic orbit L. Let D be a small open disk transverse to X, meeting L at its centre C. Let φ_t be the 1-parameter group of diffeomorphisms generated by X. If E is a sufficiently small concentric open subdisk of D, then standard results on solutions of differential equations (see for example Coddington and Levinson [C-L]) give us an open embedding, called *the Poincaré map* $\theta\colon E\to D$ with the following properties:

i) $\theta(C)=(C)$,

ii) for $x\in E$, there exists a $\tau(x)>0$ such that $\theta(x)=\varphi_{\tau(x)}(x)$ and such that $\tau(x)$ is the smallest value of t such that $t>0$ and $\varphi_t(x)\in D$,

iii) θ and τ are C^∞-functions.

In the example given in the introduction, $\tau(x)=1$ for $x\in D^2$ and θ is the rotation by $2\pi p/q$.

A.15 Assumption. *Now let us specialize to the case of a unit vector field with each orbit a circle.*

Let $\varrho: M' \to \mathbb{R}^+$ be the function which assigns to each point $x \in M'$ the length of the circle through x. As shown by the example in the introduction, ϱ is not necessarily continuous. In fact, ϱ is not necessarily locally bounded. However, we do have the following formula in the special case that the orbit of $x \in E$ defined by $x_0 = x$, $x_1 = \theta(x_0), \ldots, x_r = \theta(x_{r-1}), \ldots$ remains in E for every value of r. Since the circle through x meets E only a finite number of times, the orbit is finite, of order $n(x)$, say. Then

A.16 $$\varrho(x) = \sum_{i=0}^{n(x)-1} \tau(x_i) = \sum_{i=0}^{n(x)-1} \tau(\theta^i x).$$

The following proposition gives useful information.

A.17 Proposition. *Let L be a fixed orbit of length λ. Let $k>0$ and $\varepsilon>0$ be given (k large and ε small). Then there is a neighbourhood U of L such that if $x \in U$ and $\varrho(x) \leqslant k$, then*
i) *for some integer $i (0 < i \leqslant [k/\lambda]+1)$, $|\varrho(x) - i\lambda| < \varepsilon$ and*
ii) *the orbit through x lies in an ε-neighbourhood of L.*

Proof. By taking U sufficiently small, standard theorems (see [C-L]) about differential equations tell us that given $x \in U$, there exists an $x_0 \in L$ such that

$$d(\varphi_t(x), \varphi_t(x_0)) < \varepsilon' \quad \text{for} \quad 0 \leqslant t \leqslant k.$$

Taking $t = \varrho(x)$ we see that $d(\varphi_t(x_0), x_0) < 2\varepsilon'$ and hence, for ε' sufficiently small, t is near some integral multiple of λ. We may picture the situation as follows (Fig. A.17.1):

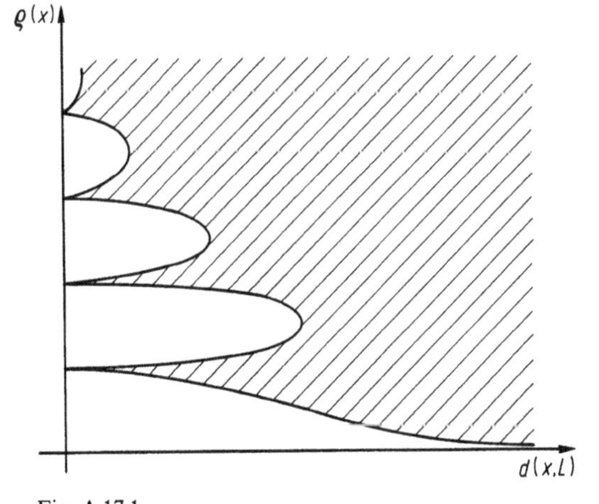

Fig. A.17.1

The unshaded region of the diagram represents pairs of real numbers which cannot be of the form $(d(x, L), \varrho(x))$ for any $x \in M'$.

A.18 Corollary. *The function ϱ is lower semicontinuous. If N is any subset of M', then the set V of points of continuity of $\varrho \upharpoonright N$ is an open subset of N. If N is locally compact, then V is dense in N.*

Proof. Only the second statement requires proof, as the first follows immediately from the proposition and the last is a standard consequence of the Baire category theorem. Let y be a point of continuity of $\varrho \upharpoonright N$. Let W be an open neighbourhood of y such that if $x \in N \cap W$, then $|\varrho(x) - \varrho(y)| < \varrho(y)/8$. Let $z \in N \cap W$. We prove that $\varrho \upharpoonright N$ is continuous at z. If $x \in N \cap W$ and $i \geqslant 2$, then

$$i\varrho(z) - \varrho(x) > i(7/8)\varrho(y) - (9/8)\varrho(y) \geqslant (5/8)\varrho(y).$$

Let $\varepsilon > 0$ and suppose that $\varepsilon < \varrho(y)/8$. In Proposition A.17, we put $k = 2\varrho(y)$ and take L to be the orbit through z. Let U be the open set given by the proposition. Then if $x \in N \cap U \cap V$, we know that $|\varrho(x) - i\varrho(z)| < \varepsilon$ for some integer $i > 0$. It follows from the preceding paragraph that $i = 1$. This proves the corollary.

A.19. Now suppose ϱ is bounded by $k > 0$ in a neighbourhood of an orbit L. Then proposition A.17 shows that there exist arbitrarily small open neighbourhoods of L which are invariant under the flow. Hence, we can find a neighbourhood E' of C in E such that $\theta E' = E'$. (E' is the intersection of E with a small invariant neighbourhood of L.) Equation (A.16) now applies. Since $\tau(C) = \varrho(C) > 0$, and since τ is continuous, we see that $n(x)$ is bounded as x varies in a neighbourhood of C in E'. Let $n(x) \leqslant m$ in this neighbourhood. Then $\theta^{m!} = id$ on this neighbourhood.

A.20. By averaging a Riemannian metric on E we may assume that θ is linear and even orthogonal near C with respect to some coordinate system on E. If r is the order of θ near C, then it follows that the complement of a finite union of proper linear subspaces consists of points of order exactly r under θ.

A.21 Proposition. *Condition* 2 *of the main Theorem A.2 implies Condition* 3.

Proof. On the double cover $\tilde{M}$ we take the unit vector field X along the leaves. Now the leaves can be consistently oriented in a small neighbourhood of a fixed leaf L. (But note that the neighbourhood itself may be non-orientable. In this case θ reverses orientations.) Therefore, each circular leaf in M is covered by two disjoint circles in $\tilde{M}$. These circles are interchanged by the covering translation σ.

We now take $M' = \tilde{M}$ and apply the analysis and notation from A.19—A.20. Let $\alpha: M' \to \mathbb{R}$ be defined by $\alpha(x) = \limsup_{y \to x} \varrho(y)$.

Since ϱ is locally bounded, the Poincaré map θ has finite order r on a sufficiently small open disk E' of codimension one. By Eq. (A.16),

$$\text{A.22} \qquad \alpha(x) = \sum_{i=0}^{r-1} \tau(\theta^i x) \quad \text{for} \quad x \in E'.$$

Since α is also invariant under the flow, we easily see that near L the map α is C^∞. It follows that α is C^∞ on the whole of M'.

The vector field X/α has the property that solution curves have period $1/s$ for some integer $s \geqslant 1$. (The integer s is equal to r on the leaf L.) Hence, we have an action of $S^1 = \mathbb{R}/\mathbb{Z}$ on $\tilde{M}$. The covering translation $\sigma: \tilde{M} \to \tilde{M}$ is an isometry which preserves the foliation. Therefore, σ preserves α and reverses X. Hence, $\sigma_*(X/\alpha) = -X/\alpha$. It follows that σ together with the circle action give rise to an $O(2)$-action on $\tilde{M}$, where σ corresponds to the matrix $\begin{bmatrix} -1 & 0 \\ 0 & 1 \end{bmatrix}$. This completes the proof of Proposition A.21.

d. Condition 3 Implies Condition 1

A.23. By integrating a metric on $\tilde{M}$ over $O(2)$, we ensure that $\sigma: \tilde{M} \to \tilde{M}$, the non-trivial covering translation, is an isometry, and that we have a one-parameter family of isometries acting on $\tilde{M}$, with every orbit a circle. The corresponding vector field X is a Killing vector field on $\tilde{M}$ with metric $\tilde{g}$. Let $g' = \tilde{g}/\tilde{g}(X,X)$. By Lemma A.13, X is a Killing vector field with respect to g' and by Lemma A.12, the orbits are geodesic with respect to g'.

Since X is preserved by σ, except for sign, and σ is a $\tilde{g}$-isometry, we see that σ is a g'-isometry. Hence, a Riemannian metric g is induced on M by g' and the image of a g'-geodesic on $\tilde{M}$ is a g-geodesic on M. Condition 1 follows.

e. Condition 1 Implies Condition 2

A.24. Without loss of generality, we may replace M by $M' = \tilde{M}$, where we have a flow as in the local analysis in c). We put $B_1 = \{x \in M' | \varrho$ is unbounded in every neighbourhood of $x\}$ and $B_2 = \{x \in B_1$ and $\varrho \restriction B_1$ is not continuous at $x\}$.

A.25 Lemma. *The function* $\alpha(x) = \limsup_{y \to x.} \varrho(y)$, *which is defined on* $M' \setminus B_1$, *is locally constant there.*

Proof. As we showed in the proof of Proposition A.21, α is a C^∞-function. Let X be the unit vector field along the leaves of M' and let φ_t be the one-parameter family generated by X.

Now $\varphi_{\alpha(x)}(x) = x$. To see this, let $\{y_i\}$ be a sequence with limit x, such that $\varrho(y_i)$ tends to $\alpha(x)$. Then we take the limit of the equation $\varphi_{\varrho(y_i)}(y_i) = y_i$. Let $\gamma: I \to M'$ be any C^∞-path. We define a one-parameter family of geodesics by

$$(s,t) \mapsto \varphi_{t\alpha(\gamma s)}(\gamma s) \; (0 \leqslant t \leqslant 1)$$

(recall that we are assuming Condition 1), where t is the parameter along the geodesic and s is the parameter varying the geodesic. By the first variation formula for the length of a C^∞-path (see 1.109 or Kobayashi and Nomizu [K-N2] p. 80), we see that the length is independent of s. But the length of such a geodesic is $\alpha(\gamma s)$.

A.26 Theorem. *Let M' be a Riemannian manifold with a C^∞-unit vector field X such that every solution curve is a geodesic circle. Then X is the derivative of a C^∞-circle action on M' (the circle being identified with $\mathbb{R}/c\mathbb{Z}$, where $c>0$ is a constant, not necessarily equal to one).*

Proof. By Lemma A.25, we need only show that $B_1=\phi$. So we suppose $B_1\neq\phi$ and prove a contradiction. Now B_1 is clearly closed in M' and, by Corollary A.18, B_2 is closed and has void interior in B_1.

Let $x\in B_1\setminus B_2$ and let U be a connected open neighbourhood of x which is disjoint from B_2, and such that

A.27 $\quad |\varrho(u)-\varrho(v)|<\varrho(w)/8 \quad \text{for} \quad u,v,w\in U\cap B_1 .$

Let V be a component of $M'\setminus B_1$ which meets U and such that

A.28 $\quad c>3\varrho(x), \quad \text{where} \quad \alpha\restriction V=c,$

where c is the constant given by Lemma A.25.

Now $\varphi_c\restriction\bar{V}$ is the identity, and $Bd V\subset B_1$. By the definition of B_1,

A.29 $\quad \operatorname{int}\bar{V}=V.$

Since U is connected and meets B_1, there is a point $y\in U\cap Bd V$. Since $\varphi_c(y)=y$, there is an integer k such that $c=k\varrho(y)$. By (A.27) and (A.28), $k\geqslant 2$.

A.30. Let D be a small open disk of codimension one, transverse to the flow, with centre y. Let E be a smaller concentric disk on which θ^i is defined for $1\leqslant i\leqslant k!$, with each $\theta^i E\subset D$. (Our notation follows that of the local analysis § c)). Since $\theta^{k!}\restriction E\cap\bar{V}$ is the identity, there is a small neighbourhood F of y in $E\cap\bar{V}$ which is invariant under θ. We take F open in $E\cap\bar{V}$. Then $F=G\cap\bar{V}$, where G is open in E. Let H be the component of y in G, and let $J=H\cup\theta H\cup\ldots\cup\theta^{k!}H$.
Then J is connected and open in E and

$$\theta(J\cap\bar{V})=J\cap\bar{V}\subset F .$$

A.31. We now prove a contradiction, by adapting an argument due to Montgomery [M-Z]. We define $\psi:J\to J$ by $\psi(x)=\theta(x)$ for $x\in J\cap\bar{V}$ and $\psi(x)=x$ otherwise. By (A.27), we see that $\theta\restriction E\cap Bd V$ is the identity, so that ψ is continuous. Since $\psi^{k!}=id$, ψ is a homeomorphism. Since $k\geqslant 2$, $\psi\restriction J\cap V\neq id$. By Newman's theorem (see Dress [DS] for a nice exposition), $J\cap V$ is dense in J. But then φ_c is the identity on J and hence on a neighbourhood of y. This contradicts the fact that $x\in B_1$. The proof is complete. □

f. Manifolds With All Geodesics Closed

A.32. Let M be a manifold such that every geodesic is an immersed circle. Let UM be the unit tangent bundle of M. Then UM has a flow which satisfies the condition of Wadsley's theorem, that each orbit is a geodesic circle. By Theorem A.26, the flow is the result of a circle action on UM. Then M must be compact, because it is the image of $S^{n-1} \times S^1$, where S^{n-1} is a single fibre in UM. See the applications of this in 7.11 and 7.12.

Results of this kind, which follow from Wadsley's theorem, are known in any case by other means. For example, the result mentioned follows from the existence of a line in a non-compact Riemannian manifold [a *line* is a geodesic $\gamma:\mathbb{R}\to M$ such that $\varrho(\gamma(t), \gamma|t')) = |t'-t|$ for every t, t' in $\mathbb{R}$].

II. Foliations With All Leaves Compact*

a. Foreword

A.33. "The forthcoming text which provides a counterexample due to W. P. Thurston closely follows the talk delivered by D. B. A. Epstein at the 'table ronde' that I organized (see my introductory letter)."

A. Besse Esq.

b. Introduction

A.34. Let M be a manifold foliated so that each leaf is compact. We study the following QUESTION:

"Is the volume of the leaves locally bounded?"

If M is compact, the word "locally" is redundant. Here we suppose for example that M has a Riemannian metric, which of course induces a Riemannian metric on each leaf. The volume of a leaf is found by integrating the volume form of the metric on the leaf. The condition that the volume of the leaves should be locally bounded can be expressed in many different ways (see [EPN 3]). For example it is equivalent to ask that the quotient space, obtained by identifying each leaf to a point, should be Hausdorff.

A.35. An affirmative answer to this question has been obtained in many different situations. When M is a compact 3-dimensional manifold foliated by circles, this is proved in [EPN 1]. Edwards, Millett and Sullivan [E-M-L] have shown that this proof can be adpated to prove the same result for any compact manifold foliated by compact manifold foliated by compact leaves of codimension two. The above proofs use the hypothesis that the foliation is differentiable, but this hypothesis can be avoided (unpublished work of Edwards and myself). In the paper already cited, Edwards, Millett and Sullivan show that if there is an r-form w, where r is the leaf dimension, such that $\int_L \omega > 0$ for each leaf L, and if M is compact, then the question has an affirmative answer. Their method is most interesting. They apply the notion of a geometric current due to Ruelle and Sullivan [R-S]. A geometric current is a current in the sense of de Rham, which is represented in each coordinate chart by a family of parallel hyperplanes and a measure transverse to these hyperplanes.

* Translated from the French by D. B. A. Epstein

A.36. One of the earliest results in foliation theory was the positive solution provided by Reeb [RB 2] in the codimension one case. Reeb also gave examples of non-compact manifolds where the answer was negative. The codimension in these examples is two. Since then it has been an open question as to whether the answer was positive whenever M was compact—this was first asked by A. Haefliger in the early 1950's.

A.37. Recently Sullivan has given a counterexample for a certain manifold of dimension five foliated by circles [SL 3]. Thurston developed Sullivan's idea further and constructed a real analytic counterexample on another compact 5-dimensional manifold. Epstein and Vogt have recently found a real analytic example in dimension four.

c. Thurston's counterexample

A.38. Let (r, v) be two real positive parameters. These parameters define a flow of speed v on a circle of radius r in the plane. We suppose that the flow is counter-clockwise. Projecting, we obtain a flow on a circle in the torus $T^2 = R^2/Z^2$ (for the moment we do not fuss about the circle intersecting itself in T^2). Now we follow Sullivan to obtain a flow on the unit tangent bundle E to T^2. If we are given the two parameters r and v and a point $e \in E$, there is a circle in the plane which is unique up to translation by elements of Z^2, of radius r and such that e is tangent to the circle in the counter-clockwise direction. This circle in the plane defines a circle in E by taking the tangent vector at each point of the circle in the plane. Obviously we get a flow on E determined by (r, v), and each orbit is an embedded circle.

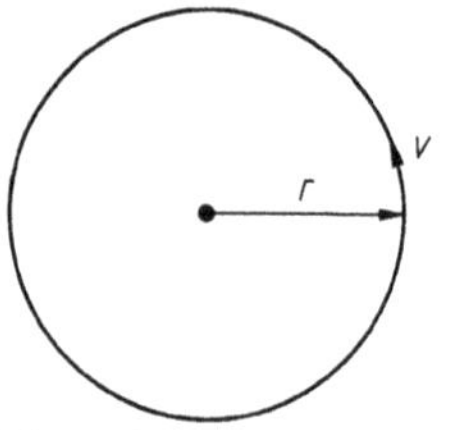

Fig. A.38

A.39. Now we want to let r tend to infinity. But when $r = \infty$ we get a straight line and not a circle. To correct this deficiency one again adapts an idea of Sullivan's. We define E', another circle bundle, with another way of lifting paths. Thurston's idea for doing this is as follows. Let N be the group of matrices of the form

$$\begin{bmatrix} 1 & 0 & 0 \\ x & 1 & 0 \\ z & y & 1 \end{bmatrix}.$$

Let Γ be the subgroup of elements of N where x, y and z are integers. Let Z be the centre of N. Then Z is the set of matrices of the form

$$\begin{bmatrix} 1 & 0 & 0 \\ 0 & 1 & 0 \\ z & 0 & 1 \end{bmatrix}.$$

We have a fibre bundle

$$\begin{array}{ccccc} Z/Z\cap\Gamma & \to & N/\Gamma & \to & N/\Gamma Z \\ \| & & & & \| \\ S^1 & & & & T^2 \end{array}$$

Here the cosets are right cosets: for example an element of N/Γ has the form Γn. The forms dx, dy and $dz-ydx$ generate the space of all left invariant forms on N. We choose such a form as a connection form on $E'=N/\Gamma\to T^2$ (for example $dz-ydx$ will do). It follows that the curvature is given by the 2-form

$$\Omega=dx\wedge dy \quad \text{on} \quad T^2.$$

A.40. It is well-known (and easy to prove by using Stoke's Theorem) that in such a situation one can calculate the holonomy over a circle of radius r in T^2 by means of the integral $\int\Omega=\pi r^2$. In other words the horizontal lift of the circle of radius r returns to the fibre where it started a distance r^2 further round the fibre.

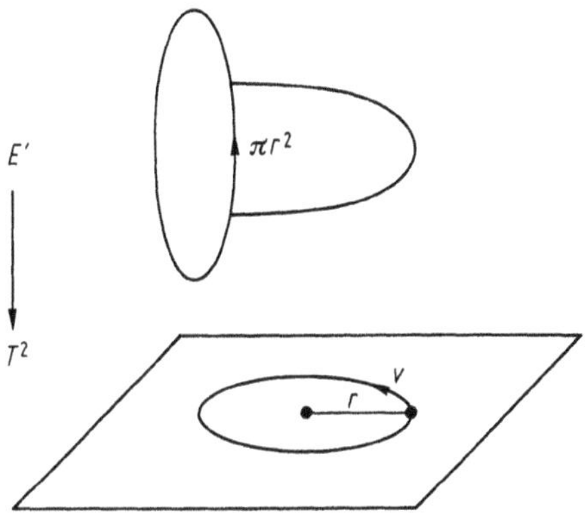

Fig. A.40

In order to arrange that the lift should become a closed curve, we correct the horizontal lift by applying a vertical vector field equal to the distance πr^2 divided by the time taken—that is to say

$$\pi r^2/(2\pi r/v)=rv/2$$

in absolute value. So the correction term is $-rv/2$.

A.41. Finally we define a vector field on $E\times_{T^2}E'$ which carries out both liftings simultaneously. We have fixed parameters (r,v). The coordinates of the vector at a point (e,e') are $(ve, v/r, -vr/2)$. The first coordinate is in the horizontal subspace of the firbre bundle $E\times_{T^2}E'\to T^2$, where we have taken the product of the trivial connection and the connection on E' described above. The second coordinate is tangent to the fibres of E and v/r is the usual formula for the derivative of a vector tangent to a circle. The third coordinate is tangent to the fibres of E' and is the correction term already discussed.

A.42. Thurston then uses the following trick on $(E \times_{T^2} E') \times S^1$ to make r tend to infinity. We take a coordinate $\theta \in R/2\pi Z = S^1$ and we define a vector field on $E \times_{T^2} E' \times S^1$ by $(e \sin 2\theta, \sin^2\theta, -2\cos^2\theta, 0)$. We have equated $v = \sin 2\theta$, $v/r = \sin^2\theta$ and $vr = 4\cos^2\theta$. We then have $|\cot\theta| = r/2$ and the length of an orbit tends to infinity when θ tends to zero or to π. However when we actually have equality, $\theta = 0$ or $\theta = \pi$, and the field is $(0, 0, 0, 2, 0)$ which has period $1/2$.

This completes the description of the counterexample.

d. Remarks

A.43. According to Wadsley's Theorem (see A.I), a necessary and sufficient condition for a foliation by circles to give a positive answer to our equation, is that there should exist a Riemannian metric such that each circle is geodesic. (We do not require the manifold to be compact in this theorem.) Therefore we have an example of a foliation by circles such that there is no metric for which each circle is a geodesic.

A.44. One could ask if Wadsley's Theorem remains true if one replaces the Riemannian metric by an affine connection. In particular we note that Thurston's example lives on flat affine manifold—that is to say one can choose coordinate charts such that the change of coordinates is always affine. Is it possible to construct a counterexample on a flat affine manifold such that each circle is locally a straight line?

A.45. We also note that in Thurston's counterexample, the fibre bundle structures on $E \times_{T^2} E'$ are not stable (where we fibre by circles). In the reverse direction Rosenberg and Langevin [R-L] have shown that one can apply Thurston's stability theorem in order to show that, for a fibre bundle in which the fibre L satisfies $H_1(L; R) = 0$, we do have stability.

Appendix B. Sturm-Liouville Equations all of whose Solutions are Periodic, after F. Neuman

By Jean Pierre Bourguignon

I. Summary

B.0. This appendix is devoted primarily to results due to F. Neuman (see [NN]) on the Sturm-Liouville equation in one variable with all periodic solutions. In the literature on differential equations there is a wide variety of books and monographs devoted to the Sturm-Liouville equation. We have selected [BA] as a reference since it inspired part of the work by F. Neuman.

In B.II we analyze how a vector-valued Sturm-Liouville equation is associated with the Jacobi equation (see 1.89), especially along a closed geodesic.

Paragraph B.III is devoted firstly to explaining why a periodic geodesic flow gives rise to a Sturm-Liouville equation with all periodic solutions and secondly, to the results of F. Neuman. We follow the exposition of N. Gontier (see [GT]). We present criteria guaranteeing that a Sturm-Liouville equation in one variable has only periodic solutions. In particular, we give the explicit form that the functions characterizing the equation must have (proving, in particular, that they depend on an almost arbitrary function).

In B.IV we come back to geometry and among the examples we previously exhibited select the ones which we can describe geometrically. We establish an inequality for the integral of the curvature along a geodesic. This gives a slightly different method of proving L. Green's theorem (see 5.59).

II. Periodic Geodesics and the Sturm-Liouville Equation

B.1. Let γ be a geodesic of a d-dimensional Riemannian manifold (M, g). We recall (see 1.89) that a Jacobi field J is a solution of the Jacobi differential equation along γ

B.2 $$J'' + R(\dot{\gamma}, J)\dot{\gamma} = 0\,.$$

This is a *vector-valued Sturm-Liouville equation.* We shall only be interested in *normal* Jacobi fields, i.e. such that $g(J, \dot{\gamma}) = 0$.

In Proposition 1.90 we saw that any Jacobi field along γ is the transverse vector field to a family of geodesics.

B.3. To simplify our study we shall consider the differential equation in $\mathbb{R}^{d-1}$ associated with (B.2) by decomposing any normal vector field Y along γ into its components $\mathscr{Y} = (Y^1, \ldots, Y^{d-1})$ in a parallel normal frame field along γ. In the new equation the independent variable will still be called s. We shall *denote by* $R(s)$ the symmetric linear operator of $\mathbb{R}^{d-1}$ associated with $R(\dot\gamma,.)\dot\gamma$ in terms of the parallel frame field. We get

B.4 $$\mathscr{Y}'' + R(s)\cdot\mathscr{Y} = 0\,.$$

Let us remark that when M is a surface, (B.4) is just the usual Sturm-Liouville equation in one variable. In this case we shall write y instead of $\mathscr{Y}$ and $r(s)$ instead of $R(s)$. We get

B.4' $$y'' + r(s)y = 0\,.$$

B.5. For a vector-valued Sturm-Liouville equation as (B.4) it is a classical result that the *Wronskian* of a family of $2d-2$ solutions is a constant. Moreover, the Wronskian is non-zero if and only if the solutions are linearly independent in which case they provide a basis for the space of solutions. Let $\langle\,,\rangle$ denote the standard scalar product in $\mathbb{R}^{d-1}$. For a pair of solutions $\mathscr{Y}_1$, $\mathscr{Y}_2$, the quantity $\langle\mathscr{Y}_1(s), \mathscr{Y}_2'(s)\rangle - \langle\mathscr{Y}_2(s), \mathscr{Y}_1'(s)\rangle$ is also a constant (these facts are, respectively, Liouville's Theorem 1.56 and the comments in 1.99 as seen from the differential equation point of view).

We also note that for $d=2$ these two properties coincide. If y_1 and y_2 are two solutions of (B.4), then $y_1(s)y_2'(s) - y_2(s)y_1'(s)$ is independent of s. This fact gives rise to *Abel's transformation*:

B.6 Lemma. *If* y_1 *is a nowhere vanishing solution of* (B.4') *on an interval* I *containing* s_0, *then*

$$y_2(s) = y_1(s)\int_{s_0}^{s}(y_1(t))^{-2}dt$$

is another solution of (B.4'), *and* y_2 *is linearly independent of* y_1.

Proof. One knows that if y_2 is another solution of (B.4'), then it satisfies

$$y_2'(s)y_1(s) - y_1'(s)y_2(s) = y_2'(s_0)y_1(s_0) - y_1'(s_0)y_2(s_0)\,.$$

This equation can be solved explicitly on any interval I where y_1 does not vanish. The solution above is the one with initial conditions $y_2(s_0)=0$, $y_2'(s_0) = 1/y_1(s_0)$. $\square$

B.7. It is also clear that the zeroes of a solution of (B.4) are simple and isolated. This follows directly from the uniqueness of the solution with given initial data $\mathscr{Y}(s_0)$, $\mathscr{Y}'(s_0)$. Moreover, when $d=2$ we can recover the function $r(s)$ from a basis of solutions y_1 and y_2 by the formula

$$r(s) = \frac{\begin{vmatrix} y_1' & y_1'' \\ y_2' & y_2'' \end{vmatrix}}{\begin{vmatrix} y_1 & y_1' \\ y_2 & y_2' \end{vmatrix}}\,.$$

B.8. Let us now suppose that γ is a periodic geodesic in M with length l. We shall still consider (B.4) as a differential equation defined on the whole of $\mathbb{R}$ (at least to be able to consider curves near γ which are not periodic). With this restriction $R(s)$ is periodic with least period l. We note that if P_m^γ denotes parallel transport once around γ starting from a point m of γ, then P_m^γ, if $N\gamma$ is the normal bundle to γ, P_m^γ is an isometry of $N_m\gamma$ onto itself. Therefore $N_m\gamma$ admits the following decomposition:

$$N_m\gamma = \Pi_1 \oplus \cdots \oplus \Pi_{j'} \oplus \Delta_1 \oplus \cdots \oplus \Delta_j$$

(with $j+2j'=d-1$), where the $\Pi_{i'}$'s are two-planes and the Δ_i's are lines invariant under P_m^γ. The parallel transport P_m^γ acts as a rotation with angle $\theta_{i'}$ in $\Pi_{i'}$ and is ε_i times the identity in Δ_i. If we want to consider a periodic vector field X along γ, its image identified through P_m^γ in $\mathbb{R}^{d-1}$ will no longer be periodic. After travelling the length L, it will turn through an angle $-\pi\theta_{i'}$ in the plane $\Pi_{i'}$ $(i'=1,\ldots,j')$ and through an angle $\varepsilon_i\pi$ along the line $\Delta_i(i=1,\ldots,j)$. For any other point m' on γ and after obvious identifications we find that P_m^γ and $P_{m'}^\gamma$ are conjugate under a rotation of $\mathbb{R}^{d-1}$. We shall drop the subscript m and consider P^γ as defined up to conjugacy by a rotation of the standard metric on $\mathbb{R}^{d-1}$. Such a behaviour will be called (l, P^γ)-periodicity.

When M is two-dimensional we have already noticed that we only have to consider an ordinary equation in one variable. Moreover, for a periodic geodesic γ, P^γ is necessarily the identity if M is orientable. In this case, to be (l, P^γ)-periodic is merely to be l-periodic.

B.9. In this context it will sometimes be interesting to think of the Sturm-Liouville operator $\frac{d^2}{ds^2}+R(s)$ as densely defined on the Sobolev space $H^1(\mathbb{R},\mathbb{R}^{d-1})$ of $\mathbb{R}^{d-1}$-valued functions. We shall especially be interested in its restriction to the subspace of $H^1(\mathbb{R},\mathbb{R}^{d-1})$ consisting of (l, P^γ)-periodic functions, i.e. which are images in $\mathbb{R}^{d-1}$ of periodic vector fields along γ. We shall call this restriction $H^1(\mathbb{R},\mathbb{R}^{d-1};\gamma)$.

III. Sturm-Liouville Equations all of whose Solutions are Periodic

B.10. We now suppose that (M,g) is a C_l-Riemannian manifold (i.e., that all of its geodesics are closed and have the same length l). This means in particular that if γ is a fixed geodesic with length l, then all variations of γ by geodesics are by periodic geodesics with the same length. Therefore, the transverse vector fields along γ of such variations, namely the Jacobi fields along γ are all periodic. If we use B.8, we find that for any geodesic γ of (M,g) the Sturm-Liouville equation transferred to $\mathbb{R}^{d-1}$ has only (l, P^γ)-periodic solutions. This can also be read off from the operator $\frac{d^2}{ds^2}+R(s)$ introduced in B.9. When $\frac{d^2}{ds^2}+R(s)$ acts on the dense subset of

$H^1(\mathbb{R}, \mathbb{R}^{d-1}; \gamma)$ on which it is defined, 0 must be an eigenvalue with maximal multiplicity, i.e. $2d-2$.

Especially when $d=2$ (i.e., M is necessarily either S^2 or $\mathbb{R}P^2$) we shall consider Sturm-Liouville equations with either l-periodic solutions alone (in the case of S^2) or with anti-periodic solutions alone (in the case of $\mathbb{R}P^2$). The results of F. Neuman ([NN]) are devoted precisely to this case. We normalize by taking $l=\pi$ and we state his results.

B.11 Proposition. (F. Neuman, [NN] p. 573). *The equation* (B.4′) *has only π-periodic (resp. π-antiperiodic) solutions if and only if there exists a non zero π-periodic (resp. π-antiperiodic) solution y_1 of* (B.4′) *such that*

B.12 $$\int_0^\pi (1/y_1^2(s) - b(s))ds = 0$$

$\left(\textit{where } b(s) = \sum_{i=1}^n A_i^2/\sin^2(s-a_i), \textit{ the } a_i\textit{'s are the zeroes of } y_1 \textit{ on } [0,\pi[\textit{ and the } A_i\textit{'s are determined by } A_i \cdot y'(a_i) = 1\right)$.

Proof. First notice that b is well defined since the zeroes of y_1 are simple and isolated (therefore, there are only finitely many a_i's in $[0,\pi[$).

On an interval $I=[s_0, s_1]$ where y_1 does not vanish we know by Abel's transformation B.6 that

$$y(s) = y_1(s) \int_{s_0}^{s} (1/y_1^2(t))dt$$

is a solution of (B.4′).

But for any set (c_i) $(i=1,\dots,n)$ of constants the function

$$y_2(s) = y_1(s)\left(\int_{a_i}^{s} (1/y_1(t))^2 dt + c_i\right)$$

defined on $\bigcup_{i=1}^{n}]a_i, a_{i+1}[$ is also a solution. We shall construct y_2 on $[0,\pi[$ by adjusting the constants c_i. We want y_2 to be at least C^2. We therefore require that hold for $i=1,\dots,n$

B.13$_i$ $$\lim_{s\to a_i^-} y_2(s) = \lim_{s\to a_i^+} y_2(s),$$

B.14$_i$ $$\lim_{s\to a_i^-} y_2'(s) = \lim_{s\to a_i^+} y_2'(s),$$

B.15$_i$ $$\lim_{s\to a_i^-} y_2''(s) = \lim_{s\to a_i^+} y_2''(s).$$

The relation (B.15$_i$) clearly follows from (B.13$_i$) since on each interval $]a_i, a_{i+1}[$ y_2 is C^2 and $y_2''(s) = -r(s)y_2(s)$ (r is continuous).

One can check directly that (B.13$_i$) is satisfied, each side of the relation converging to $1/y'(a_i)$.

By a direct check one also sees that (B.14$_i$) holds as soon as c_i has a specific value determined by c_{i-1}. We can therefore compute c_n from c_1. Now the criterion for π-periodicity (or π-antiperiodicity) will be $c_1 = c_{n+1}$. Using the values given by the induction relation for the c_i's then provides the formula (B.12). □

B.16 Theorem (F. Neuman [NN] p. 584). *The equation* (B.4′) *has only π-periodic (resp. π-antiperiodic) solutions with exactly n zeroes on $[0, \pi[$ if and only if*

i) *n is even (resp. odd);*

ii) *there exists a C^∞ π-periodic function f with $f(a_i) = f'(a_i) = 0$ for n distinct numbers (a_i) in $[0, \pi[$ such that*

B.17 $$r(s) = (\tfrac{1}{2}b'(s)/b(s) - f'(s))' - (\tfrac{1}{2}b'(s)/b(s) - f'(s))^2$$

$\left(\textit{recall that } b(s) = \sum_{i=1}^{n} A_i^2/\sin^2(s - a_i) \textit{ for non-zero numbers } A_i\right)$;

iii) *the following integral relation holds:*

B.18 $$\int_0^\pi (e^{-f(s)} - 1)b(s)\, ds = 0.$$

Moreover, one of the solutions y_1 of (B.4′) *admits the following expression*

B.19 $$y_1(s) = \operatorname{sgn}\left(\prod_{i=1}^{n} \sin(s - a_i)\right) \cdot e^{f(s)} \cdot b(s)^{-1/2}.$$

Proof. The "only if" part is obtained by a direct check using (B.19) and (B.17). The main difficulty is to verify that the objects we define are regular at the a_i's. In (B.18) we recognize condition (B.12) for the function y_1 given by (B.19).

For the "if" part periodicity is clear. What remains to be proved is that any solution has exactly n zeroes on $[0, \pi[$. Let y_2 be a solution of (B.4′) with p zeroes. We can suppose that y_2 is linearly independent of y_1, which is given by (B.19). We know that $y_1 y_2' - y_2 y_1'$ is a constant and therefore it has a definite sign ε. If a_i' is a zero of y_2, then $y_2'(a_i')$ has the same sign as $\varepsilon y_1(a_i')$. From Proposition B.11 $y_2'(a_i')$ has also the same sign as $\varepsilon'(-1)^i$. Hence, $y_1(a_1'), \ldots, y_1(a_p')$ is an alternating sequence. Since we know that y_1 changes sign only n times on $[0, \pi[$, we find that $p \leqslant n$. Exchanging the roles of y_1 and y_2, we get that $n \leqslant p$, so that $n = p$. □

B.20. The same kind of theorem in higher dimensions would be very helpful for understanding the differential equation aspect of Blaschke's conjecture. Up to now the only explicitly known examples of vector-valued Sturm-Liouville equations with only periodic solutions are those obtained by superimposing one-dimensional examples in linearly independent directions (even the equations arising from Zoll metrics on S^d are of this type, see 4.39 and what follows it).

IV. Back to Geometry with Some Examples and Remarks

B.21. Among the cases that Theorem B.16 covers, only two are geometrically realizable (think of the Betti numbers of the loop space of surfaces). The case n equals 1 for antiperiodic solutions occurs for a C_π-metric on $\mathbb{R}P^2$; the case n equals 2 occurs for periodic solutions for a C_π-metric on S^2. We saw in 5.57 that considering a Blaschke metric on S^2 is the same as considering a C_π-metric on $\mathbb{R}P^2$. We develop the first one of the above cases in more detail (it is also the simplest case) since the formulas are more reasonable.

B.22 Corollary. *The equation* (B.4′) *has only π-antiperiodic solutions with one zero on* $[0, \pi[$ *if and only if there exists a C^∞ π-periodic function f with $f(a)=f'(a)=0$ for some real number a in $[0, \pi[$ such that*

B.23 $$r(s)=1-f''(s)-(f'(s))^2-2f'(s)\cot(s-a)$$

and

B.24 $$\int_0^\pi (e^{-2f(s)}-1)/\sin^2(s-a)\ ds=0 .$$

Moreover, one of the solutions y_1 of (B.4′) *is given by*

$$y_1(s)=e^{f(s)}\sin(s-a).$$

B.25. From (B.23) one can derive an integral inequality which can be used to begin another proof of L. Green's theorem (see 5.64 and [NN] p. 588). Indeed, if we integrate r over a period, we get

$$\int_0^\pi r(s)ds=\pi-\int_0^\pi (f'(s))^2ds-\int_0^\pi 2f'(s)\cot(s-a)\,ds .$$

Therefore, after integrating by parts (the only difficulty is at $s=a$) and using the π-periodicity of f and cot, we obtain

$$\int_0^\pi r(s)ds\leqslant\pi-2\int_0^\pi f(s)/\sin^2(s-a)\,ds .$$

If we now notice that for any real number t, $e^t-1\geqslant t$, we get

$$-\int_0^\pi 2f(s)/\sin^2(s-a)\,ds\leqslant\int_0^\pi (e^{-2f(s)}-1)/\sin^2(s-a)\,ds .$$

The right-hand side is known to be zero by (B.24). Hence,

$$\int_0^\pi r(s)ds\leqslant\pi .$$

Moreover, equality occurs if and only if $f(s)=0$.

To prove L. Green's theorem (see 5.59) one has to integrate the preceding result on all geodesics, as in 5.64.

Appendix C. Examples of Pointed Blaschke Manifolds

By Lionel Bérard Bergery

I. Introduction

C.1. In this appendix, we first prove A. Weinstein's result quoted in Chapter 5 (Proposition 5.52): every manifold M which is the union of the standard disk and of a disk bundle, glued together by a diffeomorphism of their boundaries admits a Riemannian metric g such that (M, g) is a Blaschke manifold at m_0 (the center of the disk). Our proof will use the theory of Riemannian submersions. Then we examine under which conditions this construction gives SC_l^m-manifolds (see 7.7 for definitions). We also give an example of an exotic sphere (that is a manifold homeomorphic but not diffeomorphic to the standard sphere) which satisfies the previous condition.

II. A. Weinstein's Construction

C.2. We begin by fixing some notation. Let ξ $(\pi: E \to N)$ be a differentiable p-dimensional vector bundle over a differentiable n-dimensional manifold N. Let us *denote* by $D(\xi)$, $S(\xi)$ and $P(\xi)$ the total spaces of the disk bundle, the sphere bundle and the projective bundle canonically associated with ξ (with fibers D^p, S^{p-1} and $\mathbb{R}P^{p-1}$). *Let* $\tau(\xi)$ be the fixed-point free involution on $S(\xi)$ which is the antipodal map $(x \mapsto -x)$ in each fiber. Let us also suppose that

C.3. *There exists a diffeomorphism σ from $S(\xi)$ onto the canonical sphere S^{n+p-1}.*

Then we consider the d-dimensional $(d=n+p)$ differentiable manifold $M = D^d \cup_\sigma D(\xi)$ which is the union of the standard disk D^d and the disk bundle $D(\xi)$ glued together along their boundaries S^{d-1} and $S(\xi)$ by the diffeomorphism σ.

C.4 Theorem (A. Weinstein [WR 2]). *There exists a metric g on M such that (M, g) is a Blaschke manifold at m_0 the center of the disk D^d.*

Proof. For each dimension d, let us choose once and for all a Riemannian metric g_d on the standard disk $D^d = \{x | x \in \mathbb{R}^d, \|x\| \leqslant 1\}$, which is $O(d)$-invariant under the usual action of the orthogonal group $O(d)$ with m_0 fixed, and which is product near the boundary S^{d-1} of D^d. Notice that all the geodesics issuing from the center m_0 of D^d

have the same length, *say* a_d, and that the unique geodesic from m to a point p on the boundary S^{d-1} has a unit tangent vector at m_0 which is opposite to that of the geodesic from m_0 to the point $\tau(p)$ (where τ is the canonical antipodal map on S^{d-1}, see Fig. C.4 below).

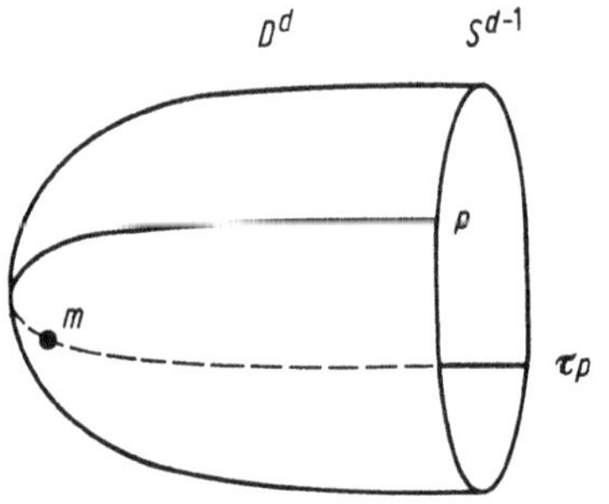

Fig. C.4

C.5. Let h be any Riemannian metric on N. By a construction due to J. Vilms ([VS]), there exists on $D(\xi)$ a Riemannian metric h' such that the fibration $(D(\xi), h') \to (N, h)$ is a Riemannian submersion with totally geodesic fibers isometric to (D^p, g_p). In the case at hand, we can describe this construction in the following way: let us choose some Euclidean metric and a *metric* linear connection on E. This connection gives a distribution on the tangent bundle TE to E, complementary to the vertical distribution (kernel of $T\pi$), and the associated parallel transport is by linear isometries. Then one constructs h' on $D(\xi)$ in the following way: one takes the vertical and horizontal distributions to be orthogonal, one lifts h from N to the horizontal distribution and one then takes the metric g_p on each fiber D^p. Parallel transport is still an isometry for these metrics on the fibers, so that the fibers are totally geodesic.

C.6. Notice that the metric h' is still product near the boundary $S(\xi)$ of $D(\xi)$, as it is for S^{p-1} in D^p. Moreover, any geodesic issuing from a point p of $S(\xi)$ orthogonal to $S(\xi)$ in $D(\xi)$ stays in the *same* fiber of the fibration and goes to the point $\tau(\xi)p$, with total length $2a_p$. In particular, all of the geodesics orthogonal to the boundary $S(\xi)$ in $D(\xi)$ have the same length. Let g_1'' be the Riemannian metric on $S(\xi)$ induced by h'.

C.7. Let us now choose a family g_t' of Riemannian metrics on S^{d-1} which are C^∞ in the parameter t for t in $[0, 1]$, constant near 0 and near 1, and such that g_0' is the restriction of g_d to S^{d-1} and g_1' is the inverse image of g_1'' by the diffeomorphism σ^{-1} from S^{d-1} to $S(\xi)$. Then we construct the metric g' on $S^{d-1} \times [0, 1]$ in such a way that the direct sum decomposition given by the projections onto the two factors is orthogonal; the metric induced on each $S^{d-1} \times \{t\}$ is g_t'; the metric induced on each $\{p\} \times [0, 1]$ is the canonical metric with length 1 on $[0, 1]$. Notice first that g' is product near the boundary $(S^{d-1} \times \{0\}) \cup (S^{d-1} \times \{1\})$.

C.8. Moreover, the curves $\gamma_p : t \mapsto (p, t)$ are geodesics in $S^{d-1} \times [0, 1]$. One way to see this fact is to notice that the projection $S^{d-1} \times [0, 1] \to [0, 1]$ is a Riemannian submersion and that the curves γ_p are the horizontal lifts of the geodesic of $[0, 1]$.

They are geodesics by [HM] (Proposition 3.1, p. 237). One can also show this directly. Let X be the vector field on $S^{d-1}\times[0,1]$ whose value at (p,t) is $\dot{\gamma}_p(t)$. For each tangent vector η to $S^{d-1}\times[0,1]$ in (p,t), there exist a vector field Y on $S^{d-1}\times[0,1]$ and a real number λ such that the value of $\lambda X+Y$ at (p,t) is η, $g'(X,Y)=0$ and Y is independent of t (that is, $[X,Y]=0$). Now $g'(X,X)=1$, so that if D is the Levi-Civita connection of g', we have

C.9
$$2g'(D_X X, \lambda X+Y)=\lambda X g'(X,X)+2Xg'(X,Y)$$
$$-Yg'(X,X)-2g'(X,[X,Y])=0,$$

whence $D_X X=0$.

C.10. We now consider the manifold M as the union of D^d, $S^{d-1}\times[0,1]$ and $D(\xi)$ glued together by the identity from S^{d-1} (boundary of D^d) onto $S^{d-1}\times\{0\}$ (a component of the boundary of $S^{d-1}\times[0,1]$) and the diffeomorphism σ from $S(\xi)$ (boundary of $D(\xi)$) onto $S^{d-1}\times\{1\}$ (the other component of the boundary of $S^{d-1}\times[0,1]$). The Riemannian metrics g_d on D^d, g' on $S^{d-1}\times[0,1]$ and h' on $D(\xi)$ are product near the corresponding boundaries. Since σ and the identity are isometries for the induced metrics on the corresponding parts of the boundaries, the metrics g_d, g' and h' may be glued together giving a Riemannian metric g on M. We also see immediately from this construction that every geodesic (for the metric g) from the center m_0 of D^d is a geodesic loop with constant length $2a_d+2+2a_p$ (see Figure C.10 below). Moreover, the cut-locus of m_0 in M is exactly the submanifold N imbedded in $D(\xi)$ as the zero section of the fibration, and the distance from m_0 to its cut-locus is a_d+1+a_p along any geodesic. Hence, (M,g) has spherical cut-locus at m_0 and by 5.43 (M,g) is a Blaschke manifold at m_0. □

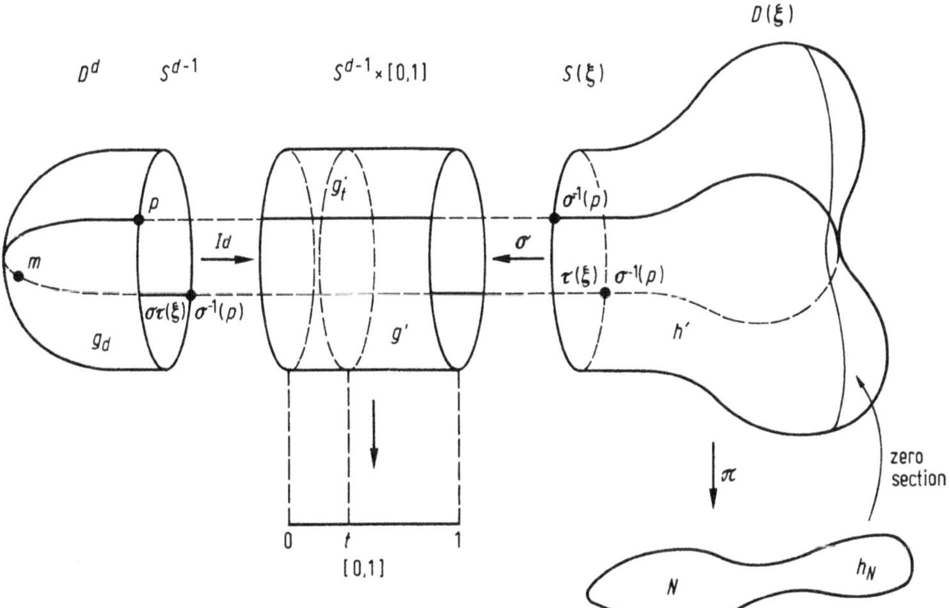

Fig. C.10

III. Some Applications

C.11. Let us now examine how the geodesics from m_0 come back to m_0. Let p be a point in S^{d-1}. Let γ_p be the unique geodesic in (D^d, g_d) from m_0 to p. Then γ_p is part of a geodesic γ issuing from m_0 in M. Let us follow the way γ goes step by step: the first part of γ is γ_p, which ends at p; the second part of γ is the curve $t \mapsto (p, t)$ in $S^{d-1} \times [0, 1]$, which ends at the point $(p, 1)$ in $S^{d-1} \times \{1\}$ (which has been identified with the point $\sigma^{-1}(p)$ in $S(\xi)$); the third part of γ is the unique geodesic of $(D(\xi), h')$ issuing orthogonally to $S(\xi)$ from $\sigma^{-1}(p)$ and it ends at the point $\tau(\xi)\circ\sigma^{-1}(p)$ of $S(\xi)$ (which has been identified with the point $(\sigma\circ\tau(\xi)\circ\sigma^{-1}(p), 1)$ of $S^{d-1} \times \{1\}$); the fourth part of γ is the curve $t \mapsto (\sigma\circ\tau(\xi)\circ\sigma^{-1}(p), 1-t)$ in $S^{d-1} \times [0, 1]$, which ends at the point $\sigma\circ\tau(\xi)\circ\sigma^{-1}(p)$ of S^{d-1}; the fifth part of γ is the unique geodesic in (D^d, g_d) issuing from $\sigma\circ\tau(\xi)\circ\sigma^{-1}(p)$ orthogonally to S^{d-1}. It crosses m_0 and ends at the point $\tau\circ\sigma\circ\tau(\xi)\circ\sigma^{-1}(p)$ of S^{d-1}. So we immediately see that the geodesic γ is *simply closed* if and only if $\tau\circ\sigma\circ\tau(\xi)\circ\sigma^{-1}(p) = p$.

We have proved the following consequence of the construction.

C.12 Corollary ([B.B 1]). *Let ξ satisfy the following assumption:*

C.13. *There exists a diffeomorphism σ from $S(\xi)$ onto S^{d-1} such that $\sigma\circ\tau(\xi) = \tau\circ\sigma$. Then $M = D^d \cup_\sigma D(\xi)$ admits a Riemannian metric g such that (M, g) is an $SC_l^{m_0}$-manifold (where m_0 is the center of D^d).* □

C.14 Remark. Another way of expressing the assumption C.13 is to say that there exists a diffeomorphism from $P(\xi)$ onto $\mathbb{R}P^{d-1}$. (Notice that $P(\xi)$ is the quotient of $S(\xi)$ under $\tau(\xi)$).

C.15. We now note that not all manifolds $M = D^d \cup_\sigma D(\xi)$ can be represented with a ξ satisfying C.13. The problem of determining such M precisely is a difficult one. It involves the classification of differentiable involutions of the canonical sphere (see [B.B 1] for a discussion). Here we shall only give a non-trivial example. Notice first that it is known from differential topology that (for $d \neq 3, 4$) any homotopy sphere (that is, a compact manifold which has the homotype type of the standard sphere) is the union of two disks glued together by an orientation-preserving diffeomorphism of S^{d-1} (see [SE]). That is, they can be written as $D^d \cup_\sigma D(\xi)$, where ξ is a vector space of dimension d considered as a vector bundle over a point. Also, the assumption C.13 is satisfied if and only if $\sigma\circ\tau = \tau\circ\sigma$.

C.16. Let Θ_d be the set of orientation-preserving diffeomorphism classes of d-dimensional oriented homotopy spheres. The set Θ_d is a finite abelian group under the connected sum ([K-M]). Let $2\Theta_d$ be the subgroup of elements which are two times an element in Θ_d. Notice that if η is a diffeomorphism of S^{d-1}, then $M(\eta\circ\eta) = D^d \cup_{\eta\circ\eta} D^d$ is the double in Θ_d of $M(\eta) = D^d \cup_\eta D^d$. We then have the following beautiful

C.17 Proposition (F. Laudenbach [B.B 1]). *If d is even ($d \neq 4$), then any element of $2\Theta_d$ may be represented by some $M(\sigma) = D^d \cup_\sigma D^d$ with $\tau\circ\sigma = \sigma\circ\tau$.*

Proof. It is known ([K-M]) that, if η is any orientation-preserving diffeomorphism of S^{d-1} and if ζ is the restriction to S^{d-1} of an orientation-preserving diffeomorphism of D^d, then $M(\eta)$ is diffeomorphic to $M(\eta\circ\zeta)$. One may choose ζ such that $\eta'=\eta\circ\zeta$ has its support (i.e., the set of points where $\eta'(p)\neq p$) in one of the two hemispheres of S^{d-1}. Then $\tau\circ\eta'\circ\tau$ has its support in the other hemisphere, so that η' and $\tau\circ\eta'\circ\tau$ commute. One now considers $\sigma=\tau\circ\eta'\circ\tau\circ\eta'$. Since d is even, τ is orientation-preserving and it is the restriction of an orientation-preserving diffeomorphism of D^d. Hence, $M(\sigma)$ is the double of $M(\eta)$ in Θ_d, and one has

C.18 $$\tau\circ\sigma=\tau\circ\tau\circ\eta'\circ\tau\circ\eta'=\tau\circ\eta'\circ\tau\circ\eta'\circ\tau=\sigma\circ\tau\,. \quad \square$$

C.19. The groups Θ_d have been computed by Kervaire and Milnor (see [K-M]). We extract from their work the example $\Theta_{10}=\mathbb{Z}_6$. Then there exists at least one exotic sphere of dimension 10 (corresponding to the elements 2 or 4 in $\mathbb{Z}_6$ according to which orientation has been chosen) which admits an SC_l^m-metric.

Appendix D. Blaschke's Conjecture for Spheres

By Marcel Berger*

I. Results

In this appendix we prove the following results (cf. 5.37, 5.57, and 5.F):

D.1 Theorem. *Let* (S^d, g) *be a Blaschke Riemannian structure on the d-dimensional sphere. Then* (S^d, g) *is (up to some constant) isometric to the standard sphere* (S^d, can).

D.2 Corollary. *Let* $(\mathbb{R}P^d, g)$ *be a Blaschke Riemannian structure on d-dimensional real projective space. Then* $(\mathbb{R}P^d, g)$ *is (up to some constant) isometric to the standard* $(\mathbb{R}P^d, \mathrm{can})$.

This follows directly from D.1 and 5.57. □

D.3 Corollary. *If* (S^d, g) *(resp.* $(\mathbb{R}P^d, g)$*) is globally harmonic (see* 6.10*), then it is isometric (up to some constant) to* (S^d, can) *(resp.* $(\mathbb{R}P^d, \mathrm{can})$*).*

This follows directly from D.1, D.2, and 6.82. □

Theorem D.1 is a direct consequence of Weinstein's Theorems 2.21, 2.24, C. T. C. Yang's result 2.24a) and of the following:

D.4 Theorem. *Let* (S^d, g) *be a Blaschke manifold with diameter equal to* π. *Then* $\mathrm{Vol}(g) \geqslant \beta(d)$, *the volume of the standard sphere* (S^d, can) *(cf.* (1.119)*). Moreover* $\mathrm{Vol}(g) = \beta(d)$ *only if* (S^d, g) *is isometric to* (S^d, can).

D.5 Remarks. The condition $\mathrm{Diam}(g) = \pi$ can always be achieved by a trivial normalization.

Theorem D.1 and Corollary D.2 solve Blaschke's conjecture (cf. 5.F) when the underlying manifold is either the sphere or real projective space. It extends L. Green's Theorem 5.59.

The method of proof is basically that suggested in 5.78; it involves the injectivity radius in disguise (see D.23) and has some connection with appendix B. The idea of the proof is the following. For a Sturm-Liouville system (SL) *denote* by A_t its resolvent endomorphism with initial conditions $A_t(t) = 0$ and $A_t'(t) = \mathrm{Id}$. If all

---

* Work partially done while at the IMPA of Rio de Janeiro and at the University of Warwick.

solutions of (SL) are π-antiperiodic then the following is true: the double-mean value

$$\int_{t=0}^{t=\pi}\left(\int_{s=t}^{s=t+\pi} \operatorname{Det}(A_t(s))ds\right)dt$$

is always bigger than or equal to the value it takes when (SL) is that of a standard sphere.

The above inequality rests eventually on an integral inequality for numerical functions whose proof was found by Jerry L. Kazdan and is given in Appendix E. We are happy to thank here Kazdan for having communicated this proof to us.

II. Some Lemmas

D.6. *In Sections II and III (with the exception of D.23, D.24 at the very end)* $(M,g)=(S^d,g)$ *will always be a Blaschke manifold with diameter equal to* π. *Hence* $\operatorname{Diam}(g)=\operatorname{Inj}(g)=\pi$ by 5.42.

We will *denote* by $\sigma: M\to M$ the antipodal involution of M; the map σ is an isometry of (M,g), see 5.57.

D.7. Let us *introduce the function*

$$f: UM\times\mathbb{R}_+\to\mathbb{R}$$

defined for u in UM and t in $\mathbb{R}_+$ by

$$f(u,t)=t^{d-1}\theta(tu),$$

where θ is the function on UM introduced in 6.3. In fact f is nothing but the volume element of (M,g) in polar coordinates. More precisely, let m be some fixed point in M and σ be the canonical measure on the unit sphere U_mM (cf. 1.M). Then $d\mu=f(u,t)dt\otimes d\sigma$ for the canonical measure μ of (M,g), cf. 1.M.

In particular, since $\operatorname{Diam}(g)=\operatorname{Inj}(g)=\pi$, we have

D.8 $$\operatorname{Vol}(g)=\operatorname{Vol}(B(m,\pi))=\int_{u\in U_mM}\left(\int_{x=0}^{x=\pi} f(u,x)dx\right)d\sigma$$

for every m in M, where $\operatorname{Vol}(g)$ is defined in 1.118.

We recall now [cf. (6.4)] that one can compute f by means of Jacobi fields as follows: let $\gamma: x\mapsto\exp(xu)$ be the geodesic with initial velocity vector u, let $\{u_i\}$ $i=2,\dots,d$ be an orthonormal basis of $u^\perp$ and let Y_i $(i=2,\dots,d)$ be the Jacobi field along γ with initial conditions $Y_i(0)=0$, $Y_i'(0)=u_i$. Then, for every t,

D.9 $$f(u,t)=|\operatorname{Det}(Y_2(t),\dots,Y_d(t))|.$$

The function f satisfies the two following lemmas:

D.10 Lemma. *For any u in UM and any s in $\mathbb{R}_+$ one has $f(-u, s) = f(u, \pi - s)$.*

Proof. We remark first that the proof of 1.91 implies that (with the notations of 1.87) any Jacobi field Y along γ such that $Y(0)=0$ can be written

$$Y = T_{(s,t)}\Gamma\left(\frac{\partial}{\partial t}\right)$$

with $\Gamma_0 = \gamma$ and $\Gamma(0, t) \in \gamma(0) = m$ for every t. The antipodal map σ yields $\Gamma(s+\pi, t) = \sigma(\Gamma(s, t))$ for every s and t, hence $Y(s+\pi) = (T\sigma)(Y(s))$ for every s. Differentiating the last equality at $s=0$ yields

D.11 $$Y'(\pi) = (T\sigma)(Y'(0)).$$

Of course

D.12 $$Y(\pi) = 0.$$

Consider now the Jacobi field Y_i $(i = 2, \ldots, d)$ figuring in (D.9). Along the geodesic $\tilde{\gamma}: s \mapsto \exp_{m'}(-s\dot{\gamma}(\pi))$, the vector field $\tilde{Y}_i: s \mapsto Y_i(\pi - s)$ is a Jacobi field with initial conditions $\tilde{Y}_i(0) = Y_i(\pi) = 0$ by (D.12) and by (D.11):

$$\tilde{Y}_i'(0) = -Y_i'(\pi) = -(T\sigma)(Y'(0)) = -(T\sigma)(u_i).$$

Since σ is an isometry, $\{-(T\sigma)(u_i)\}_{i=2,\ldots,d}$ is an orthonormal basis of $(-\dot{\gamma}(\pi))^{\perp}$; we then can apply (D.9) again to get $f(-\dot{\gamma}(\pi), s) = f(u, \pi - s)$.

Now, since f is a Riemannian invariant and σ an isometry such that $(T\sigma)(-\dot{\gamma}(\pi)) = -u$, we have

$$f(-\dot{\gamma}(\pi), s) = f(-u, s) = f(u, \pi - s). \quad \square$$

We recall (cf. 1.46) that ζ *denotes the geodesic flow on* UM. *We set*

D.13 $$\varphi = f^{1/(d-1)}.$$

D.14 Lemma. *For u in UM, x in $[0, \pi]$ and z in $[0, \pi - x]$, one has*

$$\varphi(\zeta^x(u), z) \geqslant \varphi(u, x)\varphi(u, x+z) \int_x^{x+z} \frac{dr}{\varphi^2(u, r)}.$$

Proof. By (1.89) Jacobi fields along γ are solutions of the Sturm-Liouville system

D.15 $$J'' + R(J) = 0,$$

where R is a function $t \mapsto R(t)$ with values in the space of endomorphisms of $u^{\perp}$ equal to $R(t): v \mapsto R(\dot{\gamma}(t), v)\dot{\gamma}(t)$.

Looking at $f(u, t) = |\mathrm{Det}(Y_2(t), \ldots, Y_d(t))|$, we are led to introduce the resolvant equation of (D.15), namely

D.16 $$Z'' + R \circ Z = 0$$

where Z takes values $Z(t)$ in the space of endomorphisms of $u^\perp$. We see that $f(u, t) = \mathrm{Det}((A(t))$ where A is the solution of (D.16) with initial conditions $A(0) = 0$, $A'(0) = \mathrm{Id}$. More generally

D.17 $$f(\zeta^x(u), z) = \mathrm{Det}(A_x(x+z))$$

where A_x is the solution of (D.16) such that $A_x(x) = 0$ and $A'_x(x) = \mathrm{Id}$.

We eventually pay no attention to the absolute value in determinants because working within the injectivity radius.

In the euclidean space $u^\perp$ let us *denote by* * the transpose of endomorphisms with respect to this euclidean structure. Then, because $R(t)$ is self-adjoint, the solution A_x of (D.16) can be expressed by the formula

D.18 $$A_x^*(y) = A(x) \left(\int_x^y A^{-1}(t) A^{-1*}(t) dt \right) A^*(y),$$

provided A is non singular on $[x, y]$, which is certainly the case, by assumption, if $y \in [x, \pi]$. The formula (D.18) figures, for example, in [GN 1], p. 31. Taking determinants we get

$$f(\zeta^x(u), z) = f(u, x) f(u, x+z) \,\mathrm{Det} \left(\int_x^{x+z} A^{-1}(t) A^{-1*}(t) dt \right).$$

The endomorphisms $A^{-1} A^{-1*}$ are always self-adjoint and positive-definite. We use the elementary fact (see for example [BR 10], 11.8.9.5) that the function $\xi : X \mapsto (\mathrm{Det}(X))^{-1/2}$ is strictly convex on the space of self-adjoint endomorphisms. Applying the classical inequality of convexity (see for example [H-L-P], p. 150—151, No. 204):

$$\xi \left(\frac{\int_a^b \lambda(t) p(t) dt}{\int_a^b p(t) dt} \right) \leqslant \frac{\int_a^b \xi(\lambda(t)) p(t) dt}{\int_a^b p(t) dt}$$

with $\xi = (\mathrm{Det}(\,.\,))^{-1/2}$, for the measure

$$p(t) dt = f^{-2/(d-1)}(u, t) dt = \varphi^{-2}(u, t) dt$$

and the function $\lambda(t) = \varphi^2(u, t) A^{-1}(t) A^{-1*}(t)$, we get straightforwardly

$$\mathrm{Det} \left(\int_x^{x+z} A^{-1}(t) A^{-1*}(t) dt \right) \geqslant \left(\int_x^{x+z} \frac{dt}{\varphi^2(u, t)} \right)^{d-1}$$

which yields D.14. □

D.19 Lemma. *If μ_1 denotes the canonical measure on UM (see 1.121) one has*

$$\pi \operatorname{Vol}^2(g) = 2 \int_{UM} \left(\int_{x=0}^{x=\pi} \left(\int_{z=0}^{z=\pi-x} f(\zeta^x(u), z) dz \right) dx \right) d\mu_1 .$$

Proof. We integrate (D.8) with m running through M, to get (using 1.123)

$$\begin{aligned}\operatorname{Vol}^2(g) &= \int_{m\in M} \left(\int_{u\in U_mM} \left(\int_{z=0}^{z=\pi} f(u, z) dz \right) d\sigma \right) dm \\ &= \int_{UM} \left(\int_{z=0}^{z=\pi} f(u, z) dz \right) d\mu_1 .\end{aligned}$$

We integrate this inequality with respect to x running through $[0, \pi]$ endowed with the Lebesgue measure dx. Using the invariance of μ_1 under the geodesic flow ζ (Liouville's Theorem, see 1.125) we get

$$\begin{aligned}\pi \operatorname{Vol}^2(g) &= \int_{x=0}^{x=\pi} \left(\int_{UM} \left(\int_{z=0}^{z=\pi} f(u, z) dz \right) d\mu_1 \right) dx \\ &= \int_{UM} \left(\int_{x=0}^{x=\pi} \left(\int_{z=0}^{z=\pi} f(\zeta^x(u), z) dz \right) dx \right) d\mu_1 .\end{aligned}$$

Now the idea is to split the above integral into two parts, by means of:

$$\int_{z=0}^{z=\pi} = \int_{z=0}^{z=\pi-x} + \int_{z=\pi-x}^{z=\pi} .$$

Because $u \mapsto -\zeta^\pi(u)$ is a diffeomorphism of UM which preserves μ_1 (Liouville's Theorem again) we have, using also D.10:

$$\begin{aligned}&\int_{UM} \left(\int_{x=0}^{x=\pi} \left(\int_{z=0}^{z=\pi-x} f(\zeta^x(u), z) dz \right) dx \right) d\mu_1 \\ &= \int_{UM} \left(\int_{x=0}^{x=\pi} \left(\int_{z=0}^{z=\pi-x} f(\zeta^x(-\zeta^\pi(u)), z) dz \right) dx \right) d\mu_1 \\ &= \int_{UM} \left(\int_{x=0}^{x=\pi} \left(\int_{z=0}^{z=\pi-x} f(-\zeta^{\pi-x}(u), z) dz \right) dx \right) d\mu_1 \\ &= \int_{UM} \left(\int_{x=0}^{x=\pi} \left(\int_{z=0}^{z=\pi-x} f(\zeta^{\pi-x}(u), \pi - z) dz \right) dx \right) d\mu_1 .\end{aligned}$$

Changing (x, z) into $(\pi - x, \pi - z)$ in this last integral yields for it the value

$$\int_{UM} \left(\int_{x=0}^{x=\pi} \left(\int_{z=\pi-x}^{z=\pi} f(\zeta^x(u), z) dz \right) dx \right) d\mu_1$$

which proves the lemma. The geometric significance is the equality of the two shaded areas in Figure D.19. □

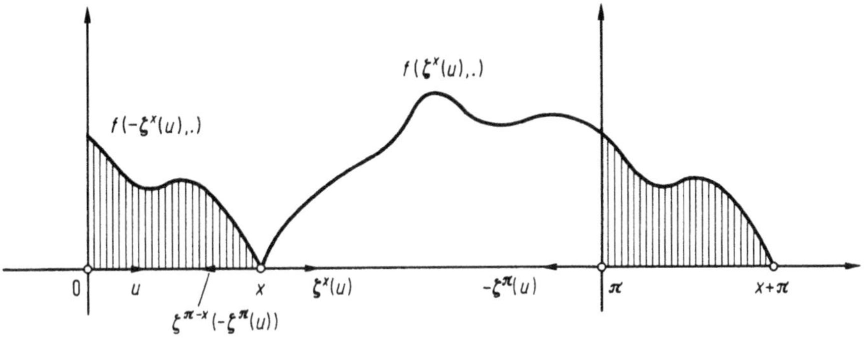

Fig. D.19

III. Proof of Theorem D.4

We set $y=x+z$ and apply in succession the Hölder Inequality and Lemma D.14 to get

D.20
$$\begin{aligned}
I(u) &= \int_{x=0}^{x=\pi}\left(\int_{z=0}^{z=\pi-x} f(\zeta^x(u), z)dz\right)dx \\
&= \int_{x=0}^{x=\pi}\left(\int_{y=x}^{y=\pi} f(\zeta^x(u), y-x)dy\right)dx \\
&= \int_{x=0}^{x=\pi}\int_{y=x}^{y=\pi} \frac{\varphi^{d-1}(\zeta^x(u), y-x)}{\sin^{d-1}(y-x)}\sin^{d-1}(y-x)dydx \\
&\geqslant \frac{\left(\int_{x=0}^{x=\pi}\int_{y=x}^{y=\pi} \varphi(\zeta^x(x), y-x)\sin^{d-2}(y-x)dydx\right)^{d-1}}{\left(\int_{x=0}^{x=\pi}\int_{y=x}^{y=\pi}\sin^{d-1}(y-x)dydx\right)^{d-2}} \\
&\geqslant \frac{\left(\int_{x=0}^{x=\pi}\int_{y=x}^{y=\pi}\int_{t=x}^{t=y} \frac{\varphi(u, x)\varphi(u, y)}{\varphi^2(u, t)}\sin^{d-2}(y-x)dtdydx\right)^{d-1}}{\left(\int_{x=0}^{x=\pi}\int_{y=x}^{y=\pi}\sin^{d-1}(y-x)dydx\right)^{d-2}}.
\end{aligned}$$

We now apply the inequality due to J. Kazdan (Theorem E.2) in the particular case where the weight function $\varrho(y-x)$ is equal to $\sin^{d-2}(y-x)$ and get

$$I(u) \geqslant \int_{x=0}^{x=\pi}\int_{y=x}^{y=\pi}\sin^{d-1}(y-x)dydx = c(d),$$

valid for every u in UM.

By D.19 and (1.124) we have

$$\pi\,\mathrm{Vol}^2(g) \geqslant 2c(d)\,\mathrm{Vol}(UM) = 2c(d)\beta(d-1)\,\mathrm{Vol}(g)$$

with the result that

$$\operatorname{Vol}(g) \geqslant \frac{2c(d)\beta(d-1)}{\pi}.$$

But

D.21 $$\frac{2c(d)\beta(d-1)}{\pi} = \beta(d),$$

which proves the inequality of D.4. To obtain (D.21), one could proceed by explicit computation of $c(d)$. Alternatively suppose that (M, g) is the standard sphere (S^d, can). Since the sectional curvature is constant and equal to 1, the endomorphisms $R(t)$ in (D.15), (D.16) are equal to the identity Id. This implies $A_x(y) = \sin(y-x)\,\text{Id}$, hence $f(\zeta^x(u), y-x) = \sin^{d-1}(y-x)$ for every x and y. In particular equality is attained in D.14 and in (D.20) for every u, which proves (D.21).

Suppose, conversely, that (M, g) is such that $\operatorname{Vol}(g) = \beta(d)$. This implies that we must have equality in D.14 and in E.2 for every u. The function $(\operatorname{Det}(\,.\,))^{-1/2}$ being strictly convex, the function $t \mapsto \lambda(t)$ has to be a constant map. Since we have $A(0)=0$ and $A'(0)=\text{Id}$ then the constant value of λ has to be the identity endomorphism. Hence $A^*(t)A(t) = \varphi^2(u, t)\text{Id}$ for every t and u. Now the equality has to be attained in E.2 for the function φ. That function φ belongs to the set T (see the proof E.2) because $\varphi(0) = \varphi(\pi) = 0$ and $\varphi'(0) = \varphi'(\pi) = 1$. So for every t and u:

D.22 $$A^*(t)A(t) = \sin^2 t \,\text{Id}.$$

Differentiating (D.22) four times at $t=0$ and using (D.16), we see that $R = R(u, \cdot)u = \text{Id}$ for every u, which says exactly that (M, g) has constant sectional curvature equal to 1. □

D.23 Remarks. Lemma D.14 is clearly valid for every Riemannian manifold (M, g) whose injectivity is greater than or equal to π. But for such manifolds Lemma D.10 no longer holds in general and in Lemma D.19 we can get only

$$\pi \operatorname{Vol}^2(g) > \int_{UM} \left(\int_{x=0}^{x=\pi} \left(\int_{z=0}^{z=\pi-x} f(\zeta^x(u), z)dz \right) dx \right) d\mu_1,$$

missing a factor 2. The rest of the proof applies and we finally prove (cf. 5.78):

D.24 Proposition. *For every Riemannian manifold (M, g) of dimension d one always has the inequality*

$$\operatorname{Vol}(g) > \frac{1}{2}\frac{\beta(d)}{\pi^d} \operatorname{Inj}^d(g).$$

Appendix E. An Inequality Arising in Geometry

By Jerry L. Kazdan*

E.1. In the preceeding appendix, Marcel Berger has proved the Blaschke's conjecture for spheres. His proof relies on an inequality which we prove here. *Let S denote* the set of functions

$$S=\{\varphi\in C[0,\pi]:\varphi(x)=x^{\alpha}(\pi-x)^{\beta}\psi(x) \text{ for some } 0\leqslant\alpha,\beta<2 \text{ and some } \psi\in C[0,\pi] \text{ with } \psi>0 \text{ on } [0,\pi]\}.$$

The main interest is for $\varphi\in C^{\infty}$, in which case α and β are either 0 or 1, so φ has a zero of a most order 1 at 0 and π.

Let

$$F(\varphi)=\int_0^{\pi}\left[\int_x^{\pi}\left(\int_x^{y}\frac{\varphi(x)\varphi(y)}{\varphi^2(t)}\varrho(y-x)dt\right)dy\right]dx,$$

where $\varrho\in C[0,\pi]$ is a given nonnegative function. To avoid trivialities, assume ϱ is positive somewhere. One can check that this improper integral converges if φ is in S.

Here is the desired inequality:

E.2 Theorem. *If* $\varrho(\pi-t)=\varrho(t)$ *for all* $0\leqslant t\leqslant\pi$, *then* $F(\varphi)\geqslant F(\sin)$.

E.3 Remark. It is probably true that equality holds only if $\varphi(x)=c\sin x$, where $c>0$ is any constant. The plausibility of this is clear from our proof below, see the very end of this Appendix.

Proof. Let $T\subset S$ *denote* those functions where $\alpha=\beta=1$. We shall prove the theorem for φ in T–and in this case our proof shows that equality holds only for $\varphi=c\sin$. Assuming for the moment that this case is done, the general case is treated by the following limiting argument. If $\varphi\in S$ and $\varepsilon>0$, *let*

$$\varphi_{\varepsilon}(x)=\begin{cases}\varphi(\varepsilon)\,x/\varepsilon & 0\leqslant x\leqslant\varepsilon\\ \varphi(x) & \varepsilon\leqslant x\leqslant\pi-\varepsilon\\ \varphi(\pi-\varepsilon)\,(\pi-x)/\varepsilon & \pi-\varepsilon\leqslant x\leqslant\pi\,.\end{cases}$$

Thus $\varphi_{\varepsilon}\in T$. It is tedious, but straightforward, to check that

$$\lim_{\varepsilon\to 0}F(\varphi_{\varepsilon})=F(\varphi),$$

* Supported in part by N. S. F. Grant MPS 74-06319-A03

which will complete the proof of E.2 for all φ in S. This limiting argument obscures the case of equality in E.2.

To prove E.2 for φ in T, write φ as

$$\varphi(x) = \sin x \exp u(x)$$

for some $u \in C[0, \pi]$. Then the functional $F(\varphi)$ becomes

E.4 $$J(u) = \int_{\Omega} \exp[u(x) + u(y) - 2u(t)] d\mu(x, y, t),$$

where $\Omega \subset \mathbb{R}^3$ is the region of integration in $F(\varphi)$ and μ is the measure

$$d\mu(x, y, t) = \frac{\sin x \sin y}{\sin^2 t} \varrho(y - x) dt\, dy\, dx.$$

Inequality E.2 then reads

E.5 $$J(u) \geqslant J(0).$$

Because exp is a convex function, Jensen's inequality (see [H-L-P], p. 150—151, No. 204) applied to (E.4) yields

E.6 $$J(u) \geqslant \mu(\Omega) \exp[(\mu(\Omega))^{-1} K(u)],$$

where K is the functional

E.7 $$K(u) = \int_{\Omega} [u(x) + u(y) - 2u(t)] d\mu(x, y, t).$$

Equality holds in (E.6) only if u is constant.

Clearly $\mu(\Omega) = J(0)$, so that (E.6) is precisely the desired inequality if we can prove that $K(u) = 0$ for all $u \in C[0, \pi]$, an identity that is certainly not obvious. In fact we will prove a stronger statement.

E.8 Lemma. $K(u) = 0$ *for all* $u \in C[0, \pi]$ *if and only if* $\varrho(\pi - t) = \varrho(t)$ *for all* $0 \leqslant t \leqslant \pi$.

Proof. We compute the three integrals in (E.7). Since

$$\int_x^y \sin^{-2} t\, dt = \frac{\sin(y - x)}{\sin x \sin y},$$

we have

E.9 $$\int_{\Omega} u(x) d\mu(x, y, t) = \int_0^{\pi} u(x) \int_x^{\pi} \sin(y - x) \varrho(y - x) dy\, dx$$

and

E.10 $$\int_{\Omega} u(y) d\mu(x, y, t) = \int_0^{\pi} u(y) \int_0^{y} \sin(y - x) \varrho(y - x) dx\, dy.$$

Let $y-x=\pi-s$ in (E.9) and, after interchanging the roles of x and y in (E.10), let $x-y=s$ there. This gives

E.11 $$\int_{\Omega}[u(x)+u(y)]d\mu(x,y,t)=\int_0^{\pi}u(x)a(x)dx\,,$$

where

$$a(x)=\int_0^x \sin s\varrho(s)ds+\int_x^{\pi}\sin s\varrho(\pi-s)ds\,.$$

For the last term in (E.7), reverse the order of integration to obtain

E.12 $$\int_{\Omega}u(t)d\mu(x,y,t)=\int_0^{\pi}u(t)b(t)\sin^{-2}t\,dt\,,$$

where

E.13 $$b(t)=\int_0^t \sin x\int_t^{\pi}\sin y\,\varrho(y-x)dydx\,.$$

Combining (E.11) and (E.12) we find that

$$K(u)=\int_0^{\pi}u(t)f(t)\sin^{-2}t\,dt\,,$$

where

E.14 $$f(t)=\sin^2 t\,a(t)-2b(t)\,.$$

To show that $K(u)=0$ for all u, we must show that $f(t)=0$. But $f(0)=0$, so we just show that the derivative is zero, $f'(t)=0$.

Now from (E.13)

$$b'(t)=\sin t\left[\int_t^{\pi}\sin y\,\varrho(y-t)dy-\int_0^t \sin x\,\varrho(t-x)dx\right].$$

Let $y-t=\pi-s$ in the first integral and $t-x=s$ in the second integral to obtain

E.15 $$b'(t)=\sin t\left[\int_t^{\pi}\sin(s-t)\varrho(\pi-s)ds+\int_0^t \sin(s-t)\varrho(s)ds\right].$$

Using (E.14), (E.15), and the formula for $a(t)$, we find–after appropriate algebra–that

E.16 $$f'(t)=\sin^2 t\left[\sin t\,\delta(t)+2\int_0^t \cos s\,\delta(s)\,ds-2c\right],$$

where

$$\delta(s)=\varrho(s)-\varrho(\pi-s),$$

and

$$c=\int_0^\pi \cos\theta\,\varrho(\theta)d\theta=\int_0^{\pi/2}\cos\theta\,\delta(\theta)d\theta.$$

It is now evident from (E.16) that if $\delta(t)=0$, then $f'(t)=0$. Thus $\varrho(t)=\varrho(\pi-t)$ is a sufficient condition that $K(u)=0$ for all functions u.

To prove that $\delta(t)=0$ is also a necessary condition for $f'(t)=0$, *let* $g(t)$ *denote* the expression in brackets in (E.16). Then $g(t)=0$ is equivalent to $f'(t)=0$. However,

$$g'(t)=\sin t\,\delta'(t)+3\cos t\,\delta(t)=\sin^{-2}t\frac{d}{dt}[\sin^3 t,\delta(t)].$$

Hence $g'(t)=0$ implies that

$$\delta(t)=\text{const}\sin^{-3}t$$

But $\delta(\pi/2)=0$. Consequently $\delta(t)=0$. In other words, the condition $\varrho(t)=\varrho(\pi-t)$ is also necessary for $K(u)=0$ for all u. □

Added in proof: As suspected, equality in E.2 holds only if $\phi(x)=c\sin x$ for any constant $c>0$. To see this, write any ϕ in S (not just in T) as $\phi(x)=\sin x\exp u(x)$, where $u\varepsilon\ (0,\pi)$ but u may now possibly blow up at the endpoints. In any case, the improper integral (E.4) still converges absolutely; thus the stated proof of ϕ in T remains valud for any ϕ in S.

Bibliography

[A-B] Artin, E., Braun, H.: Leçons de Topologie Algébrique. (Translated from the German.), Montréal: Presses de l'Université du Quebec 1973.

[AD] Adem, J.: Relations on iterated reduced powers. Proc. Natl. Acad. Sci. USA **39**, 636—638 (1953).

[AEL] Abel, N. H.: J. f. Math **I**, 311—339 (1826).

[AL] Ayel, M.: Le spectre des espaces symétriques de rang 1. Preprint, Université de Chambéry (1974).

[AM] Abraham, R.: Foundations of Mechanics. New York: Benjamin 1967.

[AN] Allamigeon, A.: Propriétés globales des espaces de Riemann harmoniques. Ann. Inst. Fourier **15**, 91—132 (1965)

[A-P-S] Ambrose, W., Palais, R. S., Singer, I. M.: Sprays. Acad. Brasil. Ciencas 32, 163—178 (1960).

[AR] Artin, E.: Algèbre Géométrique. Paris: Gauthier-Villars (1962). Traduction Française.

[AS] Adams, J. F.: On the non-existence of elements of Hopf invariant one. Ann. Maths. **72**, 20—44 (1960).

[AT] Albert, A. A. (Ed.): Studies in Mathematics, Vol. 2: Modern Algebra. Englewood Cliffs: Prentice Hall 1963. (The Mathematical Association of America).

[BA] Boruvka, O.: Lineare Differential-transformationen 2. Ordnung. Berlin: VEB Deutscher Verlag der Wissenschaften 1967.

[B.B 1] Bérard Bergery, L.: Quelques exemples de variétés Riemanniennes où toutes les géodésiques issues d'un point sont fermées et de même longueur, suivis de quelques résultats sur leur topologie. Ann. Inst. Fourier **27**, No. 1, 231—249 (1977).

[B.B 2] Bérard Bergery, L.: Les variétés Riemanniennes homogènes simplement connexes de dimension impaire à courbure strictement positive. J. Math. Pures Appl. **55**, 47—68 (1976).

[B.B-B-L] Bérard Bergery, L., Bourguignon, J. P., Lafontaine, J.: Déformations localement triviales des variétés Riemanniennes. Proc. Symp. Pure Math. **27**, Part 1, 3—32 (1975).

[B-C] Bishop, R. L., Crittenden, R. J.: Geometry of Manifolds. Pure and Applied Mathematics. New York, London: Academic Press 1964.

[BE] Blaschke, W.: Vorlesungen über Differentialgeometrie. 3rd Ed. Berlin: Springer 1930.

[BE-EB] Berger, M., Ebin, D.: Some decomposition of the space of symmetric tensors on a Riemannian manifold. J. Diff. Geom. **3**, 379—392 (1969).

[BE-GO] Berger, M., Gostiaux, B.: Géométrie Différentielle. Paris: Armand Colin 1972.

[BE-L] Berger, M., Lascoux, A.: Variétés Kählériennes compactes. Lecture Notes in Math., Vol. 154. Berlin-Heidelberg-New York: Springer 1970.

[B-G] Brown, R., Gray, A.: Riemannian Manifold with holonomy group Spin 9. Diff. Geometry in honor of K. Yano. Kinokuniya, Tokyo 1972, 41—59.

[B-G-M] Berger, M., Gauduchon, P., Mazet, E.: Le spectre d'une variété Riemannienne. Lecture Notes Vol. 194. Berlin-Heidelberg-New York: Springer 1971

[B-K] Bruck, R. H., Kleinfeld, E.: The structure of alternative division rings. Proc. Am. Math. Soc. **2**, 878—890 (1951).

[BL 1] Borel, A.: Some remarks about transformation groups on spheres and tori. Bull. Am. Math. Soc. **55**, 580—587 (1949).

[BL 2] Borel, A.: Les bouts des espaces homogènes de groupes de Lie. Ann. Math. **58**, 443—457 (1953).

[BL 3] Borel, A.: Le plan projectif des octaves et les sphères comme espaces homogènes. C. R. Acad. Sci. Paris, Serie A **230**, 1378—1380 (1950).

[BL 4] Borel, A.: Sur la cohomologie des espaces fibrés principaux et des espaces homogènes de groupes de Lie compacts. Ann. Math. **53**, 115—207 (1953).

[BM] Blumenthal, L. M.: Theory and Applications of Distance Geometry. Bronx-New York: Chelsea Publ. Co. 1970.

[BN] Buseman, H.: The Geometry of Geodesics. Pure and applied Mathematics, Vol. 6. New York: Academic Press Inc. 1955.

[BO] Bonan, E.: Sur les G-structures de type quaternonien. Cahiers Topol. Géomét. Diff. **9**, 389—463 (1967).

[BR 1] Berger, M.: Les variétés Riemanniennes homogènes normales simplement connexes à courbure strictement positive. Ann. Scuola Norm. Super. Pisa **15**, 179—246 (1961).

[BR 2] Berger, M.: Sur les variétés à courbure positive de diamètre minimum. Comment. Math. Helv. **35**, 28—34 (1961).

[BR 3] Berger, M.: Lectures on Geodesics in Riemannian Geometry. Bombay: Tata Institute of F. R. 1965.

[BR 4] Berger, M.: Sur les variétés d'Einstein compactes. C. R. IIIème Réunion Math. Expression Latine, Namur 35—55 (1965).

[BR 5] Berger, M.: Du côté de chez Pu. Ann. Sci. École Norm. Super. **5**, 1—44 (1972).

[BR 6] Berger, M.: Sur les premières valeurs propres des variétés Riemanniennes. Compositio Mathematica **26**, Fasc. 2, 129—149 (1973).

[BR 7] Berger, M.: Sur certaines variétés Kählériennes à géodésiques toutes fermées. J. Diff. Geometry **9**, 519—520 (1974).

[BR 8] Berger, M.: Some Relations between Volume, Injectivity Radius, and Convexity Radius in Riemannian Manifolds, pages 33—42 in Differential Geometry and Relativity, edited by Cahen & Flato, D. Reidel 1976.

[BR 9] Berger, M.: Volume et rayon d'injectivé dans les variétés riemanniennes de dimension 3. Osaka J. Math. **14** (1977).

[BR 10] Berger, M.: Géométrie (5 volumes). Paris: CEDIC-Nathan 1976.

[BR 11] Berger, M.: Volume et rayon d'injectivité dans les variétés riemanniennes. C. R. Acad. Sci. Paris, Serie A **284**, 1221—1224 (1977).

[BR-GB] Becker, J. C., Gottlieb, B. H.: Applications of the evaluation map and transfer map Theorems. Math. Ann. **211**, 277—288 (1974).

[B-S] Borel, A., de Siebenthal, J.: Les sous groupes fermés de rang maximum des groupes de Lie clos. Comment. Math. Helv. **23**, 200—221 (1949).

[BT] Bott, R.: On manifolds all of whose geodesics are closed. Ann. Math. **60**, No. 3, 375—382 (1954).

[BU] Buchner, M. A.: Stability of the Cut Locus. To appear.

[C-E] Cheeger, J., Ebin, D.: Comparison Theorems in Riemannian Geometry. Amsterdam: North Holland 1975.

[CG] Cheng, S. Y.: Eigenfunctions and Nodal sets. Comment. Math. Helv. **51**, 43—55 (1976).

[C-H] Courant, R., Hilbert, D.: Methods of Mathematical Physics, Vol. 1. New York-London: Interscience Publishers 1953.

[CI] Calabi, E.: On compact Riemannian manifolds with constant curvature I. Proc. Symp. Pure Math. **3**, 155—180 (1961).

[CL] Chavel, I.: Riemannian Symmetric Spaces of Rank one. Lecture Notes in Pure and applied Math. New York: Marcel Dekker 1972.

[C-L] Coddington, E. A., Levinson, N.: Theory of Ordinary Differential Equations. New York: McGraw-Hill 1955.

[CN 1] Cartan, E.: Oeuvres complètes. Paris: Gauthier-Villars 1952.

[CN 2] Cartan, E.: Sur certaines formes Riemanniennes remarquables des géométries à groupe fondamental simple. Ann. École Norm. Super. **44**, 345—467 (1927).

[C-S] Chevalley, C., Schafer, R.: The exceptional simple Lie algebras F_4 and E_6. Proc. Natl. Acad. Sci. USA **36**, 137—141 (1950).

[CY] Chevalley, C.: Theory of Lie groups. Princeton Mathematical series no. 8. Princeton: Princeton Univ. Press 1946.

[DE 1] Dieudonné, J.: Eléments d'analyse, Tome 3. Cahiers Scientifiques 33. Paris: Gauthier-Villars 1970.

[DE 2] Dieudonné, J.: Eléments d'analyse, Tome 4. Cahiers Scientifiques 34. Paris: Gauthier-Villars 1971.

[D-H] Duistermaat, J. J., Hörmander, L.: Fourier integral operators II. Acta Math. **128**, 183—269 (1972).

[D-G] Duistermaat, J. J., Guillemin, V. W.: The spectrum of positive elliptic operators and periodic bicharacteristics. Inventiones Math. **29**, 39—79 (1975).

[DI] Dessertine, J.-Cl.: Thèse de 3ème Cycle. Université Paris VII 1971.

[DS] Dress, A.: Newman's theorems on transformation groups. Topology **8**, 203—207 (1969).

[DX] Darboux, G.: Leçons sur la Théorie générale des Surfaces. Tome 3. Ed. Chelsea (3ème édition 1972). First edition: Paris, Gauthier-Villars, Vol. 1, 2, 3, and 4 (1894 to 1915).

[E-K] Eells, J., Kuiper, N.: Manifolds which are like projective planes. Publ. Math. IHES **14**, 181—222 (1962).

[E-M] Ebin, D. G., Marsden, J.: The motion of an incompressible perfect fluid. Ann. Math. **92**, no. 4, 102—163 (1970).

[E-M-S] Edwards, R., Millet, K., Sullivan, D.: Foliations with all leaves compact. Topology **16**, 13—32 (1977).

[EN 1] Ebin, D. G.: The space of Riemannian metrics. Proc. Symp. Pure Math. **15**, 11—40 (1968).

[EN 2] Ebin, D. G.: Espace de métriques Riemanniennes et mouvement des fluides via les variétés d'applications. Cours à l'Ecole Polytechnique et à Paris VII. Paris: Publications du Centre de Maths de l'Ecole Polytechnique 1972.

[EPN 1] Epstein, D. B. A.: Periodic flows on 3-manifolds. Annals Math. **95**, 68—82 (1972)

[EPN 2] Epstein, D. B. A.: Natural tensors on a Riemannian manifold. J. Differential Geometry **10** no. 4, 631—646 (1975)

[EPN 3] Epstein, D. B. A.: Foliations with all leaves compact. Ann. Inst. Fourier **26**, 265—281 (1976)

[ES] Escobales, R. H.: Riemannian submersions with totally geodesic fibers. J. Diff. Geom. **10**, 253—276 (1975).

[FK 1] Funk, P.: Über Flächen mit lauter geschlossenen geodätischen Linien. Math. Ann. **74**, 278—300 (1913).

[FK 2] Funk, P.: Über Flächen mit einem festen Abstand der konjugierten Punkte. Math. Z. **16**, 159—162 (1923).

[F-K] Flaschel, P., Klingenberg, W.: Riemannsche Hilbertmannigfaltigkeiten. Periodische Geodedätische. Mit einem Anhang von H. Karcher. Lecture Notes in Math. 282. Berlin-Heidelberg-New York: Springer 1972.

[FL 1] Freudhental, H.: Oktaven, Ausnahmegruppen und Oktavengeometrie. Utrecht: Math. Inst. der Rijksuniversiteit 1951.

[FL 2] Freudenthal, H.: Lie groups in the foundations of geometry. Advan. Math. **1**, 145—190 (1965).

[FR] Faulkner, J.: Octonion planes defined by quadratic Jordan algebras. Memoir of the Am. Math. Soc. no. 104. Providence: AMS (1970).

[GG] Greenberg, M.: Lectures on Algebraic Topology. New York: Benjamin 1967.

[G-H-V] Graeub, W., Halperin, S., Vanstone, R.: Curvature, Connections and Cohomology, Vol. I. New York: Academic Press 1972.

[G-K-M] Gromoll, D., Klingenberg, W., Meyer, W.: Riemannsche Geometrie in Großen. Lecture Notes in Mathematics no. 55. Berlin-Heidelberg-New York: Springer 1968.

[GL] Glaeser, G.: Racine carrée d'une fonction différentiable. Ann. Inst. Fourier **13**, no. 2, 203—210 (1963).

[GN 1] Green, L. W.: A theorem of E. Hopf. Mich. Math. J. **5**, 31—34 (1958).

[GN 2] Green, L. W.: Auf Wiedersehensflächen. Ann. Math. **78**, 289—299 (1963).

[GO 1] Godbillon, C.: Géométrie différentielle et mécanique analytique. Paris: Hermann 1969.

[GO 2] Godbillon, C.: Eléments de Topologie Algébrique. Paris: Hermann 1971.

[GR] Gambier, B.: Surfaces à lignes géodésiques toutes fermées. Etude spéciale de celles qui sont de révolution. Bulletin des Sciences Math. 2ème Série Tome XLIX (Mars 1925).

[G-S 1] Gluck, H., Singer, D.: The existence of non triangulable Cut Loci. Bull. Am. Math. Soc. **82**, no. 4, 599—602 (1976).

[G-S 2] Gluck, H., Singer, D.: Deformation of geodesic fields. Bull. Am. Math. Soc. **82**, no. 4, 571—574 (1976).

[GT] Gontier, N.: Equations de Sturm-Liouville à solutions toutes périodiques. Microthèse. Paris: Ecole Polytechnique 1974.

[GU] Guillemin, V.: The Radon transform on Zoll surfaces. Preprint (1976).

[GY] Gray, A.: The volume of a small geodesic ball of a Riemannian manifold. Mich. Math. J. **20**, 329—344 (1973).

[HE] Hersch, J.: Quatre propriétés isométriques de membranes sphériques homogènes. CR. Acad. Sci., Série A, **270**, 1645—1648 (1970).

[HH] Hirsch, G.: La géométrie projective et la topologie des espaces fibrés. Colloque de topologie algébrique du CNRS. Paris: Editions du C.N.R.S. 1949.

[HL] Hall, M., Jr.: The theory of groups. New York: MacMillan 1959.

[H-L-P] Hardy, G. H., Littlewood, J. E., Polya, G.: Inequalities. Cambridge University Press 1967.

[HM] Hermann, R.: A sufficient condition that a mapping of Riemannian manifolds be a fiber bundle. Proc. Am. Math. Soc. **11**, 236—242 (1960).

[HN 1] Helgason, S.: Differential Geometry and Symmetric Spaces. Pure and applied Mathematics. New York: Academic Press 1962.

[HN 2] Helgason, S.: The Radon transform on Euclidean spaces, compact two-point homogeneous spaces and Grassmann manifolds. Acta Math. **113**, 153—180 (1965).

[HR 1] Hörmander, L.: Linear Partial Differential Operators. Springer Grundlehren, Band 116. Berlin-Heidelberg-New York: Springer 1963.

[HR 2] Hörmander, L.: Pseudo differential operators and hypoelliptic equations. Proc. Symp. Pure Math. **10**, 138—183 (1966).

[HR 3] Hörmander, L.: The spectral function of an elliptic operator. Acta Math. **121**, 193—218 (1968).

[HR 4] Hörmander, L.: Fourier integral operator I. Acta Math. **127**, 79—183 (1971).

[KG 1] Klingenberg, W.: Manifolds with restricted Conjugate Locus. Ann. Math. **78**, no. 3, 527—547 (1963).

[KG 2] Klingenberg, W.: Lectures on Closed Geodesics. To appear.

[KI] Kobayashi, S.: On Conjugate and Cut Loci in Studies in Global Geometry and Analysis. Edited by S. S. Chern. Studies in Mathematics, Vol. 4. Washigton, D. C.: the Mathematical Assn. of America 1976.

[KL] Klein, J.: Espaces variationnels et mécanique. Ann. Inst. Fourier **12**, 1—124 (1962).

[K-M] Kervaire, M. A., Milnor, J. W.: Groups of homotopy spheres I. Ann. Math. **77**, 504—537 (1963).

[K-N 1] Kobayashi, S., Nomizu, K.: Foundations of Differential Geometry, Vol. 1. New York: Interscience Publishers 1963.

[K-N 2] Kobayashi, S., Nomizu, K.: Foundations of Differential Geometry, Vol. 2. New York: Interscience Publishers 1969.

[K-W 1] Kazdan, J. L., Warner, F. W.: Curvature functions for open 2-manifolds. Ann. Math. **99**, 203—219 (1974).

[K-W 2] Kazdan, J. L., Warner, F. W.: Existence and conformal deformations of metrics with prescribed Gaussian and Scalar Curvatures. Ann. Math. **101**, 317—331 (1975).

[L-L] Landau, L., Lifchitz, E.: Physique théorique, Tome 1: Mécanique, Moscou: Ed. Mir 1966.

[LO] Lopez de Medrano, S.: Involutions on Manifolds. Ergebnisse der Mathematik 59. Berlin-Heidelberg-New York: Springer 1971.

[LS 1] Loos, C.: Symmetric Spaces and Classification I: General Theory. Math. Lecture Notes Series. New York: W. A. Benjamin 1969.

[LS 2] Loos, O.: Symmetric Spaces II: Compact Spaces and Classification. Math. Lecture Notes Series. New York: W. A. Benjamin 1969.

[LZ 1] Lichnerowicz, A.: Sur les espaces Riemanniens complètement harmoniques. Bull. Soc. Math. France **72**, 146—168 (1944).

[LZ 2] Lichnerowicz, A.: Géométrie des Groupes de Transformation. Paris: Dunod 1958.

[LZ 3] Lichnerowicz, A.: Théorie globale des connexions et des groupes d'holonomie. Roma: Ed. Cremonese 1955.

[MA] Matsuhima, Y.: On a type of subgroups of compact Lie Groups. Nagoya Math. J. **2**, 1—15 (1951).

[ME] Morse, M.: The Calculus of Variations in the Large. Am. Math. Soc. Publ. **18**, 1934.

[MG] Moufang, R.: Alternativekörper und der Satz vom Vollständigen Verseit. Abhandl. Math. Univ. Hamburg, **9**, 207—222 (1933).
[MI] Michel, D.: Comparaison des notions de variétés Riemanniennes globalement harmoniques et fortement harmoniques. C. R. Acad. Sci. Paris, Serie A **282**, 1007—1010 (1976).
[MI-ST] Milnor, J. W., Stasheff, J. D.: Characteristic Classes. Ann. of Math. Studies. Princeton: Princeton University Press and University of Tokyo Press 1974.
[ML 1] Michel, R.: Sur certains tenseurs symétriques des projectifs réels. J. Math. Pures Appl. **51**, 273—293 (1972).
[ML 2] Michel, R.: Problèmes d'analyse géométrique liés à la conjecture de Blaschke. Bull. Soc. Math. France **101**, 17—69 (1973).
[MN] Malliavin, P.: Géométrie différentielle intrinsèque. Enseignement des Sciences 14. Paris: Hermann 1972.
[MO] Moser, J.: On the volume element on a manifold. Trans. Am. Math. Soc. **120**, 286—294 (1965).
[MR 1] Milnor, J.: Morse Theory. Annals of Math. Studies 51. Princeton: Princeton University Press 1962.
[MR 2] Milnor, J.: Some consequences of a theorem of Bott. Ann. Math. **68**, 444—449 (1958).
[MS 1] Myers, S. B.: Connections between differential geometry and topology I. Duke Math. J. **1**, 376—391 (1935).
[MS 2] Myers, S. B.: Connections between differential geometry and topology II. Duke Math. J. **2**, 95—102 (1936).
[MW] Mostow, G. D.: Strong rigidity of locally symmetric spaces. Princeton: Princeton University Press no. 78: 1973.
[M-Z] Montgomery, D., Zippin, L.: Topological Transformations Groups. New York: Interscience 1955.
[NA] Nakagawa, H.: A note on theorems of Bott and Samelson. J. Math. Kyoto Univ. **7**, no. 2, 205—220 (1967).
[NG] Nirenberg, L.: Lectures on Linear Partial Differential Equations. Regional Conference Series in Math. 17. Providence: Conf. Board of A.M.S. 1972.
[NN] Neuman, F.: Linear differential equations of the second order and their applications. Rend. di Matematica **3**, Vol. 4, Serie VI, 559—616 (1971).
[N-S 1] Nakagawa, H., Shiohama, K.: On Riemannian manifolds with certain cut loci. Tôhoku Math. J. **22**, 14—23 (1970).
[N-S 2] Nakagawa, H., Shiohama, K.: On Riemannian manifolds with certain cut loci II. Tôhoku Math. J. **22**, 357—361 (1970).
[OK] Oniščik, A. L.: Transitive compact transformation groups. Am. Math. Soc. Translations (2) **55**, 153—194 (1966).
[OL] O'Neill, B.: The fundamental equations of a submersion. Mich. Math. J. **13**, 459—469 (1966).
[OM] Omori, H.: A class of Riemannian metrics on a manifold. J. Diff. Geom. **2**, 233—252 (1968).
[OT] Otsuki, T.: On focal elements and the spheres. Tôhoku Math. J. **17**, 285—304 (1965).
[PT] Pickert, G.: Projektive Ebenen. Berlin-Göttingen-Heidelberg: Springer 1955.
[QC] Queinnec, C.: Sur les surfaces de révolution à lignes géodésiques simplement fermées et de longueurs égales—avec dessins hors texte. Microthèse. Paris: Ecole Polytechnique 1975.
[RB 1] Reeb, G.: Variétés de Riemann dont toutes les Géodésiques sont fermées. Acad. Roy. Belg. Bull. Cl. Sci. (5) **36**, 324—329 (1950).
[RB 2] Reeb, G.: Act. Sci. et Ind. no. 1183. Paris: Hermann 1952.
[R-L] Rosenberg, H., Langevin, R.: On stability of compact leaves and fibrations. Topology **16**, 107—111 (1977).
[R-W-W] Ruse, H. S., Walker, A. G., Willmore, T. J.: Harmonic Spaces. Consiglio Nacionale delle Ricerche. Monographie Mathematiche 8. Roma: Edizioni Cremonese 1961.
[R-S] Ruelle, D., Sullivan, D.: Currents, flows and diffeomorphisms. Topology **14**, 319—327 (1975).
[SA] Sugahara, K.: On the Cut locus and the topology of Riemannian manifolds. J. Math. Kyoto Univ. **14**, 391—411 (1974).
[SAT] Satake, I.: The Gauss-Bonnet theorem for *V*-manifolds. J. of the Math. Soc. of Japan **9**, 464—492 (1957).

[SC] Schafer, R.: An Introduction to Nonassociative Algebras. Pure and applied Math. New York: Academic Press 1966.

[SD] Steenrod, N.: The Topology of Fiber Bundles. Princeton: Princeton Univ. Press 1951.

[SE] Smale, S.: Generalized Poincaré's conjecture in dimensions greater than four. Annals of Math. (2) **74**, 391—406 (1961).

[SG] Sternberg, S.: Lectures on Differential Geometry. Englewood Cliffs: Prentice Hall 1964.

[SH] Smith, R. T.: The spherical representation of groups transitive on S^n. Indiana Univ. Math. J. **24**, 307—325 (1974).

[SI] Sakai, T.: On eigenvalues of Laplacian and curvature of Riemannian manifolds. Tôhoku Mathematical J. **23**, 589—603 (1971).

[SK 1] Spivak, M.: A Comprehensive Introduction to Differential Geometry. Vol. 2. Berkeley: Publish or Perish, Inc. 1970.

[SK 2] Spivak, M.: A Comprehensive Introduction to Differential Geometry. Vol. 5. Berkeley: Publish or Perish, Inc. 1975.

[SL 1] Sullivan, D.: Differential Forms and the Topology of Manifolds. Preceedings of The International Conference on Manifolds and Related Topics in Topology. Tokyo 1973, 37—51, edited by Akio Mattori. Tokyo: University Press 1975.

[SL 2] Sullivan, D.: A new flow. Bull. Amer. Math. Soc. **82**, 331—332 (1976).

[SL 3] Sullivan, D.: A counter-example to the periodic orbit conjecture. I.H.E.S. Public. Math. **46**, (1976).

[SM] Semyanistyi, V. I.: Homogeneous functions and some problems of integral geometry in spaces of constant curvature. Soviet Math. Doklady **2**, 59—62 (1961).

[SN 1] Samelson, H.: Beiträge zur Topologie der Gruppen-Mannigfaltigkeiten. Ann. Math. **42**, 1091—1137 (1941).

[SN 2] Samelson, H.: On manifolds with many closed geodesics. Portugaliae Mathematicae **22**, no. 4, 193—196 (1963).

[SO] Santalo, L. A.: Integral Geometry in General Spaces. Proc. Intern. Cong. Math., Cambridge, 483—489 (1950).

[SP] Springer, T.: The projective Octave plane. Indag. Math. **22**, 74—101 (1960).

[SR] Spanier, E.: Algebraic Topology. Reading, Mass.: Mac-Graw Hill 1966.

[ST] Sergeraert, F.: Un théorème des fonctions implicites sur certains espaces de Fréchet et quelques applications. Ann. Sci. Ecole Norm. Sup. 4ème Série, Tome **5**, 599—660 (1972).

[S-T] Singer, I. M., Thorpe, J. A.: The curvature of 4-dimensional Einstein spaces. In: Global Analysis, in honor of K. Kodaira, 355—366. Princeton Mathematical Series no. 29. Princeton: Princeton University Press 1969.

[SU] Souriau, J. M.: Structure des Systèmes Dynamiques. Paris: Dunod 1970.

[SV] Skorniakov, L. A.: Metrization of the projective plane in connection with a given system of curves. Izvestiya Akad. Nauk SSSR, Ser. Math. **19**, 471—482 (1955).

[S.V] Singh Varma, H. O.: Homogeneous manifolds all of whose geodesics are closed. Indag. Math. **48**, 813—819 (1965).

[S-V] Springer, T., Veldkamp, F.: Elliptic and hyperbolic octave planes. Indag. Math. **25**, 413—451 (1963).

[SY] Seeley, R.: Complex powers of an elliptic operator. Proc. Symp. Pure Math. **10**, Providence: A.M.S., 288—307 (1966).

[SZ] Szenthe, J.: A metric characterization of symmetric spaces. Acta Math. Acad. Sci. Hungaricae **20**, 303—314 (1969).

[TN] Teleman, C.: Asupra unor spatii simetrice. Sectia se stiinte matematice si fizice Tomul VII no. 4. Bucarest 1955.

[TS] Tits, J.: Le plan projectif des octaves et les groupes de Lie exeptionnels. Bull. Acad. Roy. Belgique **39**, 309—329 (1953).

[TY] Tannery, J.: Sur une surface de révolution du quatrième degré dont les lignes géodésiques sont algébriques. Bull. Sci. Math., Paris (1892), 190—192.

[VS] Vilms, J.: Totally geodesic maps. J. Diff. Geom. **4**, 73—79 (1970).

[WE] Willmore, T. J.: 2-point invariant functions and k-harmonic manifolds. Rev. Roum. Math. Pures Appl. **13**, 1051—1057 (1968).

[WF 1] Wolf, J. A.: Spaces of Constant Curvature. New York: Mc-Graw Hill Book Company 1967.

[WF 2] Wolf, J. A.: Elliptic spaces in Grassmann manifolds. Illinois J. Math. **7**, 447—462 (1963).

[WG 1] Wang, H. C.: Homogeneous spaces with non-vanishing Euler characteristic. Ann. Math. **50**, 915—953 (1949).

[WG 2] Wang, H. C.: Two-point homogeneous spaces. Ann. Math. **55**, 177—191 (1952).

[W-H] Willmore, T. J., El Hadi, K.: *k*-harmonic symmetric manifolds. Revue Roum. Math. Pures Appl. **15**, 1573—1577 (1970).

[WK] Walker, A. G.: On Lichnérowicz's conjecture for harmonic 4-spaces. J. Lond. Math. Soc. **24**, 317—329 (1948—1949).

[WN 1] Weinstein, A.: The cut locus and conjugate locus of a Riemannian manifold. Ann. Math. **87**, 29—41 (1968).

[WN 2] Weinstein, A.: The generic conjugate locus in global analysis. Proc. Symp. Pure Math. Providence: Am. Math. Soc. **15**, 299—301 (1970).

[WN 3] Weinstein, A.: On the volume of manifolds all of whose geodesics are closed. J. Diff. Geom. **9**, 513—517 (1974).

[WN 4] Weinstein, A.: Fourier Integral Operators, Quantization and the Spectra of Riemannian Manifolds. In Colloque International de Géométrie Symplectique et Physique Mathématique CNRS Aix (Juin 1974). Paris: Publication CNRS 1976.

[WN 5] Weinstein, A.: Symplectic *V*-manifolds, Periodic Orbits of Hamiltonian Systems, and the Volume of Certain Riemannian Manifolds. Communic. Pures & Appl. Math. **30**, 149—164 (1977).

[WR 1] Warner, F. W.: The conjugate locus of a Riemannian manifold. Am. J. Math. **87**, 575—604 (1965).

[WR 2] Warner, F. W.: Conjugate loci of constant order. Ann. Math. **86**, 192—212 (1967).

[WR 3] Warner, F. W.: Foundations of differentiable manifolds and Lie groups. Scott Foresman 1971.

[WY 1] Wadsley, A. W., Thesis, Ph. D.: Warwick University.

[WY 2] Wadsley, A. W.: Geodesic foliations by circles. J. Diff. Geom. **10**, no. 4, 541—549 (1975).

[YA] Yosida, K.: Functional Analysis. Springer Grundlehren Band 123. Berlin-Heidelberg-New York: Springer 1965.

[ZL] Zoll, O.: Über Flächen mit Scharen geschlossener geodätischer Linien. Math. Ann. **57**, 108—133 (1903).

Notation Index

Subject Index

Ergebnisse der Mathematik und ihrer Grenzgebiete

A Series of Modern Surveys in Mathematics

---

1. Bachmann: Transfinite Zahlen
2. Miranda: Partial Differential Equations of Elliptic Type
4. Samuel: Méthodes d'algèbre abstraite en géométrie algébrique
5. Dieudonné: La géométrie des groupes classiques
6. Roth: Algebraic Threefolds with Special Regard to Problems of Rationality
7. Ostmann: Additive Zahlentheorie. 1. Teil: Allgemeine Untersuchungen
8. Wittich: Neuere Untersuchungen über eindeutige analytische Funktionen
10. Suzuki: Structure of a Group and the Structure of its Lattice of Subgroups
11. Ostmann: Additive Zahlentheorie. 2. Teil: Spezielle Zahlenmengen
13. Segre: Some Properties of Differentiable Varieties and Transformations
14. Coxeter/Moser: Generators and Relations for Discrete Groups
15. Zeller/Beekmann: Theorie der Limitierungsverfahren
16. Cesari: Asymptotic Behavior and Stability Problems in Ordinary Differential Equations
17. Severi: Il teorema di Riemann-Roch per curve, superficie e varietà questioni collegate
18. Jenkins: Univalent Functions and Conformal Mapping
19. Boas/Buck: Polynomial Expansions of Analytic Functions
20. Bruck: A Survey of Binary Systems
21. Day: Normed Linear Spaces
23. Bergmann: Integral Operators in the Theory of Linear Partial Differential Equations
25. Sikorski: Boolean Algebras
26. Künzi: Quasikonforme Abbildungen
27. Schatten: Norm Ideals of Completely Continuous Operators
28. Noshiro: Cluster Sets
30. Beckenbach/Bellman: Inequalities
31. Wolfowitz: Coding Theorems of Information Theory
32. Constantinescu/Cornea: Ideale Ränder Riemannscher Flächen
33. Conner/Floyd: Differentiable Periodic Maps
34. Mumford: Geometric Invariant Theory
35. Gabriel/Zisman: Calculus of Fractions and Homotopy Theory
36. Putnam: Commutation Properties of Hilbert Space Operators and Related Topics
37. Neumann: Varieties of Groups
38. Boas: Integrability Theorems for Trigonometric Transforms
39. Sz.-Nagy: Spektraldarstellung linearer Transformationen des Hilbertschen Raumes
40. Seligman: Modular Lie Algebras
41. Deuring: Algebren
42. Schütte: Vollständige Systeme modaler und intuitionistischer Logik
43. Smullyan: First-Order Logic
44. Dembowski: Finite Geometries
45. Linnik: Ergodic Properties of Algebraic Fields
46. Krull: Idealtheorie
47. Nachbin: Topology on Spaces of Holomorphic Mappings
48. A. Ionescu Tulcea/C. Ionescu Tulcea: Topics in the Theory of Lifting
49. Hayes/Pauc: Derivation and Martingales
50. Kahane: Séries de Fourier absolument convergentes
51. Behnke/Thullen: Theorie der Funktionen mehrerer komplexer Veränderlichen
52. Wilf: Finite Sections of Some Classical Inequalities
53. Ramis: Sous-ensembles analytiques d'une variété banachique complexe
54. Busemann: Recent Synthetic Differential Geometry
55. Walter: Differential and Integral Inequalities
56. Monna: Analyse non-archimédienne
57. Alfsen: Compact Convex Sets and Boundary Integrals

58. Greco/Salmon: Topics in m-adic Topologies
59. López de Medrano: Involutions on Manifolds
60. Sakai: C^*-Algebras and W^*-Algebras
61. Zariski: Algebraic Surfaces
62. Robinson: Finiteness Conditions and Generalized Soluble Groups, Part 1
63. Robinson: Finiteness Conditions and Generalized Soluble Groups, Part 2
64. Hakim: Topos annelés et schémas relatifs
65. Browder: Surgery on Simply-Connected Manifolds
66. Pietsch: Nuclear Locally Convex Spaces
67. Dellacherie: Capacités et processus stochastiques
68. Raghunathan: Discrete Subgroups of Lie Groups
69. Rourke/Sanderson: Introduction of Piecewise-Linear Topology
70. Kobayashi: Transformation Groups in Differential Geometry
71. Tougeron: Idéaux de fonctions différentiables
72. Gihman/Skorohod: Stochastic Differential Equations
73. Milnor/Husemoller: Symmetric Bilinear Forms
74. Fossum: The Divisor Class Group of a Krull Domain
75. Springer: Jordan Algebras and Algebraic Groups
76. Wehrfritz: Infinite Linear Groups
77. Radjavi/Rosenthal: Invariant Subspaces
78. Bognár: Indefinite Inner Product Spaces
79. Skorohod: Integration in Hilbert Space
80. Bonsall/Duncan: Complete Normed Algebras
81. Crossley/Nerode: Combinatorial Functors
82. Petrov: Sums of Independent Random Variables
83. Walker: The Stone-Čech Compactification
84. Wells/Williams: Embeddings and Extensions in Analysis
85. Hsiang: Cohomology Theory of Topological Transformation Groups
86. Olevskii: Fourier Series with Respect to General Orthogonal Systems
87. Berg/Forst: Potential Theory on Locally Compact Abelian Groups
88. Weil: Elliptic Functions according to Eisenstein and Kronecker
89. Lyndon/Schupp: Combinatorial Group Theory
90. Edwards/Gaudry: Littlewood-Paley and Multiplier Theory
91. Gunning: Riemann Surfaces and Generalized Theta Functions
92. Lindenstrauss/Tzafriri: Classical Banach Spaces I
93. Besse: Manifolds all of whose Geodesics are Closed
94. Heyer: Probability Measures on Locally Compact Groups
95. Adian: The Burnside Problem and Identities in Groups

www.ingramcontent.com/pod-product-compliance
Lightning Source LLC
Chambersburg PA
CBHW051718090225
21648CB00002B/22
9783540081586